Lecture Notes in Computer Science 8658

Commenced Publication in 1973
Founding and Former Series Editors:
Gerhard Goos, Juris Hartmanis, and Jan van Leeuwen

Editorial Board

T0212685

Nelson Baloian   Frada Burstein   Hiroaki Ogata
Flavia Santoro   Gustavo Zurita (Eds.)

Collaboration and Technology

20th International Conference, CRIWG 2014
Santiago, Chile, September 7-10, 2014
Proceedings

 Springer

Volume Editors

Nelson Baloian
Universidad de Chile, DCC, Beauchef 851, Santiago, Chile
E-mail: nbaloian@dcc.uchile.cl

Frada Burstein
Monash University, Caulfield Campus, Melbourne, VIC 3145, Australia
E-mail: frada.burstein@monash.edu

Hiroaki Ogata
Kyushu University, 744, Motooa, Nishi-ku, Fukuoka 819-0395, Japan
E-mail: hiroaki.ogata@gmail.com

Flavia Santoro
Universidade Federal do Estado do Rio de Janeiro
Avenida Pasteur, 458, Urca, 22245-040 Rio de Janeiro, RJ, Brazil
E-mail: flavia.santoro@uniriotec.br

Gustavo Zurita
Universidad de Chile, Diagonal Paraquay 257, 8330015 Santiago, Chile
E-mail: gzurita@fen.uchile.cl

ISSN 0302-9743   e-ISSN 1611-3349
ISBN 978-3-319-10165-1   e-ISBN 978-3-319-10166-8
DOI 10.1007/978-3-319-10166-8
Springer Cham Heidelberg New York Dordrecht London

Library of Congress Control Number: 2014945635

LNCS Sublibrary: SL 3 – Information Systems and Application, incl. Internet/Web
and HCI

Typesetting: Camera-ready by author, data conversion by Scientific Publishing Services, Chennai, India

Printed on acid-free paper

Springer is part of Springer Science+Business Media (www.springer.com)

Preface

This volume contains the papers presented at CRIWG 2014: the 20th International Conference on Collaboration and Technology held during September 6–9, 2014, in Santiago, Chile.

The conference is supported and governed by the Collaborative Research International Working Group (CRIWG), an open community of collaboration technology researchers. Since 1995, conferences supported by CRIWG have been focused on collaboration technology design, development, and evaluation. The background research is influenced by a number of disciplines, such as computer science, management science, information systems, engineering, psychology, cognitive sciences, and social sciences.

The 49 submitted papers were carefully reviewed in a double-blind review process involving at least two reviewers appointed by the program chairs (on average, there were 2.9 reviews per paper). Of these, 16 were selected as full papers and 17 were selected as work in progress. Thus, this volume presents the most relevant and insightful research papers carefully chosen among the contributions accepted for presentation and discussion at the conference.

We believe that papers published in the proceedings of this year's and past CRIWG conferences reflect the trends in collaborative computing research and its evolution. We have seen a growing interest in social networks analysis, crowdsourcing, and computer support for large communities in general. A special research topic that has been traditionally present in the CRIWG proceedings has been collaborative learning. This time there were seven papers selected touching this topic in one way or another.

As usual, we saw a strong participation from South American countries with authors from Brazil, Mexico, Argentina, Peru, Cuba, and Chile. There were also a good number of European authors from Germany, Norway, Portugal, and the UK and contributions from Australia, New Zealand, and China.

As editors, we would like to thank everybody who contributed to the content and production of this book, namely, all the authors and presenters, whose contributions made CRIWG 2014 a success; the Steering Committee, members of the Program Committee, and the additional reviewers. Last but not least, we would like to acknowledge the effort of the organizers of the conference, without whom this conference would not have been realized so effectively. Our thanks also go to Springer, the publisher of the CRIWG proceedings, for their continuous support.

June 2014

Frada Burstein
Hiroaki Ogata
Flavia Santoro
Gustavo Zurita
Nelson Baloian

Organization

Program Committee

Pedro Antunes	Victoria University of Wellington, New Zealand
Renata Araujo	Universidade Federal do Estado do Rio de Janeiro, Brazil
Nelson Baloian	Universidad de Chile
Lars Bollen	Universiteit Twente, The Netherlands
Frada Burstein	Monash University, Australia
Luis Carriço	University of Lisbon, Portugal
Cesar A. Collazos	Universidad del Cauca, Colombia
Gerard de Leoz	University of Omaha, USA
Marco de Sá	Yahoo! Research, USA
Gert-Jan De Vreede	University of Nebraska at Omaha, USA
Dominique Decouchant	UAM Cuajimalpa, Mexico DF, Mexico and LIG de Grenoble, France
Alicia Diaz	Universidad Nacional de La Plata, Argentina
Yannis Dimitriadis	University of Valladolid, Spain
Jesus Favela	Centro de Investigación Científica y de Educación Superior de Ensenada, Mexico
Benjamim Fonseca	Universidade de Trás-os-Montes e Alto Douro, Portugal
Kimberly Garcia	Centro de Investigación y de Estudios Avanzados del Instituto Politécnico Nacional, Mexico
Marco Gerosa	Universidade de São Paulo, Brazil
Adam Giemza	Universität Duisburg-Essen, Germany
Eduardo Guzmán	Universidad de Málaga, Spain
Andreas Harrer	Clausthal University of Technology, Germany
Valeria Herskovic	Pontificia Universidad Católica de Chile
Ulrich Hoppe	Universität Duisburg-Essen, Germany
Indratmo Indratmo	Grant MacEwan University, Canada
Tomoo Inoue	University of Tsukuba, Japan
Marc Jansen	University of Applied Sciences Ruhr West, Germany
Ralf Klamma	Rheinisch-Westfälische Technische Hochschule Aachen, Germany
Michael Koch	Bundeswehr University Munich, Germany
David Kocsis	University of Nebraska at Omaha, USA

Benjamin Weyers	Rheinisch-Westfälische Technische Hochschule Aachen, Germany
Lung Hsiang Wong	Nanyang Technological University, Singapore
Jürgen Ziegler	Universität Duisburg-Essen, Germany
Gustavo Zurita	Universidad de Chile

Additional Reviewers

Adewoyin, Oluwabunmi
Aniche, Mauricio
Araujo, Renata

Coelho, José
Duarte, Luis

Keynote Talks

Current Challenges
in Business Process Modelling

Pedro Antunes

School of Information Management, Victoria University of Wellington
Wellington, New Zealand
pedro.antunes@vuw.ac.nz

Abstract. The Business Process Management (BPM) method has been increasingly adopted by organisations seeking to improve their business processes. BPM is seen as an enabler of business innovation, fostering change and flexibility, increasing productivity and responsiveness, and leveraging operational intelligence, while maintaining the impact of organisational complexity within reasonable bounds. Analysing these trends, we recognise that a particular IT artefact has gained unexpected importance: process models. Process models support the BPM method in two different ways. In the one hand, by providing formalised, standardised operational rules required by process-aware information systems. In the other hand, process models also influence several activities required by the BPM method such as process elicitation, documentation, analysis, and visualisation. Nowadays business process modelling represents an important market for information systems vendors, software systems developers and integrators, and consulting services providers. This market covers diverse areas of intervention such as supply chain management, customer relationships management, change management, enterprise resource planning, and quality management. It also covers different application areas such as manufacturing, financial services and healthcare. However, the complexity of existing process modelling approaches, along with the demanding characteristics of modelling languages and tools, and the skills required to translate organisational practices into process models, have turned process modelling into a complex practice. What problems are found when modelling business processes? How can process modellers overcome these problems? How can non-experts model business processes? Surprisingly, we do not know much about the process modelling practice. In this keynote we systematically review the current problems found in business process modelling. We show that most problems are related with inadequate conceptual foundations, lack of consensus about what modelling quality is, and also inadequate tool support. Considering in particular the conceptual foundations of BPM, we argue that many problems result from the prevailing mechanistic paradigm, which causes problems bridging business rules and process models, and bridging process execution with organisational behaviour. We suggest a conceptual change towards a more humanistic paradigm centred on ease-of-use, readiness and flexibility. Finally, we discuss several projects using the design-science methodology to investigate humanistic business process modelling.

Keywords: Business Process Modelling, Collaboration.

The Wisdom of Crowds and the Long Tail

Ricardo Baeza-Yates

Yahoo Labs Barcelona
Barcelona, Spain

Abstract. In this keynote we focus on the concept of wisdom of crowds in the context of the Web, particularly through social media and web search usage. As expected from Zipf's principle of least effort, the wisdom is heterogeneous and biased to active people, which may represent at the end the wisdom of a few. We also explore the impact on the wisdom of crowds of dimensions such as bias, privacy, scalability, and spam. We also cover an important related concept, the long tail of the special interests of people, as well as the digital desert, web content that nobody sees.

Summary

The Web continues to grow and evolve very fast, changing our daily lives. This activity represents the collaborative work of the millions of institutions and people that contribute content to the Web as well as more than two billion people that use it. In this ocean of hyperlinked data there is explicit and implicit information and knowledge. But how is the Web? What are the activities of people? What is the impact of these activities? Web data mining is the main approach to answer these questions. Web data comes in three main flavors: content (text, images, etc.), structure (hyperlinks) and usage (navigation, queries, etc.), implying different techniques such as text, graph or log mining. Each case reflects the wisdom of some group of people that can be used to make the Web better.

The wisdom of crowds [9] at work in the Web is best seen in social media as well as in social networks. It is also implicit in the usage of search engines [1] and other popular web applications. The wisdom behind web users is shaped by different complex factors such as the heterogeneity of user activity [10] and hence a heavy long tail [6]; different types of bias [3] that create problems such as the bubble effect [7]; privacy breaches coming from data [5]; too much data that endangers minorities [2]; or web spam in all possible ways [8].

The diversity of user activity implies that an elite of users represent most of the wisdom and that we should really talk about the wisdom of a few [4]. This diversity also generates another concept that we define as *digital desert*, web content that no one ever sees.

References

1. Baeza-Yates, R., Ribeiro-Neto, B.: Modern Information Retrieval: The Concepts and Technology Behind Search, 2nd edn. Addison-Wesley (2011)

2. Baeza-Yates, R., Maarek, Y.: Usage Data in Web Search: Benefits and Limitations. In: Ailamaki, A., Bowers, S. (eds.) SSDBM 2012. LNCS, vol. 7338, pp. 495–506. Springer, Heidelberg (2012)
3. Baeza-Yates, R.: Big Data or Right Data? In: AMW 2013, Puebla, Mexico (May 2013)
4. Baeza-Yates, R., Saez-Trumper, D.: Wisdom of the Crowd or Wisdom of a Few? An Analysis of Users' Content Generation (submitted, 2014)
5. Barbaro, M., Zeller, T.: A face is exposed for aol searcher no. 4417749. The New York Times (August 9, 2006)
6. Goel, S., Broder, A., Gabrilovich, E., Pang, B.: Anatomy of the long tail: ordinary people with extraordinary tastes. In: ACM WSDM 2010, New York, NY, USA, pp. 201–210 (2010)
7. Pariser, E.: The Filter Bubble: What the Internet Is Hiding from You. Penguin Press (2011)
8. Spirin, N., Han, J.: Survey on web spam detection: principles and algorithms. ACM SIGKDD Explorations Newsletter Archive 13(2), 50–64 (2011)
9. Surowiecki, J.: The Wisdom of Crowds: Why the Many Are Smarter Than the Few and How Collective Wisdom Shapes Business, Economies, Societies and Nations. Random House (2004)
10. Zipf, G.K.: Human behavior and the principle of least effort. Addison-Wesley Press (1949)

Appendix: Biography

Ricardo Baeza-Yates is VP of Yahoo Labs for Europe and Latin America since 2006, leading the labs at Barcelona, Spain and Santiago, Chile. He is also part time professor at the Dept. of Information and Communication Technologies of Univ. Pompeu Fabra, in Barcelona, Spain. Until 2005 he was founder and first director of the Center for Web Research at the Dept. of Computing Science of the University of Chile (where he is a professor in leave of absence until today). He obtained a Ph.D. from the University of Waterloo, Canada, in 1989. Before he obtained two masters (M.Sc. CS & M.Eng. EE) and the EE degree as well as the CS B.Sc. from the University of Chile in Santiago. He is co-author of the best-seller Modern Information Retrieval textbook published in 1999 by Addison-Wesley, with a second enlarged edition in 2011 that won the ASIST 2012 Book of the Year award. From 2002 to 2004 he was elected to the board of governors of the IEEE Computer Society and in 2012 he was elected for the ACM Council. He has received the Organization of American States award for young researchers in exact sciences (1993), the Graham Medal for innovation in computing given by the University of Waterloo to distinguished ex-alumni (2007), the CLEI Latin American distinction for contributions to CS in the region (2009), and the National Award of the Chilean Association of Engineers (2010), among other distinctions. In 2003 he was the first computer scientist to be elected to the Chilean Academy of Sciences and since 2010 is a founding member of the Chilean Academy of Engineering. In 2009 he was named ACM Fellow and in 2011 IEEE Fellow.

Collaboration and Critical Thinking

Miguel Nussbaum, Daniela Caballero, Macarena Oteo,
and Damián Gelerstein

Pontificia Universidad Católica de Chile, School of Engineering
Santiago, Chile
mn@ing.puc.cl

Extended Abstract

In recent years, increasing importance has been placed on teaching and assessing 21st Century Skills such as critical thinking and collaboration [1]. In fact, the 2015 PISA will include a large scale assessment of Collaborative Problem Solving [2], where each student will be expected to be proficient in skills such as communicating, managing conflict, organising a team, building consensus and managing progress, among others [1].

In order to bring collaboration to the classroom, design guidelines are needed to help implement such activities. To do so, the following parameters can be used to define a collaborative activity: learning objectives, task type and level of pre-structuring [3]. The learning objectives can range from open skills, such as negotiation of meaning, to closed skills, such as knowledge specification. Task type is defined by the structuring of the activity and ranges from well-structured, with limited solutions, to ill-structured, with no clear solutions. Finally, the level of pre-structuring is defined by the level of scripting that is used [4, 5, 6] and can go from high to low.

By following the outline proposed by [3], activities can be designed that are defined by these three parameters. The learning objectives can be defined using seven collaborative work structures: Identification/Exclusion, Categorizing Elements, Forming Sequences, Completing Sequences, Establishing Exact Associations, Establishing Multiple Associations and Construction [7]. Task type depends on how structured the activity is, i.e. whether it is well-structured (only one possible solution) or ill-structured (multiple solutions) [8]. Finally, the level of pre-structuring can be defined according to the pre-structuring of the roles of each member of the group. This can either be through pre-defined roles[5], [9], where the actions carried out by each student are determined by the activity, or undefined roles, where the students must negotiate the actions to be carried out by each member of the group in order to meet the objective.

The combination of these three variables will determine the degree of collaboration that is required by the activity. These activities can be classified as one of three types: Weak, Medium and Strong. Weak activities can have any type of Learning Objective, where there is only one solution (well structured activity) and the roles are defined (high level of pre-structuring). Medium activities can also have any type of Learning Objective, with two possible combinations of the remaining variables. The first is where there is only one solution (well structured activity) and the roles are not defined (low level of pre-structuring); the second is where there are multiple solutions (ill structured activity) and the roles are defined (high level of pre-structuring). Strong activities can once again have any type of Learning Objective, where there are multiple solutions (ill structured activity) and the roles are not pre-defined (low level of pre-structuring).

It has been proven experimentally that the degree of collaboration required by an activity determines the interaction between students when working collaboratively, also defined by three variables. The first variable is the existence (or lack) of shared leadership; the second is the amount of dialogue referring to coordination and the third is the level of difficulty as perceived by the students.

Considering that studies which focus on the assessment of collaborative problem solving skills are less developed than studies on collaborative learning [10], it is essential that the effectiveness of collaborative work is also taken into consideration [1]. In order to assess the effectiveness of collaborative work, it is important to bear in mind the following dimensions: Learning about the perspectives and skills of each member of the group, Building a shared representation and negotiating the general sense of the problem (together), Communicating with the members of the group about the actions to be carried out, Monitoring and repairing shared comprehension, Discovering the type of collaborative interaction required to solve the problem according to the goals/objectives, Identifying and describing the tasks that are to be completed, Publishing plans, Monitoring the results of the actions and assessing how successfully the problem has been solved, Under-standing the roles that are required in order to solve the problem, Describing roles and organising the group (communicating protocol/ground rules), Following the ground rules (encouraging peers to carry out their tasks) and Monitoring, giving feedback and adapting the organization and roles within the group [1]. These dimensions were implemented using an instrument which looked for the members of a group to agree on a process for collaborative problem solving. This consisted of identifying geometric shapes, where each of the students played a specific role. The students could only communicate with one another via a chatroom in order to use their messages to work out how to solve the problem and thus determine the presence of the above dimensions.

Finally, it is important for every activity to consider critical thinking in its design, which involves the following abilities interpretation, analysis, evaluation, inference, explanation and self-regulation [11]. It is possible to develop critical thinking through collaboration [12], normally through the means of a specific subject matter such as mathematics [13].

Acknowledgement. This research is supported by CONICYT-FONDECYT 1120177.

References

1. OECD, PISA 2015 Collaborative Problem Solving Framework. (2013), http://www.oecd.org/pisa/pisaproducts/DraftPISA2015Collaborative ProblemSolvingFramework.pdf (retrieved)
2. De Jong, J.H.: Framework for PISA 2015: What 15-years-old should be able to do. In: 4th Annual Conference of Educational Research Center, Broumana, Lebanon (2012)
3. Strijbos, J.W., Martens, R.L., Jochems, W.M.G.: Designing for interaction: Six steps to designing computer-supported group-based learning. Computers & Education 42(4), 403–424 (2004)
4. Dillenbourg, P. (n.d.).: Over-scripting CSCL: The risks of blending collaborative learning with instructional design
5. Strijbos, J.-W., Weinberger, A.: Emerging and scripted roles in computer-supported collaborative learning. Computers in Human Behavior 26(4), 491–494 (2010)
6. Weinberger, A., Ertl, B., Fischer, F., Mandl, H.: Epistemic and social scripts in computer-supported collaborative learning. Instructional Science 33(1), 1–30 (2005)
7. Nussbaum, M., Rosas, R., Peirano, I., Cardenas, F.: Development of intelligent tutoring systems using knowledge structures. Computers & Education 36, 15–32 (2001)
8. Cohen, E.: Restructuring the Classroom: Conditions for Productive Small Groups. Review of Educational Research 64(1), 1–35 (1994)
9. Martel, C., Vignollet, L., Ferraris, C., David, J.P., Lejeune, A.: Modeling Collaborative Learning Activities on e-Learning Platforms. In: ICALT, pp. 707–709 (2006)
10. Looi, C.K.: Testing collaboration at school (2013), https://cerp.aqa.org.uk/ perspectives/testing-collaboration-school (accessed September 9, 2013)
11. Facione, P.A.: Critical Thinking: A Statement of Expert Consensus for Purposes of Educational Assessment and Instruction. Research Findings and Recommendations (1990), http://assessment.aas.duke.edu/documents/Delphi_Report.pdf (retrieved)
12. Johnson, R.T., Johnson, D.W., Stanne, M.B.: Effects of Cooperative, Competitive, and Individualistic Goal Structures on Computer-Assisted Instruction. Journal of Educational Psychology 77(6), 668–677 (1985)
13. Willingham, D.T.: Critical thinking: Why is it so hard to teach? Arts Education Policy Review 109(4), 21–32 (2008)

Table of Contents

Understanding How Network Performance Affects User Experience of Remote Guidance

Angus Donovan[1], Leila Alem[1], Weidong Huang[2], Ren Liu[1], and Mark Hedley[1]

[1] CSIRO
[2] University of Tasmania

Abstract. Much research has been done to support remote collaboration on physical tasks. However, the focus of the research has been mainly on system and interface design and their impact on collaboration. Relatively less attention has been paid to investigating how network performance can affect user experience and task performance. In this paper, we present a preliminary user study on this issue in which participants were asked to work collaboratively in pair using a remote mobile tele-assistance system we developed. In this study, five network scenarios were examined and network performance (QoS) was measured using four metrics including delay, jitter, bandwidth and packet loss. User experience (QoE) was measured using both objective and subjective metrics. The formal included time taken and number of instructions repeated for task performance while the latter included user ratings of quality of audio experience, quality of video experience and overall quality of experience. The results indicated that the packet loss rate in QoS is the biggest contributor to loss in QoE. We also discuss implications of the study and possible directions of future work.

1 Introduction

Nowadays technologies are becoming increasingly ubiquitous and complex. As a result, expertise is often required for performing physical tasks on these technologies. Physical tasks are ones that require collaborators working on physical objects such as equipment maintenance. However, it is common that users do not have the required skills set and help is required to, for example, fix the technology when it breaks down. This can be potentially a big issue for many users as access to sound expertise and guidance is often lacking, particularly in rural and remote areas and in emergency situations. In response to the demand and in order to make expertise more accessible to users, a number of systems have been developed (e.g., [7, 12, 13]).

These systems typically have two network-connected units: one helper unit and one worker unit, and are constructed in a way that the remote helper is enabled to guide the local worker performing collaborative psychical tasks using both audio and visual communications just like they were co-located. Recently, we have developed a remote mobile tele-assistance tool called ReMoTe [9, 15]. The tool's worker unit is a wearable system that supports mobility of the worker

N. Baloian et al. (Eds.): CRIWG 2014, LNCS 8658, pp. 1–12, 2014.

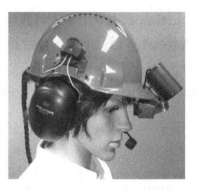

Fig. 1. Helper Interface   **Fig. 2.** Worker Interface

and frees his two hands for manipulation of physical objects, while the helper unit is a tabletop system on which the helper can perform hand gestures for guiding purposes. ReMoTe allows two people to collaborate on a "shared visual space" [7] through audio and video communication over an internet connection in real time. The shared visual space is a main feature of the system interface: as depicted in Figure 1, the helper's hand movements are detected and augmented with the work-space of the worker, which is then displayed on a near-field display as visual aid to the worker who is completing the task as shown in Figure 2.

Much research has been done to investigate usability of these systems and their impact on users' collaboration behavior with an assumption of guaranteed network availability (e.g., [10,11,14,16]). However, little attention has been paid to the impact of varying network conditions on user experience and task performance [8,19]. ReMoTe is essentially a network based distributed system, in which the quality of network is a key to its usability. Therefore there is the value of exploring the use of ReMoTe in different network settings. This will provide a basis for us to understand when a drop of QoS (quality of service) may lead to a drop in QoE (quality of experience), which in turn may lead to a wish to drop the real time remote assistance session and reverse to an offline remote guiding session. As part of effort towards this end, we are in the process of conducing a series of studies. In this paper, we present a preliminary investigation of this issue to understand how network performance affects user experience of remote guidance.

2   Background

The specific goal of this particular investigation was to map the relationship of performance between QoS and QoE. To achieve the goal, we tested five network scenarios: Ethernet, Satellite, 3G mobile technology, Wifi and Fibre optic and determined how specific QoS measures within the scenario affected the QoE of an end user.

2.1   Network QoS Metrics

QoS is defined by the ITU (International Telecommunication Union) as the "Totality of characteristics of a telecommunications service that bear on its ability to satisfy stated and implied needs of the user of the service" [3]. The QoS for this study is measured in four particular ways specified below. The performance limits for relevant applications are also given.

Delay. Delay is a measure of the interval between the system transmission and when the user receives the information. For typical personal computers or workstations as end-systems, the delay should not exceed 400ms for telephone quality speech. For interactive applications with real time sound transmission the maximum delay is of the order of 0.1 to 0.5 seconds [1,2]. Beyond 100 ms speech is still comprehensible although users may become irritated with the service.

Jitter. Jitter occurs when received signals have varying latency times on arrival. It is an error that occurs when transmitted data packets have variable queuing delays. Jitter requirements are essential to transmitting data at a constant, reliable rate. Real time sound is the most sensitive data transmission type to jitter [6]. As a result, the bounds for acceptable jitter are determined by audio quality. The jitter for videoconferencing applications should not exceed 400ms [1].

Bandwidth. Bandwidth is the average rate of successful message delivery over a communication channel measured in bits/s. The video codec selected for ReMoTe is the H.264 codec for video encoding and the Opus codec for audio encoding. Using the H.264 codec and ReMoTe's standard resolution of 640x480 (4:3), the associated bandwidth that is required is between 300kbps and 800kbps [4]. The Opus audio codec requires an average bandwidth of 25kbit/s [5].

Packet Loss. Packet loss is the amount of packets lost per time slot. This can be in seconds to mirror the bit/s throughput rate. Packet loss can occur because of signal degradation, multipath fading, channel congestion and faulty networks or hardware.

2.2   User QoE Meterics

QoE is defined as "the overall acceptability of an application or service, as perceived subjectively by the end-user [3]. QoE measures are subjective in nature, however they can be measured quantitatively by asking the following questions:

- Time taken to complete the task?
- How many instructions had to be repeated?
- Video quality rating?

– Audio quality rating?
– Overall quality rating?

Examining the time taken is a key indicator of how well the network is performing as it implies how clear the instructions and communications were to both the Helper and the Worker. A metric on the amount of instructions that have to be repeated gives an indication of where the ReMoTe system wasn't functioning in a seamless fashion due to network conditions or otherwise. The video, audio and overall quality ratings are determined using a scale from 1 (bad) to 5 (excellent).

3   Method

3.1   Design

Seven university students were recruited to participate in the study. Four of them were male and the rest are female. One of them was selected to play the role of Helper for all test sessions. The remaining six were asked to play the role of Worker. Both the Helper and the Worker were situated in the same room with both facing away from each other so that they cannot see any of the construction being undertaken on the other side. The room setup is shown in Figure 3. In this figure, the Helper is in the foreground with hands placed over the screen to convey gestures. During testing sessions, the Helper also had a headset (microphone and headphones) linked to the Helper computer that allows audio communication with the Worker. The Worker laptop is narrowly shown on the left side of the figure, in front of the Worker.

Fig. 3. Testing setup

Fig. 4. Loose blocks   **Fig. 5.** Assembled blocks

A standard task was created involving large Lego 'Duplo' blocks for each of the test subjects. The Helper constructs a random arrangement of five Duplo blocks and keeps this arrangement in front of him. The Duplo blocks used are shown in Figures 4 and 5. The Worker sits at table and has five identical un-arranged Duplo blocks. The Helper must use ReMoTe to aid the Worker in copying the random construction.

Having the same Helper throughout the testing sessions allows consistency in process and instruction. Having blocks of the same color, shape and size increases the reliance of visual communication. Several practice runs were used in the beginning to ensure that both the Helper and the Worker were familiar with each other's communication styles and the testing in general before the test was formally started.

The specifications of the QoS metrics we used for each of the five network conditions are shown in Table 1. In order to gain a baseline test of human performance with ReMoTe, an optimal network condition (no QoS impairments) was added. This means that there were six network conditions in total and each subject was required to completed the task for each condition. It should be noted that the order of the testing conditions was determined randomly.

Table 1. QoS Metrics for Network Scenarios

	Delay	Jitter	Bandwidth	Packet Loss
Ethernet	10ms	0.2ms	100Mbit/s	0.5%
Satellite	600ms	7ms	1Mbit/s	2%
3G	50ms	$\approx$0	9Mbit/s	2%
Wi-Fi	10ms	30ms	30Mbit/s	5%
Fibre	190ms	40ms	1000Mbit/s	0.5%

3.2   Testing Protocol

The following test protocol was used:

1. Welcome the subject into the testing room and explain how testing will be completed.
2. Allow the Worker to set up the helmet as well as their microphone and speaker head set. Ensure that the Worker is as comfortable as possible before undertaking testing.
3. Complete 3 initial practice tests. The results of these tests are not recorded.
4. Complete the baseline test (no impairments). Ask for video, audio and overall quality ratings.
5. Complete the 5 different network scenario tests. Ask for video, audio and overall quality ratings at the end of each network scenario test.
6. Ask for additional comments and feedback in relation to the overall testing experience.

The vocal instructions used when undertaking testing were fairly consistent. Testing began with the Duplo blocks disassembled in front of the Worker. The set of instructions would usually follow in the style shown below:

- "Start with the Yellow block as the base"
- "Place the small green block on the 4 dots of the right hand side of the yellow block"
- "Place the blue block directly on top of the green block"
- etc.

4   Results and Discussion

Each of the six subjects completed six tasks (five network scenarios plus a baseline condition). In order to easily interpret the data obtained, the experimental data for each condition were aggregated and displayed in column graphs as shown in Figures 6 to 10. Note that the height of each column gives the combined total of the test scores attained by each subject and the score of each subject is differentiated by color.

4.1   Quality of Audio Experience

As can be seen from Figure 6, the audio quality remained largely the same quality throughout the testing. The audio experience is highly important to the QoE as the user becomes exponentially dissatisfied with phone calls or video conferencing when audio transmissions experience poor QoS [17]. There was significant delay of 600ms on the audio in the Satellite scenario however, as communication in the ReMoTe system is largely one-way (Helper instructing the Worker), there is minimal dialogue. It is predominantly in communications involving heavy dialogue (phone calls, videoconferencing etc.) that latency becomes a chief factor

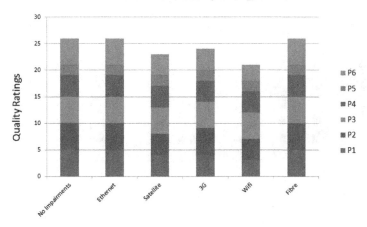

Fig. 6. Combined Audio Quality Ratings

in determining QoE. It is speculated that this is reason why QoE in the Satellite scenario was not at a lower score.

In the Wifi scenario, the high packet loss and noise created occasional 'breaking up' of the audio signal. This made the audio transmission prone to clicks and a small amount of distortion which affected the QoE ratings.

Jitter didn't seem to have any impact on the audio QoE within the experiment environment. This is confirmed by the high jitter in the fibre scenario but no loss of QoE ratings.

Bandwidth didn't seem to have an impact on the audio QoE within the experiment environment. This is supported by [18] which states that "for many applications high data rates do not translate to improved user experience unless the latency is low." However this isn't conclusive as the Satellite scenario had the lowest bandwidth of 1Mbit/s but the decrease in bandwidth may have been 'masked' by the high packet loss rate of 5%. Similarly for the 3G scenario, the comparatively low bandwidth of 9Mbit/s may have been 'masked' by the packet loss rate of 2%.

4.2 Quality of Video Experience

As can be seen from Figure 7, the quality of experience for video diminished with higher packet loss. This is clearly evident in the Satellite, 3G and Wifi scenarios where the packet loss was set to greater than 1%. This impacted video experience as the video feed would intermittently become 'garbled' where nothing was discernible on the screen for a period of approximately half of a second. For a 2% packet loss, this happened approximately every 18 seconds but for Wifi packet loss this happened more frequently at approximately once every 5 seconds. This has large implications for the video streaming in the ReMoTe system. If a hand

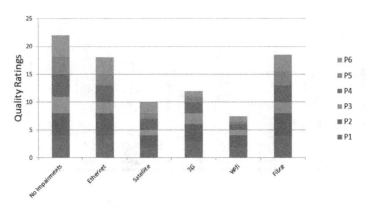

Fig. 7. Combined Video Quality Ratings

gesture is made and the video stream encounters noise at the same time, then the gesture will have to be repeated. It also reduces the QoE for the worker as it can be unclear exactly which information needs to be re-transmitted and the Worker must therefore try to describe what needs to be re-transmitted.

The delay of the Satellite at 600ms was detrimental to the video's quality of experience. When coupled with high noise, this made the ReMoTe system's operation highly unappealing to the user. It didn't however, have a particularly large effect on the ability of the Worker to complete the Duplo task. This indicates that whilst there was a poor quality of video, the audio instructions were the most effective means of communication to complete the task.

With the increased packet loss present in the Wifi scenario, it was also noted that increased pixilation and blurriness occurred in the video image. This has consequences for the quality of experience if ReMoTe is being used in a situation requiring higher precision visual information.

Jitter didn't seem to have any impact on the audio QoE within the experiment environment. This is confirmed by the high jitter in the Fibre scenario but no loss of QoE ratings.

The results suggest that the packet loss rate in a network QoS metrics is the biggest contributor to loss in QoE. Both the 3G and Wifi secnarios have a 2% and 5% packet loss rate and both these scenarios received the worst QoE scores. The Satellite scenario is not a reliable measure of the effect of packet loss on QoE because of it's high delay.

Bandwidth didn't seem to have an impact on the audio QoE within the experiment environment however this isn't conclusive for the reasons discussed in Subsection 4.1. The Satellite scenario had the lowest bandwidth of 1Mbit/s but the decrease in bandwidth may have been 'masked' by the high packet loss rate of 5%. Similarly for the 3G scenario, the comparatively low bandwidth of 9Mbit/s may have been 'masked' by the packet loss rate of 2%.

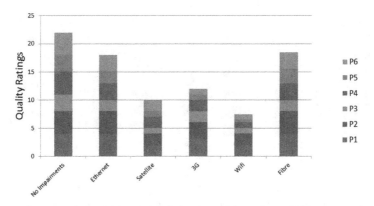

Fig. 8. Combined Overall Quality Ratings

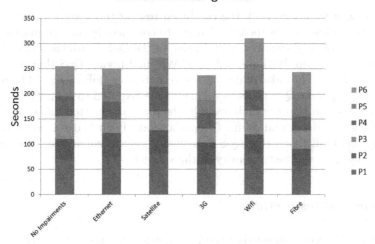

Fig. 9. Combined Testing Times

4.3 Overall Quality of Experience

As can be seen from Figure 8, the overall quality ratings show that 'Wifi' and 'Satellite' had the lowest QoE overall. This is primarily due to the larger differences in ratings that were scored in video quality ratings. The quality of audio experience contributes fairly evenly to the score of the overall experience.

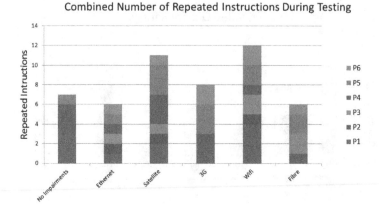

Fig. 10. Combined Number of Repeated Instructions During Testing

4.4   Testing Times

As can be seen from Figure 9, the combined testing times show that Satellite and Wifi network scenarios took the longest to complete by a small margin. The longer amount of time required is indicative of the QoS factors associated with each network scenario. Both Satellite and Wifi had poorer QoS than the other network scenarios. They also had comparatively poorer QoE ratings. The results suggest that a poorer QoS leads to a longer time required to complete tasks.

Further, it seems that the main two QoS factors that effected this timing were Satellite's 600ms delay and Wifi's 5% packet loss. Each had a heavier impact on timing and QoE than other QoS measures. This was confirmed in comments about the QoE by the vast majority of the subjects.

4.5   Repeated Instructions

The graph in Figure 10 shows that the QoS metrics was indicative of the number of mistakes made whilst carrying out testing. By comparison with similar results mentioned in previous sections, the Wifi and Satellite scenarios had the most mistakes as over 10 instructions needed to be repeated for these scenarios.

It can be argued that having to repeat instructions is the sole reason why the Satellite and Wifi network scenarios took a longer amount of time to complete than other scenarios. Therefore, the relationship between QoS and QoE is largely dependent on how many mistakes are made whilst using the ReMoTe system. Of course, additional testing with a high volume of subjects would be required in order to accurately examine this relationship.

Repeating instructions whilst carrying out a task also has implications for QoE as it can be irritating to repeat instructions.

4.6 Summary of Results

In summary, the results suggest that the packet loss rate in a network QoS metrics is the biggest contributor to loss in QoE. This has been determined by both the Wifi and Satellite scenarios which have a 2% and 5% packet loss rate and both these scenarios received the worst QoE scores. Satellite is highly likely to have performed poorly due to additional delay of 600ms. Both Satellite and Wifi also took longer times for task completion.

5 Conclusion

This paper presents a preliminary study investigating the impact of QoS on QoE using ReMoTe. In our study, QoS was measured in terms of delay, jitter, bandwidth and packet loss in five network scenarios including Ethernet, Satelite, 3G, Wifi and Fibre. QoE was measured in terms of time taken, number of instructions repeated, video quality rating, audio quality rating and overall quality rating.

ReMoTe was tested with the focus on gaining subjective results yielding the most useful information about ReMoTe and its QoE based on QoS. The results suggest that Satellite and Wifi network scenarios provide the worst QoE primarily due to packet loss but also due to signal delay.

The work presented in this paper provides the foundation for further study so that in future, the proposed measurements can be used to determine where QoS can be leveraged to provide the best QoE. More realistic testing can be completed which involves more reliance on the visual system of ReMoTe. High volume testing of numerous subjects would assist finding reliable results. These results would aid in optimizing ReMoTe's QoE over various network QoS conditions.

References

1. Fluckiger, F.: Understanding Networked Multimedia. Prentice Hall (1995)
2. Szuprowicz, B.: Multimedia Networking. McGraw-Hill Inc. (1995)
3. International Telecommunications Union, Definitions of terms related to quality of service. ITU Standardizatioin Sector (September 2008)
4. Microsoft, Network Bandwidth Requirements for Media Traffic, Lync TechNet Library (2013)
5. Sarwar, G., et al.: On the Quality of VoIP with DCCP for Satellite Communications. Internationl Journal of Satellite Communications (2000)
6. Chen, Y., Farley, T., Ye, N.: QoS Requirements of Network Applications on the Internet. Department of Industrial Engineering, Arizona State University (2004)
7. Fussell, S., Setlock, L., Yang, J., Ou, J., Mauer, E., Kramer, A.: Gestures over video streams to support remote collaboration on physical tasks. In: Hum.-Comput. Interact., vol. 19(3), pp. 273–309 (September 2004)
8. Gergle, D., Kraut, R., Fussell, S.: The impact of delayed visual feedback on collaborative performance. In: Proceedings of the SIGCHI Conference on Human Factors in Computing Systems (CHI 2006), pp. 1303–1312 (2006)

9. Huang, W., Alem, L.: HandsinAir: a wearable system for remote collaboration on physical tasks. In: Proceedings of the 2013 Conference on Computer Supported Cooperative Work Companion (CSCW 2013), pp. 153–156 (2013)
10. Huang, W., Alem, L.: Supporting hand gestures in mobile remote collaboration: a usability evaluation. In: Proceedings of the 25th BCS Conference on Human-Computer Interaction (BCS-HCI 2011), pp. 211–216 (2011)
11. Kuzuoka, H., Yamazaki, K., Yamazaki, A., Kosaka, J., Suga, Y., Heath, C.: Dual ecologies of robot as communication media: thoughts on coordinating orientations and projectability. In: Proceedings of the SIGCHI Conference on Human Factors in Computing Systems (CHI 2004), pp. 183–190 (2004)
12. Kuzuoka, H., Kosaka, J., Yamazaki, K., Suga, Y., Yamazaki, A., Luff, P., Heath, C.: Mediating dual ecologies. In: Proceedings of the 2004 ACM Conference on Computer Supported Cooperative Work (CSCW 2004), pp. 477–486 (2004)
13. Kuzuoka, H.: Spatial workspace collaboration: a SharedView video support system for remote collaboration capability. In: Proceedings of the SIGCHI Conference on Human Factors in Computing Systems (CHI 1992), pp. 533–540 (1992)
14. Kirk, D., Rodden, T., Fraser, D.: Turn it this way: grounding collaborative action with remote gestures. In: Proceedings of the SIGCHI Conference on Human Factors in Computing Systems (CHI 2007), pp. 1039–1048 (2007)
15. Tecchia, F., Alem, L., Huang, W.: 3D helping hands: a gesture based MR system for remote collaboration. In: Proceedings of the 11th ACM SIGGRAPH International Conference on Virtual-Reality Continuum and its Applications in Industry (VRCAI 2012), pp. 323–328 (2012)
16. Yamashita, N., Kaji, K., Kuzuoka, H., Hirata, K.: Improving visibility of remote gestures in distributed tabletop collaboration. In: Proceedings of the ACM 2011 Conference on Computer Supported Cooperative Work (CSCW 2011), pp. 95–104 (2011)
17. Rajkumar, R., et al.: A resource allocation model for QoS management. In: IEEE Real Time Systems Symposium (1997)
18. Mohan, S., Kapoor, R., Mohanty, B.: Latency in HSPA Data Networks. Qualcomm Research Papers (2003)
19. Wu, W., Arefin, A., Rivas, R., Nahrstedt, K., Sheppard, R., Yang, Z.: Quality of experience in distributed interactive multimedia environments: toward a theoretical framework. In: Proceedings of the 17th ACM International Conference on Multimedia (MM 2009), pp. 481–490 (2009)

Requirements for Ad-hoc Geo-referenced BPM with Microblogging

Pedro Antunes[1], Gustavo Zurita[2], and Nelson Baloian[3]

[1] School of Information Management, Victoria University of Wellington
Wellington, New Zealand
pedro.antunes@vuw.ac.nz
[2] Department of Information Systems and Management, Business and Economics Faculty,
Universidad de Chile,
Diagonal Paraguay 257, Santiago, Chile
gnzurita@facea.uchile.cl
[3] Department of Computer Science, Universidad de Chile
Av. Blanco Encalada 2120, Santiago, Chile
nbaloian@dcc.uchile.cl

Abstract. There are many scenarios in which business processes will benefit from the integration of geographical information for its management. In this paper we discuss a set of requirements for ad-hoc geo-referenced Business Process Management (BPM), noting in particular the conflicts between spatial and task dependencies when coordinating activities. We suggest the predominance of spatial dependencies and propose the integration of process models in georeferencing tools. We analyse the communication needs of geo-referenced and BPM processes and suggest the adoption of microblogging platforms for coordination support. We also discuss the implementation of an ad-hoc georeferenced BPM tool, specify the microblogging messages needed to coordinate georeferenced activities, and discuss a preliminary formative evaluation of the proposed implementation.

Keywords: BPM, GIS, geo-referenced processes, ad-hoc BPM.

1 Introduction

Many businesses require integrated process and geographical information management. For instance, in New Zealand it is common to contract private rubbish pickup companies, so that every week or on a designated date a truck comes by and picks up rubbish bins. Of course these contracts are constantly being created and cancelled, which means that collection routes have to be redefined on a weekly basis. Besides, the collection routes have to be permanently optimized. Geographical Information Systems (GIS) may help defining and optimizing routes in a visual way. But a route can also be seen as an ad-hoc business process [1] comprising a set of tasks. The term ad-hoc emphasizes the volatile nature of the business process, which requires constant redefinition.

N. Baloian et al. (Eds.): CRIWG 2014, LNCS 8658, pp. 13–22, 2014.

A system integrating the functionality provided by GIS and Business Process Management (BPM) would provide support to people managing the scenario described above and many other similar scenarios such as infrastructure maintenance, processes for fire fighting, repair of street lighting, police rounds, and forest management, just to mention few. In this paper we discuss a set of requirements for ad-hoc geo-referenced BPM, noting in particular the conflict between spatial and task dependencies in coordination. We suggest the predominance of spatial dependencies and propose the integration of process models in geo-referenced tools. We analyse the communication needs of integrated GIS and ad-hoc BPM processes and also suggest the adoption of microblogging platforms for coordination support.

The paper is organized as follows. In Section 2 we discuss the requirements for geo-referenced BPM, focusing on the conflicts between spatial and task dependencies. In Section 3 we analyse the related research. Section 4 discusses the implementation of a geo-referenced BPM tool and specifies the set of microblogging messages necessary to coordinate geo-referenced activities. Finally, in Section 5 we present preliminary results from a formative evaluation action.

2   Requirements

BPM presupposes the existence of two core constituents: Process Aware Information Systems (PAIS) and process models [2]. The acronym PAIS does not refer to a particular system but to a category of systems that adopt a process view where business goals are decomposed in a discrete number of activities. This process view also introduces the notion of coordination through task-dependencies, i.e. activities are coordinated through precedence rules expressed by workflow patterns [3]. Process models specify the tasks and dependencies that are prototypical for a particular business process. They decouple process specification from enactment and allow implementing information systems based on high-level models that are easier to specify than low-level software code.

In turn, GIS presuppose two core constituents: data transformation and visualization [4]. Data transformation concerns the acquisition and representation of geographical data in the computer domain, while visualization supports spatial reasoning and decision-making. GIS induce spatial dependencies, i.e. work activities tend to be centred on geographical places or areas and coordination is implicitly related to changing the spatial focus of attention. Users do some reasoning activities related with a certain area and, when they "move" to another area, they implicitly start a different activity.

This dichotomy between task and spatial dependencies leads us to interrogate how work could be coordinated in an ad-hoc geo-referenced BPM system. Considering the rubbish collection example, a rubbish collection process can be modelled either by selecting a set of places and associating tasks to those places, or it can be modelled with a set of consecutive tasks, each one performed at a different place. Both models lead to the same result, but this example hints that the first choice is probably the best one, as tasks seem to be more contingent to spatial relationships. Effectively, we have been experimenting geo-collaborative tools in various scenarios [5-7] and have been observing that in most scenarios coordination is centred on spaces and not tasks, i.e. task dependencies are secondary to spatial dependencies. This leads to our first requirement:

R1. In an ad-hoc geo-referenced BPM system, coordination should be primarily determined by spatial dependencies and only secondarily determined by task dependencies.

This provides some flexibility as at least two different options may be considered regarding the structure of spatial data: either 1) the spatial model defines a path between different places, and therefore there is a sequence of points to traverse; or 2) there is no such path and users may select which region may be more convenient to work on. In any case we can view a business process as a collection of places or regions, each one having associated activities. GIS usually do not impose many task-dependencies and therefore we suggest that ad-hoc geo-referenced BPM should also not impose constraints on how users interact with spatial elements. This reasoning leads to the following two requirements.

R2. An ad-hoc geo-referenced BPM system should be regarded as an ad-hoc process where users determine the order of activities and places/regions provide context.

R3. Each place/region should have an associated sub-process.

R3 accommodates the combination of spatial reasoning and ad-hoc workflow, and can actually be modelled with current process modelling languages. For instance, in BPMN (Business Process Management Notation) it means having a parent ad-hoc process with several sub-processes, one for each place/region, and where the affiliated sub-processes enclose a set of activities and dependencies related to that place/region.

Having suggested that a geo-referenced process should be an ad-hoc process, we have not yet committed our judgment about the sub-processes. In our rubbish collection example, we could consider several options. One is not detailing the activities related to collecting a rubbish bag, while a more extreme approach would define a very detailed sequence of activities related with stopping the vehicle on the road, deciding about collecting a bag or bin, ringing the bell if necessary (e.g. people with special needs), using the truck lift, and moving the vehicle. This later example is certainly an exaggeration but is presented here for illustrative purposes.

The research literature identifies categories of processes that allow us discussing the issue in more detail. These include tightly, loosely, ad-hoc and unframed processes [1]. Tightly framed processes have a model specifying all details about what and how activities should be accomplished. Loosely framed processes have a model describing normal behaviour but accept deviations such as skipping and repeating activities. Ad-hoc framed processes are unique, in the sense that the model may be constantly redefined. And finally, unframed processes do not specify any model and rely on collaboration to carry out the process.

These categories present very different requirements for geo-referenced BPM. For instance, tightly framed processes require mechanisms to reuse small process pieces, named worklets [8]. Loosely framed processes do not enforce control flow, i.e. all activities are available in the users' pools [9]. Ad-hoc framed processes require determining at each step what to do next. And unframed processes require informal communication support as a minimum basis for collaboration. We suggest the following requirements to address these issues.

R4. It should be possible to associate an unframed sub-process to a place/region where informal messages would be exchanged between the participants in that sub-process.

R5. Certain types of framed sub-processes may evolve according to tasks and control flow specified in runtime.

R4 and R5 highlight different messaging needs. In certain cases semi-structured messages with control flow events and constraints have to be exchanged, while in other cases unstructured messages have to be supported. Some messages involve communication between users, while others involve communication between users and system. We assert that the required types of communication can be supported through a microblogging platform, which main characteristics concern the capacity to send and receive short messages to and from a variety of destinations using a simple addressing mechanism [10]. In our case, a critical requirement is related with geographical referencing. This is expressed in our last requirement.

R6. An ad-hoc geo-referenced BPM system can be implemented on microblogging platform provided the exchanged messages are geographically referenced.

In Section 4 we discuss an implementation based on these requirements.

3   Related Work

3.1   Social BPM

The intersection between BPM and microblogging started receiving attention very recently and has been coined Social BPM [11,12]. An example is Tweetflows [13], a lightweight platform supporting the coordination of business processes using Twitter. The platform is aimed at crowdsourcing tasks and services using an ad-hoc approach where there is no process model and control flows are determined at runtime every time an activity finishes. The authors identify a set of primitives that support activity initiation and termination (called service request and response), besides other interesting primitives like delegating, which provide unusual flexibility to process enactment. The platform uses the typical "@" symbol to identify message receivers and hashtags to identify services. One problem pointed out by the authors is a lack of privacy/security, since messages are visible by all followers. More recently, the platform has been extended to support mobile workflow [14]. It does not support geo-referenced activities.

Böhringer [15] also addressed BPM support through microblogging, focusing again on ad-hoc processes and suggesting a close relationship between this type of process and several characteristics of social platforms such as a high degree of freedom and a more proactive approach to activity selection and execution. The author argues that by definition ad-hoc processes should not be modelled since modelling one single work instance creates unnecessary burden without benefits. The author presented the general concept of a prototype using hashtags to reference complete processes and the "@" symbol to represent human and automated activities. Following the principles suggested by case management, the proposed system does not include control flow. Instead users coordinate themselves by exchanging messages about a set of activities. Using hashtags it is possible to retrieve all message exchanged about a process.

Adaptive Case Management (ACM) has been suggested as the codename for the research area specifically concerned with the adoption of ad-hoc processes [16]. An example of an ACM system integrating Social BPM is Casebook [17]. Unlike the two systems mentioned above, Casebook provides a more heavyweight approach, structuring ad-hoc activities around cases and providing specific tools for case planning, case measuring, learning, and catalogue management.

3.2   GIS and BPM

Excluding cases where workflow techniques have been used to coordinate the computation of geographical information (e.g. [18]), which are out of our scope, the research literature on the integration of BPM and GIS is very scarce. Kaster et al. [19] and Weske et al. [20] developed a GIS with integrated decision support adopting a process view, but again this approach is out of scope since it does not address business processes in general and R1-3 in particular. Walter [21] suggests some potential advantages of using both types of systems for decision-making, for instance in the area of incident management. Incident management is an application area where the combination of framed and unframed processes could be beneficial, as it often requires combining planning and improvisation. Though we did not find examples in the literature explicitly implementing R1-3.

4   Implementation

4.1   Control Flow

Van der Aalst et al. [3] suggested 20 patterns covering most BPM control flow needs. Some of these patterns are very complex and have not yet been adequately supported by current BPM systems, while others seem to be consensually implemented by most BPM systems. Considering the exploratory nature of this work, we opted to work with a minimal set of consensual patterns: Pattern 1 (Sequence), Pattern 2 (Parallel split), Pattern 3 (Synchronization), and Pattern 11 (Implicit termination).

These patterns can be implemented differently depending on the type of process framing considered (cf. Section 2). Consider for example that user U1 has completed task T1 and a sequence control flow is to be followed with T2 done by U2. Several possibilities can be discussed:

1. U1 notifies the workflow engine that T1 was completed. A worklet determines that T2 should be done by U2.
2. U1 notifies the workflow engine that T1 was completed. The workflow engine enables T2, which can be executed by U2 or any other allowed users.
3. U1 notifies a privileged agent (moderator) that T1 was completed. The agent determines that T2 should be done by U2.
4. U1 decides that T2 should be done by U2 and notifies U2.
5. U1 decides that T2 should be done next and notifies users that T2 is enabled, which U2 may offer to execute.
6. U1 notifies users that T1 was completed. Users may discuss the issue to determine that T2 should be done next by U2, or maybe U2 offers to execute T2.

Option 1 reflects the typical behaviour of a tightly framed process. Option 2 implements a loosely framed process. Options 3-5 reflect different alternatives to implement the typical behaviour of ad-hoc framed processes. And option 6 is associated with unframed processes.

Analysing the above options in more detail, we decided not to implement option 3, since centralizing control flow decisions would require too much burden from the privileged agent. Option 5 has also been discarded as too similar to the unframed strategy.

The possibilities discussed above, which consider a sequence pattern, could also be extended to the other patterns without many surprises. The exception concerns synchronization in the context of ad-hoc framed processes. The question is who synchronizes tasks in an ad-hoc context. Let us consider an example where T1 done by U1 and T2 done by U2 should be synchronized before starting T3 done by U3. Several possibilities can be considered for control flow in an ad-hoc framed process:

1. U1 and U2 notify a privileged agent that the respective tasks were completed. The agent determines that T3 should be done by U3.
2. U1 decides that T2 should be merged and notifies U2. After receiving a notification from T2, U1 decides that T3 should be done by U3 and notifies U3. Symmetrically the same decision can be made by U2.
3. The last alternative is slightly more complex as it requires considering the parallel split that originated the two parallel flows that will be synchronized. We can consider that the system may require the user specifying a parallel split should also specify the corresponding synchronization.

Table 1. Adopted control flow mechanisms. U means "user"; U+ means "one or more users"; G means "group"; and → means "transition". E.g. U → U means "passing control from one user to another".

Control flow	Ad-hoc framed process	Unframed process
Sequence	U → U	G → U
Parallel split	U → U+	G → U+
Synchronization	U → U+ - U	U+ → G

Analysing these three alternatives, we note that having a privileged agent making control flow decisions is against the "spirit" of a pure ad-hoc approach. We can make the same comment about the last alternative, which requires users to pre-determine synchronizations when defining splits. Such an approach violates the principle that an ad-hoc framed process can be redefined as the process enfolds. We therefore decided to only implement the option 2 mentioned above. Table 1 summarizes the adopted control flow mechanisms.

4.2   Messaging for Ad-hoc and Unframed Process

As previously discussed, informal communication and control flow require informal and semi-formal messaging, respectively. We adopted Twitter to implement messaging. We use the typical "#hashtag" to refer to specific business process instances, and the also typical "@user" to refer to participants, including the workflow engine. The "%number" tag was adopted to refer to van der Aalst's workflow patterns, e.g. %1 refers to sequence and %2 to parallel split. The message specification is provided in Table 2 using regular expressions. Note that geographical locations are specified with Twitter's location application interface and therefore do not appear in the message.

Table 2. Twitter messages necessary to implement the ad-hoc geo-referenced BPM tool

```
twitter message = code? comment          termination = %11 process flow?
code = sequence | and-split | and-join-start |   process = #\w
      and-join-end | termination          flow = \^\w
sequence = %1 process agent?              agent = @\w
and-split = %2 process flow agent         comment = .*
and-join-start = %3a process flow
and-join-end = %3b process flow agent
```

One interesting characteristic of the adopted messaging scheme is that we do not explicitly refer to tasks but to users. A task is implicitly defined by sending a message to a user. The second characteristic is that parallel flows initiated by splits are explicitly named using the "^flow" tag. Let us illustrate how the messaging scheme works with the fire-fighting example shown in Table 4.

Table 3. Exchanged messages for a fire fighting ad-hoc process. Ln means a "n" geo-referenced location, Lnm is a "m" geo-referenced location/region close to "n". Ln(^flow) is a "n" geo-referenced location of a flow "flow". Ln(msge) is a geo-referenced location associated to a message.

	Sent from	Message	Locations
1	FFOfficeC1	%1 #Ruta68 @FFC1Captain Auto driver reports fire on route 68, 10 km east from Valparaiso	L1(msge)
2	FFC1Captain	#Ruta68 On our way	L1(msge)
3	FFC1Captain	#Ruta68 Arrived to fire site	L1(msge)
4	FFC1Captain	%2 #Ruta68 ^FireCloseRoad @Firefighter1C1 Attack fire close to the road	L1.1(^FireCloseRoad)
5	FFC1Captain	%2 #Ruta68 ^BuildFirewall @Firefighter2C1 Build firewall between road and fire	L1.2(^BuildFirewall)

Table 3. (*continued*)

6	FFOfficeC2	%1 #Ruta68 @FFC2Captain House owner reports trees are catching fire near house before arriving to Valparaiso near route 68	L2(msge)
7	FFC2Captain	#Ruta68 We are on our way	L2(msge)
8	FFC2Captain	#Ruta68 Arrived to fire site	L2(msge)
8	FFC2Captain	%2 #Ruta68 ^FireOnTrees @Firefighter1C2 Extinguish fire from trees	L2.1(^FireOnTrees)
9	FFC2Captain	%2 #Ruta68 ^BuildFirewall2 @Firefighter2C2 Build firewall between trees and house	L1.2(^BuildFirewall2)
10	Firefighter1C1	%1 #Ruta68 @FFC1Captain ^FireCloseRoad Fire extinguished	L1.1(^FireCloseRoad)
11	FFC1Captain	%3a #Ruta68 ^FireCloseRoad Stand by	L1.1(^FireCloseRoad)
12	Firefighter1C1	%3b #Ruta68 ^FireCloseRoad @FFC1Captain OK	L1.1(^FireCloseRoad)
13	Firefighter2C1	%1 #Ruta68 @FFC1Captain ^BuildFirewall Captain, firewall done	L1.2(^BuildFirewall)
14	FFC1Captain	%3a #Ruta68 ^BuildFirewall Stand by	L1.2(^BuildFirewall)
15	Firefighter2C1	%3b #Ruta68 ^BuildFirewall @FFC1Captain OK	L1.2(^BuildFirewall)
16	Firefighter1C2	%1 #Ruta68 @FFC2Captain Need help fighting fires on trees	L3(msge)
17	FFC2Captain	%1 #Ruta68 @FFC1Captain Need help fighting fires on trees	L3(msge)
18	FFC1Captain	%1 #Ruta68 @Firefighter1C1 Help fire on trees	L3(msge)
19	FFC1Captain	%1 #Ruta68 @Firefighter2C1 Help fire on trees	L3(msge)
20	Firefighter1C2	%1 #Ruta68 @FFC2Captain Fire on trees extinguished	L3(msge)
21	Firefighter2C2	%1 #Ruta68 @FFC2Captain Firewall done	L1.2(^BuildFirewall2)
22	FFC2Captain	#Ruta68 Going to check people in house	L2(msge)
23	FFC2Captain	#Ruta68 Everyone ok, returning to headquarters	L2(msge)
24	FFC1Captain	#Ruta68 Ok, returning to headquarters	L1(msge)
25	FFOfficeC1	%11 #Ruta68 Action completed	L1(msge)

In this example, fire appears near the road and two companies are called to extinguish the fire. Tasks are assigned to fire fighters as described in Table 3. In this case, locations are associated to flows or places where sub-processes are performed (see column "Locations" in Table 3). They represent the spatial dependencies of tasks. Messages 1-3, 6-8, 16-20 and 22-24 require the association of the location information. For example, in message 1, it is necessary to inform where the fire is taking place; in message 6, the exact place where a fire is notified should be indicated. The messages that have no associated task or that do not require changing locations are associated to the locations mentioned in the previous messages (e.g., messages 2 and 3 take place in the same location as message 1; messages 7 and 8 take place in the same location as message 6). If a change of location is needed, the new message has to explicitly reference the new location context (e.g., message 6 changes the location context to L2). When tasks need to be geo-referenced, they should be associated to a location in the map. In messages 4 and 5 of our example, tasks ^FireCloseRoad and ^BuildFirewall take place in locations L1.1 and L1.2, respectively, near location L1 were fire was reported for the first time. In this way fire fighters know exactly where they have to move to perform the task.

5   Preliminary Evaluation and Discussion

A scenario based evaluation technique was adopted to assess the viability of the fire-fighting example described in Section 4. Running the scenario with several users allowed understanding the system's perceived usability and utility, the communication and coordination problems users would have using the system, and the design options that should be further explored at the next development stages. We evaluated the fire-fighting scenario with four users. The users walked through the messages detailed in Table 3 considering a realistic fire-fighting setting. We then conducted post-session interviews with the users, focussing especially on communication and coordination breakdowns.

The formative evaluation allowed a better understanding of various design issues: a) It is better using a minimal set of workflow patterns, to keep the system simple and understandable by people unfamiliar with BPM. In particular, the users referred that the approach was simple enough, while having a high potential for describing ad-hoc activities; b) The association of messages to locations may not be required when tasks take place in the same location, but are very useful when tasks are distributed; c) There are multiple ways to specify the coordination of the fire-fighting process. Some specifications use simple, informal messages, like in a normal conversation. Though the users agreed that often some explicit coordination is needed; d) The messaging mechanism affords users with the desired level of detail necessary to accomplish different tasks. Though users suggested that each message could be expanded to allow a deeper level of detail, especially regarding the task description.

References

1. van der Aalst, W.: Business Process Management: A Comprehensive Survey. ISRN Software Engineering (2013)
2. Weber, B., Reichert, M., Rinderle, S.: Change patterns and change support features – Enhancing flexibility in process-aware information systems. Data & Knowledge Engineering 66(3), 438–466 (2008)
3. Van der Aalst, W., Hofstede, A., Kiepuszewski, B.: Workflow Patterns. Distributed and Parallel Databases 14, 5–51 (2003)
4. Goodchild, M.: Twenty years of progress: GIScience in 2010. Journal of Spatial Information Science 1, 3–20 (2010)
5. Antunes, P., Sapateiro, C., Zurita, G., Baloian, N.: Integrating Spatial Data and Decision Models in an E-Planning Tool. In: Kolfschoten, G., Herrmann, T., Lukosch, S. (eds.) CRIWG 2010. LNCS, vol. 6257, pp. 97–112. Springer, Heidelberg (2010)
6. Antunes, P., Zurita, G., Baloian, N., Sapateiro, C.: Integrating Decision-Making Support in Geocollaboration Tools. Group Decision and Negotiation 23(2), 211–233 (2014)

7. Antunes, P., Zurita, G., Baloian, N.: Key Indicators for Assessing the Design of Geocollaborative Applications. International Journal of Information Technology & Decision Making 13(2), 361–385 (2014)
8. Adams, M., ter Hofstede, A.H.M., Edmond, D., van der Aalst, W.M.P.: Worklets: A service-oriented implementation of dynamic flexibility in workflows. In: Meersman, R., Tari, Z. (eds.) OTM 2006. LNCS, vol. 4275, pp. 291–308. Springer, Heidelberg (2006)
9. van der Aalst, W., Weske, M., Grunbauer, D.: Case handling: A new paradigm for business process support. Data & Knowledge Engineering 53(2), 129–162 (2005)
10. Honey, C., Herring, S.: Beyond microblogging: Conversation and collaboration via Twitter. In: 42nd Hawaii International Conference on System Sciences, Hawaii, pp. 1–10. IEEE (2009)
11. Brambilla, M., Fraternali, P., Vaca, C.: BPMN and design patterns for engineering social BPM solutions. In: Daniel, F., Barkaoui, K., Dustdar, S. (eds.) BPM Workshops 2011, Part I. LNBIP, vol. 99, pp. 219–230. Springer, Heidelberg (2012)
12. Erol, S., Granitzer, M., Happ, S., Jantunen, S., Jennings, B., Johannesson, P., Schmidt, R.: Combining BPM and social software: contradiction or chance? Journal of Software Maintenance and Evolution: Research and Practice 22(6-7), 449–476 (2010)
13. Treiber, M., Schall, D., Dustdar, S., Scherling, C.: Tweetflows: flexible workflows with twitter. In: Proceedings of the 3rd International Workshop on Principles of Engineering Service-Oriented Systems, pp. 1–7. ACM (2011)
14. Treiber, M., Schall, D., Dustdar, S., Scherling, C.: Creating mobile ad hoc workflows with Twitter. In: Proceedings of the 27th Annual ACM Symposium on Applied Computing, pp. 1998–2000. ACM (2000)
15. Böhringer, M.: Emergent case management for ad-hoc processes: A solution based on microblogging and activity streams. In: Muehlen, M.z., Su, J. (eds.) BPM 2010 Workshops. LNBIP, vol. 66, pp. 384–395. Springer, Heidelberg (2011)
16. Motahari-Nezhad, H., Swenson, K.: Adaptive Case Management: Overview and Research Challenges. In: IEEE 15th Conference on Business Informatics (CBI), pp. 264–269. IEEE (2013)
17. Motahari-Nezhad, H., Spence, S., Bartolini, C., Graupner, S., Bess, C., Hickey, M., Rahmouni, M.: Casebook: A Cloud-Based System of Engagement for Case Management. IEEE Internet Computing 17(5) (2013)
18. Chen, Q., Wang, L., Shang, Z.: MRGIS: A MapReduce-Enabled high performance workflow system for GIS. In: IEEE Fourth International Conference on eScience, pp. 646–651. IEEE (2008)
19. Kaster, D., Medeiros, C., Rocha, H.: Supporting modeling and problem solving from precedent experiences: the role of workflows and case-based reasoning. Environmental Modelling & Software 20(6), 689–704 (2005)
20. Weske, M., Vossen, G., Medeiros, C., Pires, F.: Workflow management in geoprocessing applications. In: Proceedings of the 6th ACM International Symposium on Advances in Geographic Information Systems, pp. 88–93. ACM (1998)
21. Walter, M.: Situational Awareness for Enhanced Incident Management (SAFE-IM). In: Military Communications Conference, pp. 1–6. IEEE (2007)

Construction and Evaluation
of a Collaboration Observation Model

Paula Ballard da F. Gentil, Maria Luiza M. Campos, and Marcos R.S. Borges

Graduate Program in Informatics, Federal University of Rio de Janeiro, Rio de Janeiro, Brazil
{paula.fonseca,mluiza,mborges}@ppgi.ufrj.br

Abstract. This article proposes a Collaboration Observation Model (COM) to allow a systematic observation of aspects involved in collaboration, in order to facilitate coordination of observation activities in the group as well as to improve collaboration, interaction and communication among participants. Effective collaboration is a key factor of success in teamwork activities to achieve shared goals, because it can improve the quality of interactions and reduce tasks execution time. To evaluate the proposed model, a real experiment was conducted to verify its adequacy to collaborative aspects and to refine the model.

Keywords: collaboration, observation model, conceptual model.

1 Introduction

Collaboration is an important factor for organizations to achieve their goals of productivity, quality and knowledge sharing [16]. To work collaboratively, individuals negotiate and exchange information relevant to achieve common goals.

A conceptual model is a high level description of a problem domain, contemplating some aspects of the physical and social world around us for purposes of understanding and communication [17]. It represents an application domain through the perception of the users and the development team [12]. It is essential that conceptual models have a good level of expressivity so that they can capture as much as possible the intended meaning of Universe of Discourse [1]. They should also facilitate communication between the stakeholders in a project, regardless platform of hardware and software.

This article proposes a Collaboration Observation Model, using the concepts of the collaboration domain ontology (CONTO) proposed by Oliveira [18]. The CONTO ontology is based on the Unified Foundational Ontology (UFO) [9, 10] and on the 3C Model (Communication, Coordination and Cooperation) [6]. The Collaboration Observation Model allows a more expressive and accurate representation of the model elements, their relations and rules, to better support applications and other initiatives using this model.

---

[1] Represents all the knowledge under consideration in a particular domain studied.

N. Baloian et al. (Eds.): CRIWG 2014, LNCS 8658, pp. 23–37, 2014.
© Springer International Publishing Switzerland 2014

The observation of people work activities in the organizations is a useful way to understand the interactions, practical skills and tacit knowledge developed by the work teams [22]. When we observe how people carry out their activities, it is possible to understand their reasons and motivations, as well as the problems or difficulties related with the execution of tasks. An experiment was performed and observation results were evaluated to verify the effectiveness of the proposed Conceptual Model and to enable the refinement of the model.

In the next section the related work is presented. In section 3, the main elements of CONTO ontology and UFO are explained. The process used for the construction of the Conceptual Model and the Collaboration Observational Model are detailed in 4. Section 5 shows the real experiment conducted and the evaluation of the Conceptual Model. The conclusion and future work are in section 6.

2   The Need for an Observation Model

Observation is a means of discovering actual working knowledge and practices: how people interact with systems and technologies; how they cooperate with each other; how they use cognition to deal with complex or unanticipated situations, detecting and solving problems as they happen. Moreover, subsequent communication problems may be reduced, since observation allows better knowledge of a situation.

Anguera [1] proposes a classification used to categorize observation techniques, considering different aspects:

- The means: unstructured (unsystematic) or structured observation (systematic). An observation is unstructured when the researcher collects and records facts of reality, in an informal way, without using special technical means. An observation is structured when one needs a structured description of a task or checks hypotheses for the causes of certain phenomena. In systematic observation the observer knows which aspects of group activity are important to the research objective and creates a specific script before starting the observation.

- The participation of the observer: non-participant or participant observation. In non-participant observation the observer doesn't integrate the observed group, watching developed tasks without participating of them. In participant observation, the observer participates of the observed group and performs roles within it.

- The number of observers: individual or team, according to those who are involved in the observation.

- The place where it is performed: real settings (fieldwork) or performed in the laboratory. Usually the observations are made in real environment, where events can be observed as they actually happen. However, the observation also can be made in laboratory.

The degree of structuring and participation varies according to the goal of the observation. In exploratory studies, observations are usually unstructured and the observer tends to participate in the observed group.

Another approach that should be taken into consideration is the Ethnography. This is a methodology derived from Social Anthropology that provides detailed descriptions of human activity, along with its behavior, social interactions, techniques and cultural practices. It results from an observer point of view, participant or not, situated in the environment for prolonged periods [13]. Ethnographic methods are completely or partially based on participant's observations [2].

Observation data are typically structured and are usually described in an ad-hoc way. The task of working with models for this domain is hard, because it is necessary to deal with a large amount of data. Besides the discovery, organization and integration of collected data are very difficult. As a consequence, the definition of schemas for observation data sets is essential to represent them with an expressive and effective semantics [3].

The need to define a mechanism for describing the observation data can be found in some studies that proposed models of observational data [4] [24] and ontologies [14, 15] [25]. However, these approaches for modeling observational data semantic are not generic and extensible. They provided specific domain vocabularies for describing data models intended for storage of certain types of observation data.

The Open Geospatial Consortium (OGC) has an International Standard of Observations and Measurements [4] which defines a conceptual schema for observations and for features involved in sampling when making observations. The standard was developed in the context of geographic information systems and can be extended for other domains. An Observation Data Model (ODM) was presented in [24]. The observations data model was built to store hydrologic observations and information about the data values allowing them to be unambiguously interpreted and used.

In [15] an extensible and reusable ontology for solar-terrestrial physics was designed. The structure of the ontology supports reuse in multiple virtual observatory projects. An overview of some ontologies developed for an ecological project, called SEEK ("Science Environment for Ecological Knowledge"), was presented in [25]. The ontologies address two broad areas, scientific observations and ecological and environmental science. In [14] a formal ontology for capturing the semantics of generic scientific observation and measurement was presented. The ontology provides a basis for adding detailed semantic annotations to scientific data, improving the meaning of observational data.

A framework for conceptual modeling was presented in [3], to capture the semantics of observational data and describe them in a flexible and effective way. The approach extends previously works in conceptual modeling adding new constructs to describe the observations, measurements and context that are suitable for data annotation.

A research was conducted in [23] to observe and understand what people do when they perform an activity. The work activity of small groups was videotaped and analyzed in order to understand collaborative work and to guide the design, development and introduction of tools to support that activity. The work has not developed a Conceptual Model.

In [11] and [20], observation was used as an evaluation method for shared-workspace groupware, to identify collaboration. In [11] a conceptual framework was defined to articulate the mechanisms of collaboration for shared-workspace groupware, including low-cost evaluation methods, such as heuristic evaluation, guidance, user observations and questionnaires, to help collaborative activities. Observational User Testing was performed so that the evaluator applied a set of criteria, such as explicit communication, coordination, planning, monitoring, assistance, and protection. The application of this method had several issues because it is harder to schedule and predict the expected group interaction. In [20] a modeling technique called Collaboration Usability Analysis (CUA) was developed to represent the variability inherent in group work, considering collaboration issues and teamwork aspects of shared activities to help task analysis for groupware. CUA was based in mechanisms to represent collaborative interactions, collected from previous research into shared workspace collaboration [5] [23], such as explicit communication, information gathering and management of shared access.

This work is different from previous ones because it studies the observation of collaborative aspects. The transfer of such knowledge to a computational model is highly desirable, providing the acquisition of knowledge involved in collaborative activity and the ability to incorporate many features and skills of experts.

3   Collaboration Ontology and Unified Foundational Ontology

The Collaboration Domain Ontology (CONTO) proposed by Oliveira [18] was developed using the SABIO (Systematic Approach for Building Ontologies) methodology [7] and the modeling language proposed by Guizzardi [9], using the Unified Foundational Ontology (UFO).

UFO is based on theories from Formal Ontology, Philosophical Logics, Philosophy of Language, Linguistics and Cognitive Psychology [9, 10]. It consists of three complementary parts: UFO-A is an ontology of enduring individuals (endurants) and is the heart of the UFO; UFO-B is an ontology of events (perdurants); and UFO-C is an ontology of social entities.

The semantic distinctions added by UFO significantly contributed to the model construction, helping to define model elements that may explain some relevant ontological distinctions already identified by UFO, reducing the ambiguity in its interpretation and increasing expressiveness and consistency of the concepts associated with the represented domain.

CONTO was developed to formalize knowledge about the collaboration domain in order to provide a common conceptualization and vocabulary to improve the reuse and sharing through applications and groups [8] [18].

Besides UFO's support, the structure defined by the 3C Model (Communication, Coordination and Cooperation) was also used in CONTO [6]. In Cooperation, a Collaboration Session is an event that is composed of the actions of its participants. These actions are instantaneous events (atomic events) and they are named Participations which are performed by Participants. A Collaboration Session has one or more objectives, which can depend on other goals and each one have a priority level according to its importance for that Collaboration Session.

According to CONTO, a Communication Action is composed of two Participations, executed by Agents. Each participation event may have an associated Message that represents the exchanged information. A Message is expressed through a Language. It also uses one Communication Media that is the instrument used to carry out communication.

In order to help the Coordination process [19], Collaborative groups are formed by Agents and tasks defined by Collaborative Agreements. These groups recognize and follow rules known as a Protocol, allowing collaboration between them to happen. This Protocol is also responsible for rules that define the Collaborative Roles performed by agents.

Further details of CONTO will be presented on Section 4.2. CONTO ontology elements, used in the Collaboration Observational Model, are represented in Figures 2, 3 and 4.

4   Collaboration Observation Model

4.1   Conceptual Model Construction Process

To define a generic Collaboration Observation Model, a process was used with necessary steps to ensure adequacy and logical consistency to related concepts. This section briefly explains the conceptual modeling process defined for the construction of the Collaboration Observational Model (Figure 1). Our aim is not to introduce a comprehensive process definition, but to explain the steps used to develop and evaluate the Conceptual Model.

Initially, the Conceptual Model was derived from the elements of CONTO Ontology [18] and complemented by activities of ontological analysis based on the Unified Foundational Ontology (UFO).

In order to evaluate the Conceptual Model, a real experiment was planned and executed. Further details will be explained in the next sections. The model is being refined iteratively. Conceptual modeling is not a one-shot process, but one that is repeated and refined a number of times during its construction [12].

Thus, as shown in Figure 1, the Conceptual Model is applied in the planning stage of the real experiment. The results are evaluated to verify if the model also contemplates fundamentals collaboration aspects like communication, coordination and cooperation. After evaluating, when the results are not satisfactory, it is necessary to perform a refinement to adjust it. If considered satisfactory, then the model has reached stability and can be used.

The Collaboration Observation Model can also be used to build a simulation model, which will produce a log file that could, ultimately, be used to help the development of learning agents [21]. However, the description of the simulation model is out of the scope of this article.

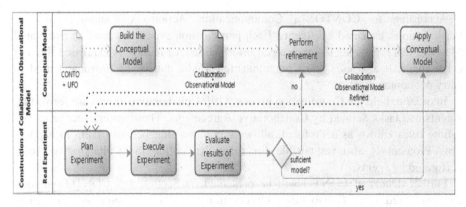

Fig. 1. Construction Process of the Collaboration Observational Model

4.2   Overview of the Collaboration Observation Model

Analogous to CONTO, the Collaboration Observation Model was built following the structure of the 3C Collaboration Model and an ontological analysis based on UFO.

CONTO ontology emphasizes communication to achieve cooperation, supporting the interactions between stakeholders and generating commitments that are managed by the coordination during the process of communication and cooperation. Coordination is responsible for organizing the cooperation processes. This paper will explain only the main concepts from CONTO used in this work. Further details about the CONTO can be found in [18].

4.2.1   Communication Package

As in CONTO, Agents participate in a Collaborative Session through intentional acts of contributions (Action Contribution) which may be material (Material Contribution) or communicative (Communicative Act). Communicative Acts have a propositional content or message (Message). A communicative interaction (Communicative Interaction) is composed of a communicative act (Communicative Act) and a perception (Perception) of this communicative act by different agents involved (Sender/ Receiver).

In Communication Package, the *Participant is* a person that takes part in Collaborative Agreements.

An *Observation Session* is a Collaborative Session and represents the event in which participants interact with the purpose of collaboration.

An *Observation Announcement* represents the message that needs to be exchanged among participants in the Collaborative Session. The message has a form of communication, which can be textual, verbal, using signals or body language. Besides, it can also use a particular mode of transmission developed by involved agents. The Observation Announcement can be an *Internal Announcement*, when the message is exchanged between agents, internally in the Collaborative Session; or

External Announcement, when the message is exchanged between agents, when the participant that sent the message is a person outside of Collaborative Session

As an example, it is possible to imagine the communication between two or more team members (Participant), in which messages (Internal Announcement/ External Announcement) can be captured and exchanged during the execution of activities. A message can result in the execution of new activities.

Figure 2 displays the Communication Package of the Conceptual Model. Concepts in gray come from UFO and CONTO.

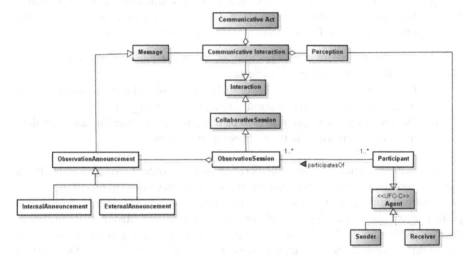

Fig. 2. Communication Package

4.2.2   Coordination Package

As in CONTO, Collaborative Groups are formed by Participants and are defined by Collaborative Agreements, which represent conditions for defining roles. These groups recognize rules known as Protocol, allowing that collaboration between them to happen. This Protocol, derived from CONTO, establishes rules, recognized by the group, that define Collaborative Roles played by Agents.

A Collaborative Group specializes in *Observation Group*, which observes the activities in a Collaborative Session, and *Operation Team*, which performs the activities in Collaborative Session, being observed according to a perspective. The *Observation Group* represents the Collaborative Group defined to observe the Collaborative Session, and consists of the set of Participants, defined by Collaborative Agreements, which have the roles of Observers. This group can be remote (*Observation Group Remote*) or local (*Observation Group Local*). The *Operation Team* represents the collaborative group defined to perform the actions in a Collaborative Session.

Collaborative Roles, which are derived from CONTO, represent the roles that each participant must perform in a group. They are featured by Closed Commitments to achieve certain goals by performing specific actions. A Closed Commitment is a concept from UFO-C and it is a commitment (intention) done by an agent. By taking a Collaborative Role, an Agent adopts the prescribed objectives for that role. *Observation Roles* represent the roles that participants must perform in a collaborative group, temporarily or permanently, which are observed.

The organization of a group involves the definition of roles assumed by its participants. Each role is associated with a set of responsibilities. The participant can play a role in a Collaborative Session, performing a task on which the amount of required energy determines his capacity. The participant also has a context that can change during the execution of tasks, indicating his situation in a Collaborative Session. If necessary, a participant, due to an unexpected fact, may need to change the execution of his/her tasks. The following roles were defined in the model:

- *Coordinator*: Person whose role is to be responsible for coordinating activities in the Collaborative Session.
- *Observed*: Person whose role is to be responsible for performing actions within the Collaborative Session, aiming to achieve a goal.
- *Observer*: Person whose role is to be responsible for observing and recording the actions performed in the Collaborative Session.

Coordination is critical in any collaboration work, in order to avoid double efforts and to ensure the integration between the individual parties involved in a group.

As an example, it is possible to observe team members (Operation Team) of a project in charge of developing software, such as: Project Manager, Requirement Analyst, Software Architect and Developer, assuming various roles (Observation Roles) for performing the relevant activities. The group can be observed (Observation Group) to verify the progress of team activities.

Figure 3 displays the **Coordination** Package of Conceptual Model.

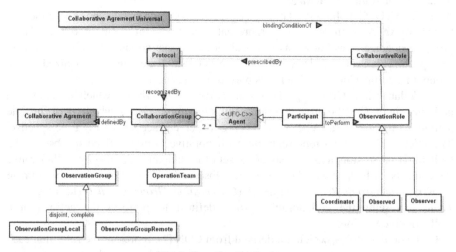

Fig. 3. Coordination Package

4.2.3   Cooperation Package

The observation activity contemplates perspectives which represent aspects to be observed, according to particular points of view, following agreed rules. The observation is associated with a *Data Log* file that records the activities being performed in the Collaborative Session, according to the perspective of observation.

In an Observation Session, the *Observation Operation* is the Collaborative Operation which is performed by the team of observers. In this session, there is always an *Operation Occurrence*, as this is the occurrence of the operation to be observed which was caused by an *Observation Operation*.

The occurrence of an action in a Collaborative Session is defined as *Activity Occurrence*. It can be a *Complex Activity Occurrence*, when may be composed of one or more atomic activities, or *Atomic Activity Occurrence*, when cannot be decomposed. The Atomic Activity has an *Observation Activity*, which represents the *Observation Activity* that occurs in a Collaborative Session. The *Observation Activity* occurs in a *Step* that defines the time interval in which each activity, and it also has an *Observation Situation*, that characterizes the occurrence of an activity (pre-state/ pos-state).

In *Observation Activity*, there is the *Human Resource Allocation*, which represents the allocation of human resources for the execution of a particular activity. A *Human Resource* can perform the activities of the Collaborative Session, using a *Material Resource*.

A *Work Plan* defines the planning of group activities and the work strategy. The complex activities are shared into smaller activities, which are allocated to different participants

An *Event* modifies the Collaborative Session, causing the execution of new activities. An event can change an operation occurrence and can be: expected, unexpected or impeditive. *Expected Events* correspond to situations that were previously planned. *Unexpected Events* are situations that were not planned and that must be solved as quickly as possible by the involved group. They have a complexity which determines the time of execution. *Impeditive Events* are those on which the execution of activities must be interrupted. Any of these events can initiate other actions.

As an example, it is possible to imagine the activities (Atomic Activity Occurrence) performed for the achievement of a goal, according to some planning (Work Plan), such as analysis of a use case, while working in requirements elicitation. For this activity, two or more professionals (Human Resource), which use/generate artifacts (Resource Material), can be allocated (Human Resource Allocation).

During the execution of activities, planned events (Expected Events), such as the lack of a business user or the initial increase in complexity provided, which can alter the performance of activities. Such events may modify time and costs relative to the activities. All these activities (Observation Activities) can be observed and recorded (DataLog) for later use.

Figures 4 and 5 display the **Cooperation** part of the Conceptual Model.

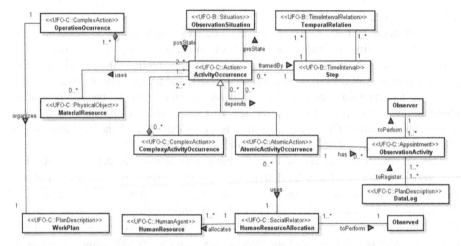

Fig. 4. Cooperation Package

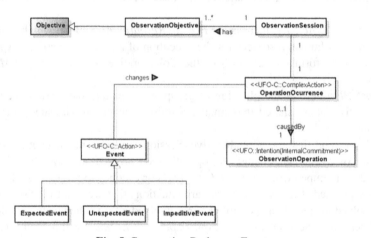

Fig. 5. Cooperation Package – Events

5   Real Experiment

To evaluate the Conceptual Model of observation, an experiment was performed with a real scenario, so that we could ensure that the developed model covers all observed aspects of collaboration. The experiment phases consisted of preparation, execution and evaluation, as detailed below.

5.1   Preparation

To conduct the real experiment, the scenario of "Bucket Brigade" was used, as explained in Figure 6. This scenario, already known in literature, was chosen because despite being simple, it allowed to observe the activities of communication, collaboration and coordination performed by the group, as well as, the observation and analysis of situations expected.

The Bucket Brigade is a method of carrying buckets of water from a source of low pressure water to a location where it is needed and the water source is not available. In the past, this method was used in fires when there was no way to transport water in pipes or trucks.

The person at the source of water fills the bucket with water and passes to the next person in line. The bucket is then transferred from person to person, until the last person in line empty the bucket. The empty buckets can be transported back to the water source.

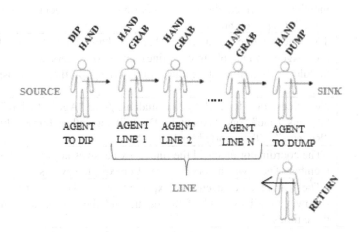

Fig. 6. Bucket Brigade

5.2   Execution

The experiment was conducted with twelve students of the Knowledge Engineering Group of Federal University of Rio de Janeiro. In this experiment, five attempts were performed, each one lasting 4 minutes, with twelve buckets without water.

Before the experiment a meeting was performed with the aim of explaining the purpose of the work and then the group members began to interact with each other to define the respective roles to be played by the team to achieve the goals.

Features of Collaboration observed during the experiment, according Table 1:

• In Communication

- The participants talked to each other to exchange experiences, during the execution of activities, and also to change roles in activities.

- Messages exchanged by the group, through spoken language and gestures, helped the group to perform activities and to keep concurrent access to the bucket (resource material), contributing to the maintenance of effective communication and collaboration.

Table 1. Result of Experiment

	Run	Description
Strategy/ Transported buckets	1^0	One coordinator, one "Agent to Dip", one "Agent to Dump", eight "Agent Line", and one "Runner", organized in a single line. Transported buckets: 86
	2^0	One coordinator, five "Agent Line" (organized in a single line), two "Agent to Dip", two "Agent to Dump" and two "Runners". Transported buckets: 111
	3^0	Two lines with five students (one "Agent to Dip", three "Agents Line", one "Agent to Dump", in each line), and two "Runners". A member performed the role of coordinator and agent at the same time. Transported buckets: 153
	4^0	Two lines of five agents (one "Agent to Dip", three "Agents Line", one "Agent to Dump", in each line), and two runners arranged in the middle of two lines to reduce the moving of them. Transported buckets: 140
	5^0	Two lines: one formed by six students (one "Agent to Dip", four "Agents Line", one "Agent to Dump") and other formed by four "Runners". Transported buckets: 110
Characteristics	1^0	- The coordinator changed the runner twice, so as not to tire this team member. There was no occurrence of unexpected events.
	2^0	- There was the occurrence of expected events: the bucket fell twice, following until the end of the queue and the runner was changed three times.
	3^0	- No runner was changed. As the Coordinator had to accumulate functions, he failed to identify problems that occurred in this execution. - The agents, with role of dipping the buckets, sometimes, were temporarily without them.
	4^0	- The students that were dipping and dumping the buckets had to be replaced.
	5^0	- The purpose of this strategy was to improve the return of the buckets to source, reducing the runner's displacement. - This strategy wasn't effective, because it was even slower than the previous one.

- In Coordination

- Each participant performed a role within the group, in which he was responsible for the implementation of some activities, in order to achieve an objective of the group. The observed group was shared in the following roles: "Coordinator", who could intervene in group activities; "Agent to Dip", who dipped buckets in water source; "Agent to Dump", who dumped the buckets; "Agent Line", who were in line to hand and to grab the bucket; and "Runner", who returned the empty buckets to water source.

- The observer group didn't interact with the group observed.
- The Coordinator determined rules to be followed by the group and performed the exchange of roles and strategies, during the experiment.

• In Cooperation

- In each run, the observed group had as objective to define an efficient strategy (work plan), to transport a larger amount of buckets in a fixed time.
- Each student (Human Resource) was allocated in an activity with the bucket (Material Resource).
- There were occurrences of expected events, such as: the participants leaving the buckets fall and also, sometimes, the bucket being inclined in a specific angle that, in a real experiment, would result in loss of some amount of water; new participants being included in the team to perform a role; students playing specific roles; participants needing to be replaced; and a group member being removed.
- The results obtained by the observer group, such as characteristics, strategies, tasks and events, were recorded in notes (Data Log).

5.3   Evaluation of Collaboration Observation Model

After running the experiment, an analysis was performed to verify that all aspects of collaboration were covered by the model, as detailed below.

• Communication analysis

The experiment indicated that each participant of the observed group must not only perform his/her specific tasks, but also must have awareness of the other participants for an effective collaboration.

• Coordination analysis

The coordinator of observed team has a key role in the performance of activities. Thus, he needs to work with some aspects observed during the execution of activities, such as the psychological part of participants or even your tiredness during the tasks, which is an expected factor and part of participant's characteristics. However, psychological part isn't expected in Conceptual Model. This factor can contribute to the execution of tasks, because when the coordinator is aware of this information before beginning tasks, he can distribute participants more appropriately.

• Cooperation analysis

It is worth noting that the occurrence of expected events can cause different kinds of actions in response. For example, if the bucket falls to the ground, it is possible to bring the bucket back to the beginning to be filled again or carry it to the end of the line, even empty. This was observed and can be attended by the Conceptual Model.

6   Conclusion and Future Work

The information model captures collaboration aspects of communication, coordination and cooperation, which can be valuable to perform works in collaboration area. The proposed Conceptual Model can be used as the basis for integrating software tools that support collaborating. The model allows observing, studying and organizing collaborative data systematically. Observing collaborative work can lead to a better understanding of a collaborative activity and can guide the design, development and introduction of new tools to support it.

In the near future, we plan to conduct another Bucket Brigade experiment, using water, in order to further verify whether the collaboration is well represented in Conceptual Model and all scheduled and unscheduled events can be observed as defined.

We also intend to use the Collaboration Observation Model as a basis to build a Simulation Model, which will be based too in models of bucket brigade. A simulator is being built to perform experiments in a simulated environment. Thus it will be possible to assess whether the constructed model is also suitable for the simulation model built.

In a separate work we will be using the model to support planning and adaptation of the plan on-the-fly during execution. We want to examine how easily a plan can adapt to changes in the context and in unpredicted situations. This situation was observed during the experiment and has motivated this separate but related research.

References

1. Anguera, M.T.: Observational Typology. Quality & Quantity. European-American Journal of Methodology 13(6), 449–484 (1979)
2. Atkinson, P., Hammersley, M.: Ethnography and participant observation. In: Denzin, N., Lincoln, Y. (eds.) Handbook of Qualitative Research, pp. 248–261. Sage, Thousand Oaks (1994)
3. Bowers, S., Madin, J.S., Schildhauer, M.P.: A conceptual modeling framework for expressing observational data semantics Conceptual Modeling. In: Li, Q., Spaccapietra, S., Yu, E., Olivé, A. (eds.) ER 2008. LNCS, vol. 5231, pp. 41–54. Springer, Heidelberg (2008)
4. Cox, S.: Observations and measurements. Technical Report 05-087r4, OGC (2006)
5. Clark, H.: Using Language. Cambridge University Press, Cambridge (1996)
6. Ellis, C.A., Gibbs, S.J., Rein, G.L.: Groupware - Some Issues and Experiences. Communications of the ACM 34(1), 38–58 (1991)
7. Falbo, R.A.: Experiences in Using a Method for Building Domain Ontologies. In: Proc. of International Workshop on Ontology In Action, Banff, Alberta, Canada (2004)
8. Gómez-Pérez, A., Fernández-López, M., Corcho-Garcia, O.: Ontological Engineering – with examples from the áreas of knowledge management, e- commerce and the semantic web, pp. 107–153. Springer (2003)
9. Guizzardi, G.: Ontological Foundations for Structural Conceptual Models. PhDthesis, University of Twente. Universal Press, The Netherland (2005)

10. Guizzardi, G., Falbo, R.A., Guizzardi, R.S.S.: A importância de Ontologias de Fundamentação para a Engenharia de Ontologias de Domínio: o caso do domínio de Processos de Software. IEEE Latin America Transactions 6(3), 244–251 (2008)

11. Gutwin, C., Greenberg, S.: The Mechanics of Collaboration: Developing Low Cost Usability Evaluation Methods for Shared Workspaces. In: IEEE 9th Int'l Workshop on Enabling Technologies: Infrastructure for Collaborative Enterprises (WET-ICE 2000) (2000)

12. Kung, C.H.: Conceptual modeling in the context of software development. IEEE Trans. on Software Eng. 15(10), 1176–1187 (1989)

13. Machado, R.G., Borges, M.R.S., Gomes, J.O.: Supporting the System Requirements Elicitation through Collaborative Observations. In: Briggs, R.O., Antunes, P., de Vreede, G.-J., Read, A.S. (eds.) CRIWG 2008. LNCS, vol. 5411, pp. 364–379. Springer, Heidelberg (2008)

14. Madin, J., Bowers, S., Schildhauer, M., Krivov, S., Pennington, D., Villa, F.: An ontology for describing and synthesizing ecological observation data. Eco. Inf. 2, 279–296 (2006)

15. McGuinness, D., et al.: The virtual solar-terrestrial observatory: A deployed semantic web application case study for scientific research. In: AAAI (2007)

16. Magdaleno, A.M., Araujo, R.M.D., Borges, M.R.S.: Designing Collaborative Processes. In: Workshop on Business Process Modeling, Development, and Support (BPMDS), Trondheim, Norway, pp. 283–290 (2007)

17. Mylopoulos, J.: Conceptual Modeling and Telos. In: Loucopoulos, P., Zicari, R. (eds.) Conceptual Modeling, Databases and CASE, ch. 2, pp. 49–68. Wiley (1992)

18. Oliveira, F., Antunes, J., Guizzardi, R.S.S.: Towards a Collaboration Ontology. In: Proceedings of the Second Brazilian Workshop on Ontologies and Metamodels for Software and Data Engineering (WOMSDE 2007), pp. 97–108 (2007)

19. Oliveira, F.F.: A Collaborative Ontology and its Applications. Dissertation (Master in Computer Science) - Federal University of Espirito Santo (2009) (in Portuguese)

20. Pinelle, D., Gutwin, C., Greenberg, S.: Task analysis for groupware usability evaluation: Modeling shared-workspace tasks with the mechanics of collaboration. ACM Trans. Comput. 10(4), 281–311 (2003)

21. Rekabdar, B., Shadgar, B., Osareh, A.: Learning teamwork behaviors approach: learning by observation meets case-based planning. In: Ramsay, A., Agre, G. (eds.) AIMSA 2012. LNCS, vol. 7557, pp. 195–201. Springer, Heidelberg (2012)

22. Silva Junior, L.C.L., Borges, M.R.S., Carvalho, P.V.R.: Collaborative Ethnography: An Approach to the Elicitation of Cognitive Requirements of Teams. In: Proceedings of The 9th International Conference on Computer Supported Work in Design (CSCW-D) (2009)

23. Tang, J.C.: Findings from Observational Studies of Collaborative Work. International Journal of Man-Machine Studies 34(2) (February 1991)

24. Tarboton, D., Horsburgh, J., Maidment, D.: CUAHSI community observations data model (ODM), version 1.0 (2007)

25. Williams, R., Martinez, N., Goldbeck, J.: Ontologies for Ecoinformatics. J. of Web Semantics 4, 237–242 (2006)

Monitoring Student Activities with a Querying System over Electronic Worksheets

Nelson Baloian[1], Jose A. Pino[1], Jens Hardings[1], and Heinz Ulrich Hoppe[2]

[1] Universidad de Chile, Department of Computer Science, Beauchef 851, Santiago, Chile
[2] University of Duisburg-Essen, Lotharstr. 63/65, 47048 Duisburg, Germany

Abstract. Monitoring students' work in the classroom has been recognized as one of the key factors for successful teaching since only a good real-time assessment enables the teacher to give proper and timely feedback. However, it is not an easy task to systematically supervise what students do in the classroom. It also might consume a considerable amount of teachers' resources. This paper presents a work in which computer technology is used in classrooms by students working on electronic worksheets on their. We explore the possibilities of assessing students' work during classroom by automatically analyzing the structure of the documents and the changes along time while students work on them. An experiment is described, showing the system is able to give the teacher valuable information. This information is intended to assess the students' performance and provide them with proper feedback.

Keywords: Monitoring students' work, automatic assessment, improving classroom teaching, architectures for educational technology systems, group workspace awareness.

1 Introduction

Despite progress in distance learning, classroom settings continue being massively used in education throughout the world. Of course, the context is not the same than even a few years ago. Students at all levels are aware of at least some Information Technology (IT) tools and thus, computer pervasiveness has made its definitive entrance in schools. Our vision for the future is The Collaborative Classroom (TCC), an evolution of our previous proposal, the Computer-integrated-Classroom (CiC) [2]. TCC includes IT hardware for all educational roles as they may be suitable [7] and recent educational software technology, implementing mechanisms supporting teachers and students' work inside the classroom. One of them is Learning Analytics (LA) [5], of which we have our own version. As it is known, most LA efforts are not for real-time decisions; the work presented in this paper concerns LA for teacher's use during classroom activity.

In particular, the focus of this article is on automatic monitoring of students' progress for quick adjustment of the teacher's work. It may also be considered as provision of teacher's awareness on the students' workspace activities. An example may clarify these statements. Suppose a high school mathematics class in which the

N. Baloian et al. (Eds.): CRIWG 2014, LNCS 8658, pp. 38–52, 2014.
© Springer International Publishing Switzerland 2014

students are supposed to individually solve sets of two linear equations; the teacher has previously explained how to do it and now the students are asked to solve their own equations. The problem we are dealing with is how to provide a suitable tool for the teacher to quickly grasp the solution progress of all students so that she can decide on whether to move ahead to more challenging equations or go back to reinforce the previous level of difficulty.

For many theories of learning and instruction, feedback is an essential part of the learning model, absolutely necessary to successfully achieve learning [3]. In [9] authors mention that feedback is one of the most powerful factors influencing learning and achievement, but this impact can vary in effectiveness depending on the type of feedback provided and the way it is given. Effective feedback requires teachers to make appropriate judgments about when, how, and at what level to provide it. Activities supplying teachers with relevant information to make these judgments are commonly known as assessment.

In general, the systematic monitoring of the students' work should be a key success factor, since teachers will be better prepared to give meaningful and timely feedback when they are aware of the students' current learning state [6]. According to [4], the existing educational research literature identifies the practice of monitoring student learning as an essential component of high-quality education. But monitoring students during in-classroom work may involve teachers moving around the classroom, being aware of how well (or poorly) students are progressing with their assignments, and working with students one-to-one as needed. These activities might be quite time consuming and sometimes difficult to perform even in classes with a reduced number of student. According to [8] due to the need to attend all students individually, teachers find it difficult to accomplish their role as facilitators in a classroom, and recommend the development of tools to support them in this task.

Some authors have developed systems intended to monitor students work mainly for the case of distance learning supported by a Learning Management System which are suitable for tracking the student's activity since most of them provide at least a low level logging which registers all students' actions. Log files can afterwards be automatically analyzed to extract high level information regarding students' progress. However, there is little literature reporting the monitoring of in-classroom students work to support assessment, although the required technology is already available [11].

In the past, computer-based learning material has been developed in the form of "electronic worksheets" in order to implement in-class learning activities for the students [12]. These materials have been called "Active Documents" [14]. These active documents provide the students with a rich environment for interaction. Also, they allow collaborative work by making use of available networks. In most cases, an XML Document Object Model has been used as a way to manipulate these documents and to store them in permanent storage devices. If students work on these electronic documents by modifying their contents, then it is possible to do an automatic - and hence systematic - analysis of their work. For example, the analysis can be used to find out how the students are advancing in the completion of the tasks described in the active document, whether or not they are filling the document with the right answers, and so forth. This constitutes the basis of our proposal.

2   Query-Based Assessment for Monitoring Students' Work

The specific goal of this work is to find out whether it is possible to develop a system in which a teacher can flexibly monitor the work of the students while working on electronic worksheets. The suitable worksheets are those which can be mapped to an XML Document Object Model (DOM) representation. Our proposal is a system that allows a teacher to "send" query agents through the network; these agents analyze the current state of students' documents and deliver information back to the teacher; she can then use this information to assess how the students are performing. In order to illustrate this we will use a very basic example: let us consider again the scenario of a mathematics class where now K8 students have to individually solve a series of exercises related to the subject being taught, e.g., arithmetic multiplication. The exercises are distributed as an electronic worksheet consisting of 3 sections, with ascending degree of difficulty. For instance, the first section may contain multiplications of positive numbers of at most 2 digits each. The second section introduces exercises with multiplication of negative numbers and the third section introduces multiplication with many digits. The structure of these worksheets clearly defines each section containing several exercises each. Each drill exercise is structured as a question and answer, and the answering part is to be modified by the students during their work. In this case, the teacher could make use of the following information:

1. **Students' progress:** the teacher wants to control how many exercises each student has answered up to now, so she can query how many answer parts have been modified. This can be presented as a total number, as a percentage or as a table specifying which exercises have been modified.
2. **Correct answers:** in this case, we need to extract the contents of the answer for each modified exercise, and compare it to the corresponding entry in a table containing the right answers. The results of the comparisons can be presented as a total number, a percentage relative to total number of exercises or number of modified exercises.
3. **Correct answers aggregated by section:** the previous information can be presented divided by section.
4. **Differences among students' progress:** we can apply query 2 for each student and present the numerical results in a table in descending order.
5. **Solving order:** using query 1 we can also determine whether the students are solving the exercises in the presented order or in another sequence.
6. **Student's work pace:** we can apply the previous queries at various times and present the differences in the resulting information. This will show the progress during that interval.

This simple list of queries shows we may consider two types of them: those gathering basic information (like the first two queries above), and those aggregating results from basic queries (like the rest of the queries above). The example describes a scenario for young children. The system we developed for this work was tested on a scenario for high school or university students. This may hint the suitability of this methodology for a wide range of learning scenarios.

3   System Description and Architecture

In order to develop a system which allows a teacher to monitor the students work in the classroom we need in the first place a software and hardware architecture enabling the communication between students' and teacher's workplaces. This architecture should also allow sending agents from teacher to students and capturing their findings with the needed information. Instead of conceiving a new one, we base the present work on a previous architecture, namely, CiC version 2 (CiCv2) [2]. Both teacher and students use computers in face-to-face sessions in this framework, allowing them to interact at various levels. The teacher can present teaching material, typically using a large interactive display, distribute assignments, exchange individual or group messages and share documents created on-the-fly or taken from an archive. Assignments can include constructive or creative tasks on the part of the students. A central repository allows users to authenticate and to access files as well as to interact with other users' applications. In the second place, we need software implementing electronic sheets allowing teachers to prepare the material students have to work on. We also opted for using an already existing tool called "**FreeStyler**" [10]. It implements a series of visual languages for modeling in a variety of subject domains such as Petri Nets and UML diagrams for computer science or system dynamics for physics, biology or economics (Fig. 1). On an abstract level, Freestyler can be seen as a graph editor, which allows the inclusion of various "palettes" defining a group of specific nodes and arcs with particular functionalities in order to model a certain system. Using FreeStyler and CiCv2 together the teacher can access documents from the repository, share them with the class, and send them to individual users or groups. By sharing a document we mean working simultaneously on the same document, propagating the changes to the participants as they occur.

The **FreeStyler modeling tool** can be used as a whiteboard application to present material and solutions to proposed problems, as well as an exercising environment for the students, who can work developing their own models according to instructions given by the teacher. FreeStyler manages content organized in pages and thus supports notebook-style usage pattern as well as page-based documents. These pages can be added, erased and copied. The **Students' FreeStyler** (a FreeStyler version customized for students) allows them to interact with the repository by sending and receiving files. Within a session, the student's application can also interact with the teacher and other students by exchanging messages in a chat, exchanging documents to be opened independently or sharing documents. In the latter case, they can modify a common document in real-time. The **Teacher's FreeStyler** (a version customized for the teacher) has the same functionalities as the student's application plus additional ones allowing monitoring the students' performance while working on assignments proposed by the teacher using FreeStyler documents (Fig. 2).

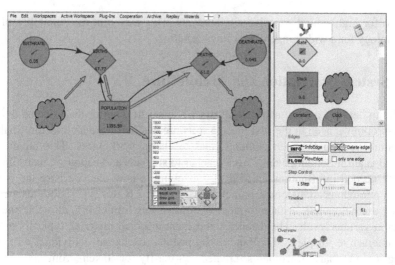

Fig. 1. Screenshot of FreeStyler. At the right hand side we see the palette. In this example, it corresponds to the System Dynamics plug-in, which allows the modeling and simulation of dynamical systems. At the center we see an already constructed model (graph).

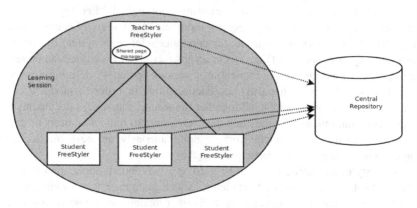

Fig. 2. CiCv2 Basic architecture

4   A Support System for Query-Based Assessment

As stated above, our interest in the plug-in inside the FreeStyler modeling tool is on the possibility for the teacher to monitor the classroom situation without cluttering the interface with external modules. This plug-in allows the teacher to access relevant information during the sessions, receiving this information from several sources. We call this plug-in the Querying System, since it is based on the visual composition of querying elements, as it will be shown below. Like other plug-ins in FreeStyler, a model consists of a graph. Nodes represent atomic queries which can be combined in a graph to form more elaborated queries. The inspiration for developing this graph comes from the "pipe" metaphor first introduced by the Unix operating system: the results of applying a program to a data set is a new data set which is input to the next program specified by the "pipe".

The teacher can access meaningful information sources available in the CiCv2 scenario by using the querying capability. A query is an object containing the specifications for gathering, composing and presenting certain information which is currently distributed in various files across the system. In order to process a query object and generate the answer, the system extracts information from files and log information from the central repository, the logs of the shared pages manager, the locally stored documents, and from the documents and logs of all student applications participating in the session.

The queries which will be used in a certain learning session are generally prepared in advance along with the designing of the learning activities and learning material that will be used. They can be specific to the activities in that particular session; they can also be general purpose queries which may be useful in any session. Either way, the query definitions are readily available as a way to minimize the teacher's involvement in technical details during the classroom session. However, it is also possible for the teacher to adjust specific parameters on queries to achieve the desired results, as it will be presented below.

When the teacher has selected a query object, she asks for its execution by pressing a button, and the result will either appear beside the graphic representation of the query or generate some changes to the currently active document, such as adding new pages with results. It is also possible to program queries to be executed periodically or triggered at a specific time, having access to updated results without any further interaction.

5   Query Implementation

A core set of basic query objects, which we call **Basic Queries**, was developed during the implementation of our system. These contain the specifications for retrieving information which is recurrently needed during the monitoring of students' work. These basic queries are the nodes contained in the palette implemented by the query plug-in. They can be directly used as they are by just dragging them from the palette to the working area or they can be combined to create new composed queries. The query composition is as follows: the output of one query object is connected as input to another query object by graphically drawing a directed arc between the nodes corresponding to those queries. The system checks the correctness of the composition by checking that the structure of the output data of the predecessor node matches the required input data structure of the subsequent node.

As an example, it is possible to obtain the difference between two documents by using the FileQuery twice for obtaining each file and connecting them to a DiffQuery. In order to hide complexity, these three queries can be encapsulated into a ComplexQuery, so the end user sees only one simple query which performs as expected. As a result, the teacher's interface shows just one complete query hiding all complexity and delivering a result when needed.

A context is defined for each query. It describes and gives access to the relevant documents it needs to perform. When a query needs to process remotely located documents (e.g., documents that are located on a student's computer) the querying engine sends the corresponding agent to the remote location. There, the context is set accordingly, so that the remotely executing agent has access to the local resources and it sends the results back to the original location.

6   Basic Queries

The basic query objects are the simplest building blocks that allow arbitrarily complex queries to be built and processed by combining them. Some queries do not have any input, and only generate output, such as a constant query always returning a fixed result and a "current document query" which always returns the current document as defined by the context where it is being executed. Other queries are terminal queries and do not provide any output, such as "Save Query", which simply saves its input into a file whose name it also fetched from its input, or "Object Creation Query", which creates a visual object being added as a new element in the modeling tool. The other queries have inputs that are processed to create a single output, such as XQuery, which executes a particular XQuery on its input, generating a single output.

Some basic queries are used to execute a particular query in a different context. As the teacher initiates the queries, the current context would always be the teacher application. In order to execute some query remotely on a student's application or at the repository, the basic queries RepositoryQuery and StudentQuery were defined. These are the objects implementing the query agents, since both take some query, send it to a remote location and trigger its execution in that remote context, receiving the result back at the original context. In the case of the StudentQuery, the remote location can be several students, either a list of predefined students, or all the students in the current session. This will trigger the execution of several agents. The results will be structured containing the result of the same query executed at the several locations.

7   Execution of Queries

Several queries can be combined by connecting the output of one to be the input of the next one, using the "pipe" metaphor mentioned above.

If we want to extract the information for the **"differences among students' progress"** example, then it is necessary to use a Complex query containing a linked list of sub-queries. The first step inside this complex query is to execute the query corresponding to **"correct answers"** in the context of each student using a StudentQuery. The result is the information for each of the students within the session in a DOM. This information is afterwards transformed by the next sub-query into a table. A last query should take this table and show it in the application user interface.

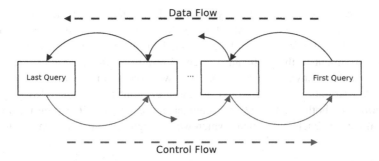

Fig. 3. Control (from left to right) and data flow (from right to left)

The "last query" is automatically identified and triggered when the teacher executes the complex query. This last query in turn triggers the previous query or queries, until the query which requests the single data from each student (StudentQuery) is reached. At that point, a query in each student environment is triggered and the resulting information is returned to the StudentQuery. From then on, the information is processed pipeline-wise in the reverse order.

The execution for the general case can be seen in Figure 3, where the control flow advances from left to right. At any point, the control flow continues its execution remotely, as in the StudentQuery, sending the query and the remaining queries to a remote location. The query will be rebuilt at the remote location, it will have access to the local context and the query will continue its control flow. This process continues with the input queries, activating all of their respective input queries, until the leaf queries are reached, which do not have any external input, beginning the data flow of results in the opposite direction.

Then, the information flows in the opposite direction (right to left in Fig. 3), delivering the results being processed in every step, until reaching the "last query". If the query was executed remotely, the result is sent back over the network, finishing the remote execution and continuing with the local processing. The "last query" receives the final result and it generally is one of the so called "terminal queries", which present or save the result in a useful way for the teacher.

8 Trials and Observations

Our goal is to explore the utility of the proposed system in practice. For that purpose, we have developed a series of three sessions involving students working while the teacher uses the proposed system in order to perform the monitoring of the activities. We intend to show in a qualitative way that the system is effective in providing useful information to the teacher in a timely manner.

We developed three sessions lasting 90 minutes each, in order to test the monitoring activities stated above. Each experiment is situated in a specific context or scenario, where students are asked to perform some activities and the teacher uses the querying system to monitor the session. The group of students for all experiments is the same, consisting of 16 university undergraduate and 2 graduate students, aged between 22 and 26 years old. All students were taking the course on distributed computing at the University of Chile, and none had previous knowledge of either the CiCv2 environment or the particular presented problems. As we intended to use the tool in as real an environment as possible, we used it in learning sessions with the same constraints a teacher would normally encounter. This included very short introductions of both the CiCv2 environment and each modeling plug-in, no more than 10 minutes in each experiment, which was enough to get the sessions started.

Session 1
In this session, the students were asked to use a collaboration framework for the Java programming language, called Matchmaker [13]. After an introduction, the students were handed out a document detailing the activities, one activity on each page, with some aspects of the activity being optional. The exercise on page 1 asked the student to create a collaborative session on a server by using some programming methods previously discussed in class. Optionally, the student can verify whether the session was successfully created, and the teacher can see which students have completed the optional part ("create + See" instead of "createOnly"). On page 2, the exercise requires the students to connect to the server and fetch all existing sessions, to print them out. On page 3, the students need to connect to a particular session, but many times the students forget to first check whether that session exists or not. On page 4, students are expected to create modifications in a session and on page 5, they are asked to fetch data from a session and print it out.

This session was set up to check the students' state while they are working on their tasks. The state is available as a set of indicators by which the teacher may identify partial progress in specific sub-tasks as well as total progress, considering each student as well as the entire group. For this purpose, information needs to be gathered from diverse sources within the system, and then aggregated into a specific output. Figure 3 presents a query which is available to the teacher (a), and the output being generated by that query (b).

In this scenario, the teacher can perform a query searching for a particular solution in order to automatically identify students who have successfully solved a problem. This can also be applied for identifying students making common errors or omissions.

Results of Session 1
Figure 4 shows the result for each exercise and each student in a matrix depicting the level of progress. It starts with a "none" value, changing as the student completes subsequent programming steps. The teacher can see students "pedro" and "juan" only do the "join" part of the exercise, without getting the list first, which is an incomplete activity on page 3. Similar partial results exist for exercises on pages 4 and 5.

The teacher can identify which students are currently working on the system and which ones are likely to need assistance by using activity levels. She thus gets awareness on the pace of both individual students and the whole group. Students progressing either slower or faster than expected will be quickly noticed by the

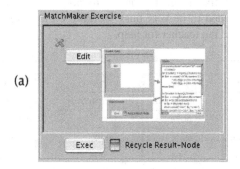

(a)

(b)

Student	Page 1	Page 2	Page 3	Page 4	Page 5
pedro	createOnly	get + print	joinOnly	join + mod	read + print
jorge	create + See	get + print	none	none	none
alberto	createOnly	getOnly	get + join	joinOnly	getOnly
juan	createOnly	get + print	joinOnly	join + mod	none

Fig. 4. Determining students' pace and specific solutions

teacher. She can take a closer look, either by directly approaching the student, or by using a new query to have a specific look at the student's work. When a student is advancing faster than his peers, the teacher might want to provide additional challenges, ideas or assignments in order to keep stimulating him. She may also share the good work with other members of the class in order to discuss the solution and its alternatives. On the other hand, when a student is having problems to solve the proposed exercise, then the teacher can take adequate measures to overcome these specific problems.

In figure 4, part (b), the teacher can see that in the various stages, students have fulfilled none, a part, or the whole exercise. The columns show each student's state on the exercise on a specific page, allowing the teacher easily compare each student's activities. For example, the student with username "jorge" has not finished exercises on pages 3 to 5, but he was the only student to achieve more than the rest in exercise on page 1. In this case, jorge actually did more than was asked in the exercise, while pedro finished all exercises, but did not complete the exercise on page 3.

Figure 5 shows the evolution of a single student's progress at different periods of time. We can clearly see how the results evolve according to the students' work. In this case, student "pedro" first solved exercise on page 1, but only partially. Afterwards, the student worked on exercise on page 2, completing the first part and then the second one. The results are presented by the system automatically, re-calculating the query according to the period of time set by the teacher (2 minutes in this case). The hints given by the teacher to the students were relevant, and they helped to guide the class in a natural way. In particular, the teacher could do early identification of at least five students who were not making any progress at all, despite the fact they seemed to be very busy working on the exercises. Only two of these students approached the teacher for help.

pedro	t1	t2	t3	t4
Page 1	none	createOnly	createOnly	createOnly
Page 2	none	none	getOnly	get + print
Page 3	none	none	none	none
Page 4	none	none	none	none

Fig. 5. Determining model complexity

Session 2

Let us consider a new scenario. We intend to show how the querying system is capable of identifying a student choosing one of several possible ways of solving a problem. University students were asked to model two stochastic processes to simulate lottery games, in order to determine which one had the highest probability of winning. The students would have to create a rather complex model by following the game descriptions literally. However, they could identify a simplification that radically reduced complexity of the model, without affecting the results. The teacher had access to a query that identified which of the paths each student seemed to have chosen. With this information, the teacher could start discussions in which students evaluate their peers' solutions and learn several ways to solve the problem.

A second aspect considered in this scenario is to determine whether the student groups are doing real collaborative work or each student is advancing on his/her own. For this purpose, the teacher has access to data of each group of students, identifying active students and passive ones. We can see the query that provides the statistics (Figure 6a), and the results within one specific group (Figure 6b).

Results of Session 2

The gathered data in our experiment did not show major changes in the distribution of activities among the participants as the session progressed. Generally it was possible to identify one or two students in a group who had a participation that was slightly over the level of the other participants. As the teacher observed these figures, it was possible to take a closer look at the groups showing a large gap in the members' participation. Perhaps the students were collaborating using direct face-to-face communication, or they were blocked because a difficult problem was encountered.

It was also possible to use queries to discover common students' errors like using the wrong element for modeling a particular stochastic scenario. As expected, several users made a quite common mistake, resulting in wrong results. When the teacher detected this kind of errors, she determined the best way to handle the situation. Some of these ways were asking for differing results, starting a discussion and either letting students find out why the results are different and which result is correct, or taking a more direct approach by telling the correct solution.

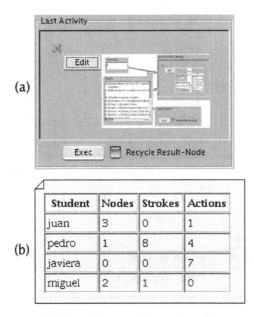

(a)

(b)

Student	Nodes	Strokes	Actions
juan	3	0	1
pedro	1	8	4
javiera	0	0	7
miguel	2	1	0

Fig. 6. Student activity within a group

Session 3
The aim of session 3 was to find out if it is possible to find out how elaborated is the answer the students are developing. This might be an indicator about the correctness of the solution the students are working on, for example, when the model a certain student is developing is much more complex than the one they are supposed to construct as result of the task given by the teacher. This is with high probability a case when the student is working on a wrong answer. In the case of complex models, it is increasingly difficult to characterize a "right" or "wrong" solution, and the teacher has to use generic information to choose where to look for problems or right answers. The queries used in this scenario are generally applicable to any situation. Fig. 7 shows the teacher has sent a query in order to obtain the number of elements (nodes, arcs, strokes, etc.) contained in each student's model. It is also possible to separate the number of nodes and arcs according to their type, which might provide even further information about the graph the student is developing. A documented example of this type of evaluation of model construction complexity can be found in [1].

Results of Session 3
During the session, it was possible to identify cases in which students were working on models that appeared extremely complex when compared to the model solution available to the teacher. In all cases the students used over twice as many nodes and edges than expected. A close look revealed they were creating a new model on the same page as the first one. Other students also decided to start a new model because the first one was unsatisfactory, creating the new model on another page, or by

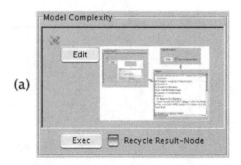

(a)

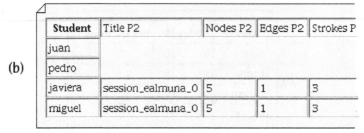

(b)

Student	Title P2	Nodes P2	Edges P2	Strokes P
juan				
pedro				
javiera	session_ealmuna_0	5	1	3
miguel	session_ealmuna_0	5	1	3

Fig. 7. Determining model complexity

deleting the previous model before starting over. These three approaches provided different results. A teacher needs to be able to interpret these results, verify that the real cause of the results effectively matches the possible interpretations (e.g. by approaching a student or using a query to view the student's model) and possibly take remedial actions.

9   Conclusions

The experiments described in the previous chapter show it is in fact possible to automatically extract valuable information from the worksheets the students are using. This information helps to assess their work and give meaningful feedback at the right moment at least in three different ways:

- Tracking the students in order to monitor their advance on the work, as seen in session 1. This gives the chance to assist students who might be working at a slower pace than the rest of the class or identify advanced students to give them positive feedback and/or provide them with additional problems to solve.
- Identify students that might have reached a correct solution in order to show them to the rest of the class in case their solution is a classic one or an unexpected one. This was the case of the session 2. This procedure can also be useful to identify cases where student are developing too complex solutions compared with the "classical" one the teacher might have in mind. In these cases, the teacher can assist the students guiding them in order to help them find simple solutions.

- Monitoring the students' level of activity of the students either during individual or collaborative learning sessions. This was explained in session 3 (Fig. 6). A low level of activity can indicate the students are not well prepared to solve the proposed task or that the task itself could be ill-designed. It can also serve as an indicator that perhaps the switching between different learning activities (e.g. from individual solving problem to collaborative work) might be not well designed [2].

The last point of the previous list indicates this tool can also be used to validate or discard a set of planned learning activities as "best practices". In fact, a teacher can test whether a certain set of learning activities the students should follow in a predetermined order can be considered as a "best practice" candidate by monitoring the students' work and checking if they advance as expected.

A very important feature of this approach is its generality. In fact, the developed query mechanism can be applied to various computer supported learning environments. The applicable environments must contain documents to be worked by students which are structured following the DOM standard and have their corresponding representation in XML. This requirement comes from the fact the query system only compares XML documents and it does not use their semantics. Therefore, this tool can be applied to monitor students' work for any other discipline, not just the one mentioned here.

We used a modeling tool as an example to test our approach. It is shown that automatic assessment by querying the documents the students are working on is very powerful. This is indeed a complex scenario since highly structured tasks can be given to the students thus making the querying much easier in many cases. For example, the working sheet might be reading a text and answering questions about its content with multiple choices or a list of mathematical exercises with unique answers, like in the case of arithmetic basic operations.

The reported research is representative of the type of services we would like to develop for TCC. It provides services that could not be offered without standards such as DOM and XML and without computer devices for each student and teacher. Furthermore, devices such as tablets allow the teacher to monitor students' activities while walking around the classroom and visiting students' workplaces.

References

[1] Antunes, P., Pino, J.A., Xiao, L.: Assessing the impact of Educational Differences in HCI Desing Practice. International Journal of Technology and Design Education (in press, 2014)

[2] Baloian, N., Pino, J.A., Hoppe, H.U.: Dealing with the Students' Attention Problem in Computer Supported Face-to-Face Lecturing. Educational Technology & Society 11(2), 192–205 (2008)

[3] Bangert-Drowns, R., Kulik, C., Kulik, J., Morgan, M.: The Instructional Effect of Feedback in Test-Like Events. Review of Educational Research 61(2), 213–238 (1991)

[4] Cotton, K.: Monitoring student learning in the classroom. School Improvement Research Series Close-Up #4. Northwest Regional Educational Laboratory, U.S. Department of Education (1988)

[5] Ferguson, R.: Learning analytics: drivers, developments and challenges. International Journal of Technology Enhanced Learning 4(5/6), 304–317 (2012)

[6] Grüntgens, W., Melzer, R.: Diagnostische kompetenz von lehrkräften im primar- und sekundarbereich (2004), http://pz.bildung-rp.de/pn/pb1_04/diagnostischekompetenz.html (retrieved on September 2008)

[7] Guerrero, L., Ochoa, S., Pino, J.A., Collazos, C.: Selecting Computing Devices to support Mobile Collaboration. Group Decision and Negotiation 15(3), 243–271 (2006)

[8] Gutierrez-Santos, S., Mavrikis, M.: Intelligent Support for Exploratory Environments: Where are We and Where Do We Want to Go Now? In: Proceedings of the 1st Int. Workshop in Intelligent Support for Exploratory Environments on EC-TEL 2008, Maastricht, The Netherlands (2008), http://ftp.informatik.rwth-aachen.de/Publications/CEUR-WS/Vol-381/ (retrieved on February 2011)

[9] Hattie, J., Timperley, H.: The power of feedback. Review of Educational Research 77, 81–112 (2007)

[10] Hoppe, H.U., Gassner, K.: Integrating Collaborative Concept Mapping Tools with Group Memory and Retrieval Functions. In: Stahl, G. (ed.) Computer Support for Collaborative Learning: Foundations for a CSCL Community. Proceedings of CSCL 2002, pp. 716–725. Lawrence Erlenbaum Associates, Inc., Hillsdale (2002)

[11] Pearce-Lazard, D., Poulovassilis, A., Geraniou, E.: The Design of Teacher Assistance Tools in an Exploratory Learning Environment for Mathematics Generalisation. In: Wolpers, M., Kirschner, P.A., Scheffel, M., Lindstaedt, S., Dimitrova, V. (eds.) EC-TEL 2010. LNCS, vol. 6383, pp. 260–275. Springer, Heidelberg (2010)

[12] Pinkwart, N.: Collaborative Modeling in Graph Based Environments. Ph.D. thesis, Universität Duisburg-Essen, Germany (2005)

[13] Tewissen, F., Baloian, N., Hoppe, H.U., Reimberg, E.: MatchMaker: Synchronizing Objects in Replicated Software-Architectures. In: Procs. of the 6th Collaboration Researchers' International Workshop on Groupware (CRIWG), pp. 60–67. IEEE Computer Society Press, Madeira (2000)

[14] Verdejo, M.F., Barros, B., Read, T., Rodríguez-Artacho, M.: A system for the specification and development of an environment for distributed CSCL scenarios. In: Cerri, S.A., Gouardéres, G., Paraguaçu, F. (eds.) ITS 2002. LNCS, vol. 2363, pp. 139–148. Springer, Heidelberg (2002)

Two Make a Network: Using Graphs to Assess the Quality of Collaboration of Dyads

Irene-Angelica Chounta[1], Tobias Hecking[2],
Heinz Ulrich Hoppe[2], and Nikolaos Avouris[1]

[1] HCI Group, University of Patras, Greece
{houren,avouris}@upatras.gr
[2] Collide, University of Duisburg-Essen, Germany
{hecking,hoppe}@collide.info

Abstract. In this paper we explore the application of network analysis techniques in order to analyze synchronous collaborative activities of dyads. The collaborative activities are represented and visualized as networks. We argue that the characteristics and properties of the networks reflect the quality of collaboration and therefore can support the analysis of collaborative activities in an automated way. To support this argument we studied the collaborative practice of 228 dyads based on graphs. The properties of each graph were evaluated in comparison to ratings of collaboration quality as assessed by human experts. The activities were also examined with respect to the solution quality. The paper presents the method and the findings of the study.

Keywords: learning analytics, collaboration, analysis, network, graph theory, SNA, CSCL.

1   Introduction

The analysis of collaborative activities is a popular research area in the interdisciplinary fields of CSCL (Computer-Supported Collaborative Learning) and CSCW (Computer-Supported Cooperative Work). It allows researchers to gain insight into the nature of collaboration, to support designers and orchestrators and to provide feedback to users with respect to their practice. CSCL constitutes an ideal application field for automated analysis techniques based on activity log files. Interaction analysis and automated metrics of interaction have been widely used for the quality assessment of collaborative activities [1, 2]. Several activity metrics, such as the number or rate of messages, the average word count, the roles alternations and the symmetry of actions, have been introduced as automated methods of analysis [3, 4]. However, the nature of data sets limitations to that kind of methods. Data-driven approaches are not stemming from a theoretical framework and it is argued that the lack of qualitative research undermines the depth of analysis and entails drawbacks [5, 6].

There are several examples of using graph representations and Social Network Analysis (SNA) analysis techniques for the analysis of learning activities [7, 8]. Most of the studies focus on individual students or groups of students of bigger size in order

N. Baloian et al. (Eds.): CRIWG 2014, LNCS 8658, pp. 53–66, 2014.

to reveal patterns of interaction and minimize the complexity of the analysis [9]. Graph theory has also been used to describe and analyze cognitive structures during collaborative learning activities [10]. In this paper, we propose the use of networks for the quality assessment of collaborative activities that involve dyads of students. Even though dyads are common in collaborative learning scenarios, they are hardly considered as "networks" or studied with SNA techniques. We argue that the interaction of dyads can be effectively represented as a network and that the network properties will reflect the quality of the collaboration. In order to support this argument, we used the log files of collaborative activities of dyads to construct networks that represent interaction structures. The graphs were analyzed and studied in comparison to the ratings of collaboration quality as assessed by human evaluators. A correlation analysis between the network properties and the ratings of the evaluators was conducted.

The paper is organized as follows: In the second section we provide an overview of related work, in section 3 we present the background of the study, the collaborative setting and the dataset used. In section 4 we describe the method of the study. The results are presented in section 5 and the paper concludes to a discussion on the findings as well as future work.

2   Related Work

2.1   Generating Networks from Event Logs

The representation of log files as networks has been part of existing and ongoing research on collaboration analysis. An advantage is the explicit representation of relations between entities that are not directly observable by looking at log files. Networks can be further analyzed and modeled by using well established techniques of graph theory [11].

A well known framework for the graph-based representation of joint activities is described in [12]. Based on tables of event logs, relations between events are extracted according to predefined contingency relations. These contingency graphs can be further processed to extract graph based models for interaction, mediation and sociograms. Nasirifard et al. [13] use subject-verb-object representations of read and write actions on documents in a shared workspace to extract multi-relational networks between users and objects, as well as relations between the users themselves.

One problem that arises with the network representation of activity logs is that one loses an explicit representation of time. While logs usually carry a timestamp, the network representation aggregates log based relations over a certain period of time. One solution can be to sample networks from action logs in subsequent time windows. This results in subsequent time slices of an evolving network. This technique has been used in [14] in order to analyze changes of the affiliations of groups of students to learning resources over time. In some cases however, networks bear an inherent notion of time. This is the case when the nodes of the network relate to events, and directed links between nodes establish a partial temporal ordering, as in citation networks [15]. The networks described later in this paper also have this property of an implicit notion of time. Consequently the applied analysis does not require sampling the networks into time slices.

2.2   Theory of Network Evolution

It could be shown that the evolution of many real world networks does not follow a random model. Moreover, there are fundamental mechanisms that govern the emergence of the given network structure. Hence, the structural characteristics that can be observed by investigating those networks are potentially meaningful. This is the basic argument to apply network analysis methods in order to reveal basic patterns in network data. In this study, networks are constructed from event logs of collaborating dyads and the assumption is that those networks reflect the quality of the collaboration. This section reviews the major research on networks models that helps to interpret the results presented in section 5. The Erdös Rényi model [16] describes the evolution of graphs when the likelihood of a connection between any two nodes is uniformly distributed. In those networks, the distribution of the degree of the nodes (number of connections) follows a Poison distribution where most nodes have a degree close to the average degree. The average path length in such networks tends to decrease when new edges are added to the network [17]. However, it can be shown that this network generation model does not apply to most real world networks. Moreover, the degree distribution often follows an inverse power law instead of a Poisson distribution [18]. This means that the majority of nodes is not well connected but, however, the probability to have a few highly connected nodes (hubs) is much greater than in random networks according to the Erdös Rényi model. These networks are scale-free because of the power law degree distribution. Although most nodes in scale-free networks are sparsely connected with the rest of the network, there are densely connected regions (clusters). The hubs often bridge between those clusters. One model that explains the emergence of scale-free networks is the preferential attachment model [18]. In this model, nodes connect more likely with nodes that already have many connections.

Recently it could be shown that scale-free networks can also emerge when nodes are subsequently added to a network. In this model new arriving nodes connect to existing nodes by chance but also to nodes that their neighbors already connect with [19]. This model comes quite close to networks that are constructed from subsequent events, as in this work.

3   Background of the Study

In this study we explore the analysis of collaborative activities through networks. We use a dataset of collaborative activities that has been previously evaluated with qualitative and quantitative methods involving human judgment, so it can serve as a reference and benchmark. The dataset consists of 228 collaborative sessions that took place during a programming course in the University of Patras (Greece). In particular, dyads of students were asked to collaborate synchronously through a groupware application in order to complete a task. The task was the collaborative construction of an algorithmic flowchart. The duration of the activity was about 90 minutes. The activity was mediated by a shared workspace groupware application [20]. The application comprises two shared spaces: (a) a common workspace for the construction of diagrammatic representations and (b) a chat tool to support the synchronous communication of users (Fig. 1). The dataset has been used in earlier studies by human experts for the empirical evaluation of a rating scheme. In addition, it was further analyzed for the training of a machine learning model for the classification of collaborative activities. The findings of earlier studies were used as valuable input in the current case.

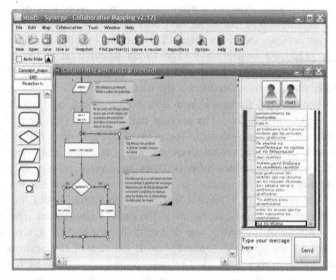

Fig. 1. The user interface of the collaborative application Synergo. It consists of the common workspace used for the construction of diagrammatic representations and the chat tool that mediates user communication.

3.1   Assessment of Collaboration Quality by Human Experts

The dataset was previously evaluated by two human experts with the use of a rating scheme [21]. The evaluators rated each collaborative session on various aspects of collaboration with respect to their quality with the support of a rating handbook and after a training phase. According to the scheme (Table 1), the quality of collaboration was rated on six dimensions that represent the four fundamental collaborative aspects: Communication, Joint information processing, Coordination and Interpersonal relationship. Each dimension was rated on a 5-point Likert scale ([-2, +2]). The general dimension of the quality of collaboration (Collaboration Quality Average, CQA) was computed as the average of all six collaborative dimensions. The ratings were successfully tested for inter-rater reliability and consistency. Depending on the collaborative dimension, ICC ranged between 0.83 and 0.95, while Cronbach's alpha ranged between 0.91 and 0.98. Therefore, the rating scheme was proved to be a successful means for the assessment of the quality of collaborative activities.

Table 1. The rating scheme used for the assessment of the quality of collaboration: the collaborative dimensions in relation to the general aspect of collaboration that they represent

General aspects of collaboration	Collaborative Dimensions
Communication	Collaboration flow
	Sustaining mutual understanding
Joint information processing	Knowledge exchange
	Argumentation
Coordination	Structuring the problem solving process
Interpersonal relationship	Cooperative orientation

3.2   Assessment of Collaboration Quality with the Use of Time Series

The current dataset was further analyzed and evaluated using time series analysis techniques [22]. In particular, the collaborative activities were represented as time series of aggregated events within certain time frames. A memory-based learning model was used for the classification of collaborative sessions with respect to the similarity of time series. The research hypothesis was that collaborative activities of similar collaboration quality will unfold in similar ways.

The memory based model was used as an automatic rater of the quality of collaboration (CQA). The ratings of the model were compared to the ratings of human experts. We conducted a correlation and error analyses in order to evaluate the efficiency and accuracy of the model. The analysis showed that the ratings of the model correlated with the ratings of expert evaluators ($\rho=0.3$, $p<0.05$), while the mean absolute prediction error MAE was less than one on a 5-point scale (MAE=0.89). It was also shown that time series of activity aggregated within small time frames (60 seconds or less) portray collaboration more effectively. The effect that the size of time frames has on the performance of the model was further explored. It was shown that the time frames ranging from 15 to 30 seconds are the most suitable for depicting the quality of collaboration, thus pinpointing that meaningful interaction for synchronous communication takes place within time frames of less than a minute. This finding comes into agreement with other related studies [1, 23].

We use the results of previous studies to reveal potential relations between the network properties and the ratings of human experts on collaborative dimensions. The research question we aim to explore is whether the quality of collaboration is depicted on the network that represents the practice of the dyad and how this information can be used to support the analysis of collaborative learning activities.

4   Method of the Study

In the present study, we aim to assess collaboration quality using network analysis. To that end, each collaborative session is represented as a network and is visualized as a graph. The nodes stand for the actions taking place in the common workspace and the chat tool. The edges of the network represent some kind of dependency between actions or relation. Therefore, in order to consider two, or more, actions as relevant and thus, connect them, the following conditions should apply:

(a) The time distance between actions must fall within a certain range. Based on the findings of previous study [22], the time range signifying relevant actions is defined from 10 to 30 seconds.
(b) Workspace actions are considered relevant, and therefore connect, only if they involve the same artifact. In the case of chat messages, only temporal proximity is taken into consideration.
(c) The identity of the actor should differ. The objective of the study is to map the interplay between users rather than the practice of a single user. Therefore we map consequent actions of different actors, aiming to reveal *reciprocal activity*.

The instance of a log file used for the construction of the networks is portrayed in Fig. 2. The log files are structured according to the OCAF format [24] as:

$$<ID><timestamp><actor><event\text{-}type><attributes>$$

where:

— <ID> is an incremental identifier, unique for each event
— <timestamp> the time that the event occurred
— <actor> the user responsible for the event
— <event-type> the type of the event, i.e. chat message or the type of the workspace action
— <attributes> a field related to additional information that can be interpreted in combination with the type of event, i.e. content of chat message or (x-y) coordinates of an artifact on the workspace.

Taking into consideration the aforementioned rules for constructing nodes and edges, we elaborate on the paradigm of Fig. 2. For the log file entries with ID number from 18 to 22, there will be five nodes, one for each entry. The node "18" will be connected through an edge with node "19" because these actions come from different actors and their time distance is less than 30 seconds. Likewise, the node "19" will be connected with nodes "20" and "21" but not with node "22" since their time distance is more than 30 seconds. The instance of the network that corresponds to the instance of the log file of Fig. 2 is presented in Fig. 3.

17	0:07:31	ece7108	Set object to front	[Alternate process (1)]
18	0:07:46	ece7108	Change Concept Entity text	[Alternate process (1), Start, (x=385,y=32,w=90,h=60)]
19	0:08:02	a08-6930	Chat message	[then we will write the contents of A and B]
20	0:08:12	ece7108	Chat message	[right]
21	0:08:28	ece7108	Chat message	[it says so in the book]
22	0:08:43	ece7108	Chat message	[that we should use this shape]
23	0:08:51	ece7108	Insert Entity	[Input-Output, (x=382,y=127,w=90,h=60), Input-Output (2)]

Fig. 2. Instance of the log file of a collaborative activity as recorded by Synergo

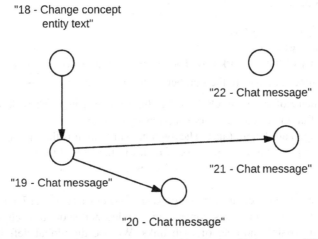

Fig. 3. Example of a network constructed by the log file instance of Fig. 2

The graphs of collaborative activities were constructed by the web-based application SiSOB Workbench [25]. Two networks, one for a collaborative session of good quality and one session of poor collaboration quality respectively, are portrayed in Fig. 4. On one hand, the session of good collaboration quality is portrayed as a big network that depicts intense activity, i.e. a large number of user actions. A number of nodes appear to be strongly connected to each other revealing some kind of activity bursts caused by key events while the rest are sparsely connected. On the other hand, the session of poor collaboration quality is portrayed as a smaller network. The instances of activity bursts are less and not so tightly structured as in the previous case.

We argue that the quality of collaboration is reflected by the characteristics of the network that represents each activity. To support this argument, we study the corresponding basic network properties and compare these to the ratings of collaboration quality, as described in section 3.1. A correlation analysis is carried out in order to establish whether the network properties correlate to the quality of collaboration.

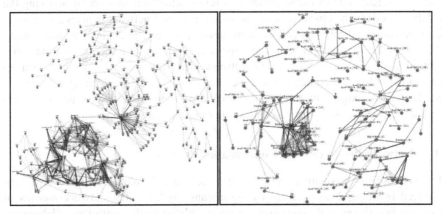

Fig. 4. Two networks representing a session that was rated as of good quality of collaboration (left) and a session of bad collaboration quality (right)

The network properties used in the study are:

- The number of nodes (N)
- The number of edges (E)
- The density of the network (D). The density of a network is defined as the ratio of the number of edges to the number of possible edges: $= \dfrac{2E}{N(N-1)}$
- The diameter of the network (d). The diameter of a network is defined as the longest of all the calculated shortest paths in a network
- The average path length (apl). The average path length is the ratio of the sum of all shortest paths between all pairs of nodes to the sum of total number of pairs. It is used to indicate the number of steps needed to move from one node to another
- The clustering coefficient average (cc_avg). The clustering coefficient is defined as the ratio of existing links connecting a node's neighbors to each other to the maximum possible number of such links. We use the global definition of the clustering coefficient for the network as a whole. This is defined as the number of closed triangles (three fully connected nodes) and open triangles (three nodes and two pairs are connected)
- The assortativity (A). Assortativity is described as the tendency of nodes to connect with other nodes, similar to themselves
- Power Law Fits (plFits). To verify if the network can be considered as a scale-free network, a power law has fitted to the degree distribution of the networks. Then a Kolmogorov-Smirnov test has been conducted in order to measure the goodness of fit. The smaller this value is, the better is the fit of the power law distribution.

5   Results

In the current study, we used the dataset of 228 collaborative activities, as described in section 3. Each one of the activities was represented as a network by applying the guidelines as described in section 4. The aforementioned network metrics and properties were computed for each collaborative session. A correlation analysis was carried out between the network properties and the ratings of human evaluators on the collaboration quality average (CQA). The network properties were also studied with respect to the individual collaborative dimensions as defined by the rating scheme and also the grades assigned by the teachers for the solution quality. Spearman's correlation coefficient was used since the data are not normally distributed.

5.1   Network Properties and Collaboration Quality

The results of the correlation analysis are presented in Table 2. All the correlations portrayed in Table 2 are statistically significant ($p<0.05$). The network metrics of clustering coefficient and assortativity were correlated neither with the collaboration quality average (CQA) nor with the solution grade on a significant level. The correlations that were not statistically significant are not presented in the results. Most of the network properties that correlate to the quality of collaboration also correlate to the individual collaborative dimensions.

Table 2. Correlations of the network properties and the ratings of collaboration quality average (CQA) and for each collaborative dimension of the rating scheme

Collaborative Dimensions	#Nodes (N)	#Edges (E)	Density (D)	Diameter (d)	(plFits)	(apl)
Collaboration flow	0.358		-0.405	0.267	-0.229	0.226
Sustaining mutual understanding	0.351	0.179	-0.275	0.246		0.205
Knowledge exchange	0.400	0.190	-0.340	0.228	-0.192	0.191
Argumentation	0.416	0.169	-0.360	0.273	-0.225	0.223
Structuring the problem solving process	0.339	0.156	-0.268	0.185	-0.208	0.154
Cooperative Orientation	0.410	0.195	-0.323	0.295	-0.196	0.257
CQA	0.446	0.180	-0.394	0.294	-0.233	0.243

The number of nodes correlates highly with the quality of collaboration (ρ_N=0.446). On the other hand, the number of edges correlates on a low level with collaboration quality (ρ_E=0.180). This shows that sessions with intense activity are also of better collaboration quality. However, the number of connections does not necessarily point to better quality. The number of nodes correlates significantly with all collaborative dimensions ratings and in particular with those that stand for the aspect of Joint information processing (Knowledge exchange and Argumentation). These dimensions assess how effectively students exchange information and give appropriate explanations to their partners. Thus, it is expected that a student who is motivated to explain and to make a valid argument will post a higher number of explanatory messages or add helpful notes/artifacts on the common workspace, resulting in intense activity.

In the case of good collaboration quality networks, there is a significant negative correlation of the collaboration quality (CQA) with the value of the Kolmogorov-Smirnov test for the power law degree distribution, where a small value indicates better fit (ρ_{plFits} = -0.233). Hence, the networks generated from good collaborative sessions appear to be scale-free networks. These networks have some nodes that are strongly connected with other nodes while the majority is only sparsely connected. Since the nodes of the network represent user activity and they can only link to events that happened later in time (condition [a]), one can conclude that good collaborations contain a few key events that cause bursts of subsequent events related to the key event. The relation of activity bursts and the scale-free property has been studied thoroughly [26].

The diameter (d) and the average path length (apl) as metrics of the linear size of a network also correlate positively with the quality of collaboration (ρ_d=0.294, ρ_{apl}=0.243). This finding indicates that good collaboration usually results in big scale-free networks with some key events. Those networks are characterized by long paths that correspond to longer uptake chains. The longer path indicates that the teams that collaborate efficiently are more influenced by previous events than those who collaborate poorly, since condition [b] requires connected events to be on the same object. Therefore, collaborative activities of good quality appear to have more key events that dominate user activity and direct future focus. Taking into account the temporal condition for the network construction, it is evident that activities of good collaboration quality unfold faster in time. The diameter and average path length correlate with the ratings of human experts on all collaborative dimensions, especially on the collaborative aspect of Cooperative orientation.

Density (D) is a metric to measure the saturation of relationships between the nodes of a network in relation to a completely connected network. One would expect that good collaboration would lead to denser networks. However the results indicate that dense networks stem from collaborative practices of bad quality (ρ_D = -0.394). Consequently, density is a negative predictor of collaboration quality in the sense that dense networks point to poor collaboration. An explanation for this counterintuitive finding is that in scale-free networks the density is anti-proportional to the number of nodes. Therefore, a higher density indicates a smaller map [27]. In addition, density correlates negatively to all individual collaborative dimensions. The highest negative correlation is scored for the dimension of Collaboration flow, which stands for the aspect of Communication (ρ = -0.405). Collaboration flow, in particular, expresses how naturally communication flows between actors and among shared resources. It is therefore expected that good communication flow is depicted by a bigger network map and thus, lower density.

5.2   Network Properties and Solution Quality

Apart from the log files of the activity, the students had to deliver the final algorithmic diagram, as it was developed in the shared workspace. The diagram was graded separately by the teacher of the course on a scale from 0 to 10. The assessments of the solution were straightforward since the task had a demonstrably correct solution with a few easily defined alternatives in parts of the diagram.

The correlations are presented in Table 3. The properties that do not appear statistically significant correlations are not included. The solution grade correlates significantly with the number of nodes and edges of the networks. Lower positive correlations also appear for the solution grade and network the diameter and average path length. The results of the correlation analysis show that thorough solutions, i.e. detailed flowcharts that result in intense activity and, therefore, bigger networks, are evaluated as good. This is expected since a correct solution requires a fully developed flowchart, accompanying notes etc.

We should note that the solution grade does not correlate either to the network density or the power law degree value. As aforementioned, these network properties characterize collaborative practice and meaningful interplay among users. In this case however, a good collaboration does not ensure a good solution quality and vice versa. For example, a detailed solution could be the result of one student's work while his partner does not contribute on the task. This would be an example of a good solution grade and bad collaboration quality. On the other hand, a good collaboration does not ensure that the students will succeed in their task. This would result in a bad solution grade but good collaboration quality. The difference between solution quality and collaboration quality is depicted in the results of the correlation analysis and the network properties.

Table 3. Correlations of the network properties and the ratings on solution quality

	#Nodes (N)	#Edges (E)	Diameter (d)	Average Path Length (apl)
solution grade	0.319	0.305	0.202	0.189

6   Conclusions and Future Work

This study explores the use of networks for the analysis and evaluation of collaborative activities of student dyads. Graphs and social network analysis techniques have been introduced for the analysis of the collaborative practice of small or bigger teams that usually work together over discussion forums or supported by various groupware applications [27, 28]. Even though dyads are a popular way of students grouping in a learning context, they are hardly considered as networks. We argue that the application of networks for the mapping of dyads interaction, can indicate the meaningful interplay and successful collaborative practice, as well as offer a valuable tool for the assessment of collaboration quality.

For the purpose of the study, we analyzed a dataset of 228 collaborative sessions where each session was represented by a network. The network properties of each session were studied in combination to ratings of human experts on collaboration quality and a correlation analysis was carried out. The results showed that the quality of collaboration reflects certain properties of the networks that represent joint activities. The size of the network, in terms of number of nodes and diameter, points to activities of good collaboration quality. Efficient collaboration is expected to result in intense activity and therefore, bigger networks. It was also shown that good practices unfold faster in time and form longer chains of actions (i.e. longer paths). The size of the network is a good indicator for the solution quality as well. Collaborative activities represented by big networks are assessed with higher grades for the quality of the solution. We should note however that a good solution does not presuppose effective collaboration. This was also portrayed by the network properties. Collaboration quality was found to correlate negatively to the density of the network as well as the power law fits criterion value, while that was not the case for the solution quality. Density is used to measure the relationships of a network and the power law fits is used to indicate scale-free networks. The negative correlation shows that good collaboration results in scale-free networks where a key action leads to reciprocal interplay among the actors. This is portrayed in the form of few central nodes that are strongly connected to others while the majority of nodes is sparsely connected leading to low density and a small value for the power law fit criterion. In the case however that a good solution grade is the result of one-sided activity, i.e. one student who takes the lead, this kind of interaction is absent.

In this study, the construction of the networks was based solely on the temporal and spatial proximity of user activity. In future studies, we plan to use content analysis techniques in order to refine the relations and connections between user actions. This is believed to lead to more accurate representations of collaborative interaction and add up semantic value to activity patterns. The application of advanced methods such as path analysis and sequential data analysis is also an interesting direction that could be pursued in order to define patterns that indicate good or bad collaboration quality.

References

1. Schümmer, T., Strijbos, J.W., Berkel, T.: A new direction for log file analysis in CSCL: Experiences with a spatio-temporal metric. In: 2005 Conference on Computer Supported Collaborative Learning (CSCL 2005). International Society of the Learning Sciences, pp. 567–576 (2005)
2. Martínez-Monés, A., Harrer, A., Dimitriadis, Y.: An interaction-aware design process for the integration of interaction analysis into mainstream CSCL practices. In: Analyzing Interactions in CSCL, pp. 269–291. Springer (2011)
3. Bratitsis, T., Dimitracopoulou, A., Martínez-Monés, A., Marcos, J.A., Dimitriadis, Y.: Supporting members of a learning community using Interaction Analysis tools: The example of the Kaleidoscope NoE scientific network. In: Eighth IEEE International Conference on Advanced Learning Technologies, ICALT 2008, pp. 809–813. IEEE (2008)
4. Kahrimanis, G., Chounta, I.A., Avouris, N.: Study of correlations between logfile-based metrics of interaction and the quality of synchronous collaboration. In: 9th International Conference on the Design of Cooperative Systems, Workshop on Analysing the Quality of Collaboration, International Reports on Socio-Informatics (IRSI), p. 24. Aix en Provence (2010)
5. Stahl, G.: Rediscovering CSCL. In: CSCL 2: Carrying Forward the Conversation, pp. 169–181. Lawrence Erlbaum Associates, Mahwah (2002)
6. Stahl, G., Koschmann, T., Suthers, D.: Computer-supported collaborative learning: An historical perspective. In: Cambridge Handbook of the Learning Sciences (2006)
7. Reffay, C., Teplovs, C., Blondel, F.-M.: Productive re-use of CSCL data and analytic tools to provide a new perspective on group cohesion. In: Connecting Computer-Supported Collaborative Learning to Policy and Practice, pp. 846–850 (2011)
8. Toikkanen, T., Lipponen, L.: The applicability of social network analysis to the study of networked learning. Interactive Learning Environments 19, 365–379 (2011)
9. Palonen, T., Hakkarainen, K.: Patterns of interaction in computersupported learning: A social network analysis. In: Fourth International Conference of the Learning Sciences, pp. 334–339 (2013)

10. Ifenthaler, D., Masduki, I., Seel, N.M.: The mystery of cognitive structure and how we can detect it: tracking the development of cognitive structures over time. Instructional Science 39, 41–61 (2011)
11. Wasserman, S.: Social network analysis: Methods and applications. Cambridge University Press (1994)
12. Suthers, D.D.: Interaction, mediation, and ties: An analytic hierarchy for socio-technical systems. In: 2011 44th Hawaii International Conference on System Sciences (HICSS), pp. 1–10. IEEE (2011)
13. Nasirifard, P., Peristeras, V., Hayes, C., Decker, S.: Extracting and utilizing social networks from log files of shared workspaces. In: Camarinha-Matos, L.M., Paraskakis, I., Afsarmanesh, H. (eds.) PRO-VE 2009. IFIP AICT, vol. 307, pp. 643–650. Springer, Heidelberg (2009)
14. Hecking, T., Ziebarth, S., Hoppe, H.U.: Analysis of dynamic resource access patterns in a blended learning course. In: Proceedings of the Fourth International Conference on Learning Analytics And Knowledge, pp. 173–182. ACM, Indianapolis (2014)
15. Halatchliyski, I., Hecking, T., Göhnert, T., Hoppe, H.U.: Analyzing the flow of ideas and profiles of contributors in an open learning community. In: Proceedings of the Third International Conference on Learning Analytics and Knowledge, pp. 66–74. ACM (2013)
16. Erdős, P., Rényi, A.: On the evolution of random graphs. Magyar Tud. Akad. Mat. Kutató Int. Közl 5, 17–61 (1960)
17. Watts, D.J., Strogatz, S.H.: Collective dynamics of 'small-world' networks. Nature 393, 440–442 (1998)
18. Barabási, A.-L., Albert, R.: Emergence of scaling in random networks. Science 286, 509–512 (1999)
19. Kumar, R., Raghavan, P., Rajagopalan, S., Sivakumar, D., Tomkins, A., Upfal, E.: Stochastic models for the web graph. In: Proceedings of the 41st Annual Symposium on Foundations of Computer Science, pp. 57–65. IEEE (2000)
20. Avouris, N., Margaritis, M., Komis, V.: Modelling Interaction during small-group Synchronous problem solving activities: the Synergo approach. In: 2nd International Workshop on Designing Computational Models of Collaborative Learning Interaction, 7th Conference on Intelligent Tutoring Systems, ITS 2004, Maceio, Brasil, pp. 13–18 (2004)
21. Kahrimanis, G., Meier, A., Chounta, I.-A., Voyiatzaki, E., Spada, H., Rummel, N., Avouris, N.: Assessing collaboration quality in synchronous CSCL problem-solving activities: Adaptation and empirical evaluation of a rating scheme. In: Cress, U., Dimitrova, V., Specht, M. (eds.) EC-TEL 2009. LNCS, vol. 5794, pp. 267–272. Springer, Heidelberg (2009)
22. Chounta, I.-A., Avouris, N.: Time Series Analysis of Collaborative Activities. In: Herskovic, V., Hoppe, H.U., Jansen, M., Ziegler, J. (eds.) CRIWG 2012. LNCS, vol. 7493, pp. 145–152. Springer, Heidelberg (2012)
23. Suthers, D.D., Dwyer, N., Medina, R., Vatrapu, R.: A framework for conceptualizing, representing, and analyzing distributed interaction. International Journal of Computer-Supported Collaborative Learning 5, 5–42 (2010)
24. Avouris, N.M., Dimitracopoulou, A., Komis, V., Fidas, C.: OCAF: an object-oriented model of analysis of collaborative problem solving. In: Proceedings of the Conference on Computer Support for Collaborative Learning: Foundations for a CSCL Community, International Society of the Learning Sciences, Boulder, Colorado, pp. 92–101 (2002)
25. Göhnert, T., Harrer, A., Hecking, T., Hoppe, H.U.: A workbench to construct and re-use network analysis workflows: concept, implementation, and example case. In: Proceedings of the 2013 IEEE/ACM International Conference on Advances in Social Networks Analysis and Mining, pp. 1464–1466. ACM (2013)

26. Barabasi, A.-L.: The origin of bursts and heavy tails in human dynamics. Nature 435, 207–211 (2005)
27. Hoppe, H.U., Engler, J., Weinbrenner, S.: The Impact of Structural Characteristics of Concept Maps on Automatic Quality Measurement. In: International Conference of the Learning Sciences (ICLS 2012), Sydney, Australia (2012)
28. Reihaneh, R., Takaffoli, M., Zaïane, O.R.: Analyzing participation of students in online courses using social network analysis techniques. In: Proceedings of Educational Data Mining (2011)

A Programming Interface and Platform Support for Developing Recommendation Algorithms on Large-Scale Social Networks

Alejandro Corbellini, Daniela Godoy, Cristian Mateos,
Alejandro Zunino, and Silvia Schiaffino

ISISTAN Research Institute - Consejo Nacional de Investigaciones Científicas y Técnicas
(CONICET), Univ. Nacional del Centro de la Provincia de Bs. As. (UNICEN), Campus
Universitario, Paraje Arroyo Seco (BBO7001B), Tandil, Buenos Aires, Argentina

Abstract. Friend recommendation algorithms in large-scale social networks such as Facebook or Twitter usually require the exploration of huge user graphs. In current solutions for parallelizing graph algorithms, the burden of dealing with distributed concerns falls on algorithm developers. In this paper, a simple yet powerful programming interface (API) to implement distributed graph traversal algorithms is presented. A case study on implementing a followee recommendation algorithm for Twitter using the API is described. This case study not only illustrates the simplicity offered by the API for developing algorithms, but also how different aspects of the distributed solutions can be treated and experimented without altering the algorithm code. Experiments evaluating the performance of different job scheduling strategies illustrate the flexibility or our approach.

1 Introduction

Friend recommendation algorithms try to infer missing edges among members of a social network that are likely to be established in the near future. For large-scale social networks such as Twitter or Facebook, these algorithms face challenges related to the amount of data to be processed. Graph-specific databases or frameworks for parallel processing of graph algorithms have arisen to address these issues. These supports do not provide by themselves means to improve different aspects of the cluster usage according to the algorithm requirements and, thus, the responsibility falls on the programmer who must modify its algorithm to handle job distribution.

In this paper, we focus on providing a simple yet powerful API to implement graph traversal algorithms. By using the API and its related abstractions, new graph-based recommendation algorithms can be quickly prototyped without thinking about distribution and parallel concerns. Also, the platform provides customizable scheduling strategies or "rules" that determine the job-node mapping that the algorithm will use, allowing a fine-grained control over the cluster usage without altering the original algorithm.

In order to illustrate and evaluate the proposed approach we present a case study in which an algorithm for followee recommendation in Twitter, described in [1], was adapted to the proposed API and executed using different scheduling strategies, namely, a Round Robin strategy, a Location Aware strategy and two Memory-based strategies. For each strategy we compared the adapted algorithm performance in terms of recommendation time, memory usage and physical network usage.

N. Baloian et al. (Eds.): CRIWG 2014, LNCS 8658, pp. 67–74, 2014.

The rest of this paper is organized as follows. Section 2 overviews of the existing work on distributed graph processing frameworks and APIs. Section 3 describes our proposed framework and the implementation of the back-end storage support. Section 4 presents the API of our adjacency Graph that allows the programmer to query the structure of the graph and set the scheduling strategy to be used. Section 5 presents a case study of developing a followee recommendation algorithm and its testing with a Twitter dataset. Finally, conclusions drawn from this work are summarized in Section 6.

2   Related Work

Similarly to the case study presented, there are many link-prediction algorithms that are based on the structure of the graph rather than its content. For example, SALSA [10] is an algorithm that uses the "hub" and "authority" notion of HITS [6], and the "random surfer" model from PageRank [13]. The Who To Follow [4] user recommendation algorithm developed by Twitter, uses PageRank to build the initial group of results and then uses that group as an input to a SALSA-based recommendation algorithm. A more specialized recommendation algorithm that recommends a person's contacts in the case of emergency scenarios based on her social network can be found in [15].

Several frameworks that support graph processing algorithms in distributed environments can be found in the literature. For example, HipG [7] allows modeling hierarchical parallel algorithms, which includes divide-and-conquer graph algorithms. Pregel [11] is a closed-source framework used at Google that provides a computational model for large-scale graph processing. Sedge [18] implements the Pregel model, but focuses on graph partitioning. Trinity [14] is a graph processing framework created by Microsoft, that provides some interesting features such as online query processing and native graph representation. Graph processing frameworks usually load data from a persistent store and then construct a distributed in-memory representation of the graph. As stated in [4], on large-scale graphs this can be a problem if the growth of the amount of physical memory to store the graph cannot keep pace with the growth of the graph.

In this work we created a distributed graph store that allow users to perform distributed computation over its data. In its current implementation the store provides a graph traversal API over an existing *back-end* store. This allowed us to focus on the development of the API and its abstractions rather than the inners of the graph storage.

There are many implementations of graph APIs over existing storage supports. For example, FlockDB [17] is a graph database that uses MySQL as an storage back-end and provides a graph API on top of SQL. Similarly, Titan [2] is a graph database that provides a graph traversal API while supporting several storage backends. On the other hand, there are graph databases that use structures specially suited for linked data. For example, Neo4j [12] is a popular graph database that uses data structures adjusted to its graph representation.

3   Distributed Execution and Storage Support

We based our proposed approach on a set of tools created to efficiently distribute code and data across a number of nodes. From the set of tools, the Distributed Execution

Framework and the Distributed Key-Value Store are the most relevant to the graph API proposed and are described below.

3.1 Distributed Execution Framework

A Distributed Job Execution Framework (DEF) was developed to simplify the creation of programs that send data and code (jobs) between computational nodes. The framework is composed of several software layers, implemented in Java, that provide abstractions at different levels to execute distributed recommendation algorithms. The first layer is the Network Module that handles networking details. The second layer is an RPC module that handles remote method calling and marshalling (conversion of objects to bytes and viceversa) of parameters and return values. The topmost layer is the Job Execution module that uses the RPC Module to communicate to other Job Execution modules. This layer sends jobs to execute on a specific node, it gathers results and handles errors.

3.2 Distributed Key-Value Store

Our graph implementation (Section 4) is built upon a distributed key-value store (KVS) that saves the target graph in a distributed manner. A KVS has, at least, two operations: a *put* operation that stores a given value under a given unique key, and a *get* operation that retrieves the value associated to a given key. Thus, the structural information of a graph (links and vertices) can be saved in a key-value store by assigning every vertex a unique key, e.g. a Twitter user identifier. Then, the list of vertices pointing to and from a vertex is stored as a value under the vertex key. This type of structure is called an Adjacency Graph.

Although there are many databases that provide a key-value API [3,16], a DEF-based implementation provide us fine-grained control over data distribution, memory consumption and peer-vertex allocation.

4 Adjacency Graph

As mentioned before, we created an Adjacency Graph for representing the graph structure and persisted it in a distributed KVS. This graph implementation uses the DEF model to perform queries and distribute data handling among a number of nodes. Such distribution can be customized using Scheduling Strategies that map a group of vertices being queried to an specific node. Next, we will describe the API of the Adjacency Graph and the Scheduling Strategies that will be used for the experiments.

4.1 Adjacency Graph API

The starting point to obtain information about the graph is the AdjacencyGraph object and its getVertex(id:int) method. This method creates a VertexQuery, i.e., a query that represents a vertex or group of vertices. This query corresponds to a ListQuery type, which means it is a query that returns a list of vertices. A ListQuery contains a handful of methods to query connected vertices:

– *incoming/outgoing:ListQuery*
Returns a ListQuery representing the list of vertices that have *outgoing* (or *incoming*) connections to the current list of vertices.
– *countIncoming/countOutgoing:CountQuery*
Counts the amount of unique vertices with incoming (or outgoing) connections to the current list of vertices. The CountQuery object represents a dictionary of vertices and number of occurrences of each vertex.
– *union/intersect(q:ListQuery):ListQuery*
Executes both *q* and the current query and intersects (or unites, if union is used) their results, creating a new ListQuery.
– *remove(q:ListQuery):ListQuery*
Removes the list of vertices resulting from executing *q* from the current list.

Regarding the CountQuery object, a user can call the *top(n:integer):TopQuery* method to obtain the top *n* vertices with most appearances. Both ListQuery and CountQuery create new queries for each of the mentioned methods to programmatically concatenate an arbitrary number of queries.

4.2 Scheduling Strategies

When a user requests the incoming or outgoing vertices of a given group of vertices, a scheduling strategy is used to divide the request in several jobs and each job is assigned to a specific node. Indeed, many job allocation strategies can be used for this purpose. Then, we implemented and compared four scheduling strategies:

– Round Robin: This strategy simply divides the given list of vertices by the amount of nodes and assigns a sublist to each of the nodes.
– Available Memory: This strategy checks the memory currently available on each node and divides the given list of vertices accordingly.
– Total Memory: The Total Memory strategy uses the maximum amount of memory that a peer can use to divide the list of requests.
– Location Aware: This strategy takes advantage of the *locality* of the vertices, dividing the input into different lists of vertices grouped by their storage location.

The selection of strategies depends on the type of improvement that the user wants. For example, a Location Aware policy is the less network-consuming strategy. This is because the request for each incoming or outgoing list of a vertex is made on the node that stores the vertex data. However, if the vertices are not well balanced among nodes, a node may receive many requests to be locally processed, degrading its performance. Moreover, the Location Aware strategy must process the given vertices list and calculate the location of each vertex to make the division.

5   Case Study

The algorithm used as case of study, proposed in [1], is based on the characterization of Twitter users made in several studies [5,8], by which users are mainly divided in two categories: information sources and information seekers. Information sources tend

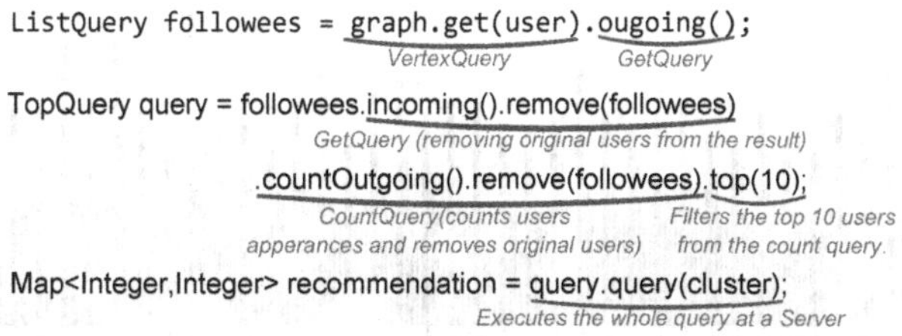

```
ListQuery followees = graph.get(user).ougoing();
                      VertexQuery        GetQuery
```

TopQuery query = followees.incoming().remove(followees)
                           GetQuery (removing original users from the result)

.countOutgoing().remove(followees).top(10);
CountQuery(counts users   Filters the top 10 users
appearances and removes original users)   from the count query.

Map<Integer,Integer> recommendation = query.query(cluster);
                                      Executes the whole query at a Server

Fig. 1. The studied recommendation algorithm expressed using the DEF Query API

to have a large amount of followers as they are actually posting useful information or news. In turn, information seekers follow several users to obtain the information they are looking for, but rarely post a tweet themselves. The aim of the algorithm is finding candidate information sources to recommend to a target information seeker. To do so, it makes the asumption that those followees of the target's followers are similar to the target, and then, the users that they follow may be of interest to the target.

Next, we will describe the adaptation of the algorithm to our API and then the experiments' layout and results.

5.1 Followee Recommendation Algorithm Adaptation

The original recommendation algorithm was adapted to execute on top of our Adjacency Graph. Using the query API presented above, we expressed the algorithm as shown in Figure 1, which resulted in the following steps:

1. Get the followees (*outgoing*) of a user.
2. Get the followers (*incoming*) of the followees and remove the original followees from the list.
3. Count the followees (*countOutgoing*) of the followers and remove the original followees from the map.
4. Obtain the top *n* elements with most appearances ($n = 10$ in this example).

Essentially, the algorithm collects lists of users at different steps, and finally joins the final list to elaborate a ranking of candidate information sources.

5.2 Experiments

The experiments were carried out using a Twitter dataset[1] consisting of approximately 1,400 million relationships between 40 million users [9]. For selecting a representative test user list, we first filtered the list of users using the information source ratio [1], denoting to what extent the user can be considered as an information source. This ratio is defined as follows:

$$IS = \frac{followers(u)}{followers(u) + followees(u)} \quad (1)$$

---

[1] http://an.kaist.ac.kr/traces/WWW2010.html

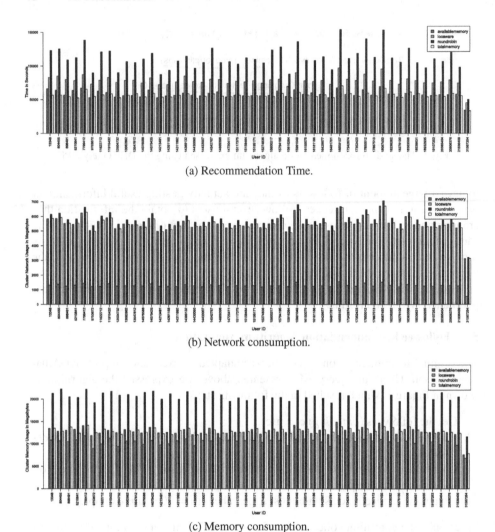

(a) Recommendation Time.

(b) Network consumption.

(c) Memory consumption.

Fig. 2. Experiment results

By this definition, an information seeker is a user that has an IS lesser than 0.5, i.e. it has more followees than followers. Moreover, we wanted to stress the platform by generating big intermediate results. To do so, we selected from the list of information seekers the top-50 users that had the longest followees lists.

Regarding the cluster characteristics, we used an heterogeneous cluster of 8 nodes. 5 of them had a 6-core AMD Phenom II CPU with 8GB of RAM, and 3 of them had a 6-core AMD FX 6100 CPU with 16 GB of RAM. Other characteristics, like network speed and hard disk, were similar for all of them.

The experiments were carried out as follows. For each user, we ran the algorithm 10 times for each strategy. For every run we measured the bytes sent over the network, the maximum memory consumed (the biggest memory spike) and the total recommendation time. Then we calculated the average for the whole execution.

The results obtained from the experiments are depicted in Figure 2. In Figure 2a we show the average time for recommendation. The memory-based strategies yielded the best recommendation times. As the algorithm consumes a lot of memory, a distribution of work based on memory results in better balance. In Figure 2b it can be seen that the amount of network consumed by the Location Aware strategy is, as we initially expected, far less than the consumption of other strategies. Finally, 2c shows the average of the maximum memory consumed on the cluster. The memory usage pattern of the Round Robin strategy yielded bigger memory spikes on every node of the cluster whereas the Location Aware strategy used the least RAM memory, which was expected as less memory was needed to make remote database requests.

6 Conclusions

In this paper we presented a graph API for developing distributed link prediction algorithms without concerning about the inner job-node distribution mechanisms. For testing this approach we used an existing recommendation algorithm for the Twitter social network as case study, with data of a complete Twitter graph of 2009. The objective of the algorithm is to recommend users that share the same followers than the followees of the target user.

The proposed API hides most of the distributed processes involved in the adjacency Graph implementation. Moreover, it also offers customizable Scheduling Strategies, i.e. the way jobs are created and assigned to nodes, to be used when querying the graph API. The election of the strategy may have an impact on the execution of the algorithm and the way it uses the cluster resources. To show the effect of strategy selection, we experimented with four types of scheduling strategies: Round Robin, Location Aware, Available Memory and Total Memory.

In our experiments, the Location Aware strategy showed approximately 84% less network bandwidth consumption than the other strategies and the least overall memory usage. Despite the time for recommendation was acceptable in comparison to the Round Robin strategy (about 50% less), it was 15% slower than the Memory-based approaches, which yielded the fastest recommendations. As expected, the Round Robin strategy yielded the worst results.

Future work includes a) the improvement of the underlying graph storage, b) the implementation of new strategies and c) the testing of this approach with other graph-based recommendation algorithms.

References

1. Armentano, M., Godoy, D., Amandi, A.: Topology-based recommendation of users in micro-blogging communities. Journal of Computer Science and Technology 27(3), 624–634 (2012)

2. Aurelius. Titan (2014), http://thinkaurelius.github.io/titan/ (accessed April 14, 2014)
3. Cattell, R.: Scalable SQL and NoSQL data stores. ACM SIGMOD Record 39(4), 12–27 (2011)
4. Gupta, P., Goel, A., Lin, J., Sharma, A., Wang, D., Zadeh, R.: WTF: The who to follow service at Twitter. In: Proceedings of the 22th International World Wide Web Conference (WWW 2013), Rio de Janeiro, Brazil (2013)
5. Java, A., Song, X., Finin, T., Tseng, B.: Why we Twitter: Understanding microblogging usage and communities. In: Proceedings of the 9th WebKDD and 1st SNA-KDD 2007 Workshop on Web Mining and Social Network Analysis, San Jose, CA, USA, pp. 56–65 (2007)
6. Kleinberg, J.M.: Authoritative sources in a hyperlinked environment. Journal of the ACM 46(5), 604–632 (1999)
7. Krepska, E., Kielmann, T., Fokkink, W., Bal, H.: HipG: Parallel processing of large-scale graphs. ACM SIGOPS Operating Systems Review 45(2), 3–13 (2011)
8. Krishnamurthy, B., Gill, P., Arlitt, M.: A few chirps about Twitter. In: Proceedings of the 1st Workshop on Online Social Networks (WOSP 2008), Seattle, USA, pp. 19–24 (2008)
9. Kwak, H., Lee, C., Park, H., Moon, S.: What is Twitter, a social network or a news media? In: Proceedings of the 19th International Conference on World Wide Web (WWW 2010), Raleigh, NC, USA, pp. 591–600 (2010)
10. Lempel, R., Moran, S.: SALSA: the stochastic approach for link-structure analysis. ACM Transactions on Information Systems 19(2), 131–160 (2001)
11. Malewicz, G., Austern, M.H., Bik, A.J.C., Dehnert, J.C., Horn, I., Leiser, N., Czajkowski, G.: Pregel: A system for large-scale graph processing. In: Proceedings of the 2010 International Conference on Management of Data (SIGMOD 2010), Indianapolis, IN, USA, pp. 135–146 (2010)
12. Inc. Neo Technology. Neo4J (2013), http://www.neo4j.org/ (accessed August 5, 2013)
13. Page, L., Brin, S., Motwani, R., Winograd, T.: The PageRank citation ranking: bringing order to the web, pp. 1–17 (1999)
14. Shao, B., Wang, H., Li, Y.: The Trinity Graph Engine. Technical Report MSR-TR-2012-30, Microsoft Research (March 2012)
15. da Silva, S.T.F., Oliveira, J., Borges, M.R.S.: Contextual analysis of the victims' social network for people recommendation on the emergency scenario. In: Herskovic, V., Hoppe, H.U., Jansen, M., Ziegler, J. (eds.) CRIWG 2012. LNCS, vol. 7493, pp. 200–207. Springer, Heidelberg (2012)
16. Strauch, C., Sites, U.L.S., Kriha, W.: NoSQL databases. Lecture Notes, Stuttgart Media University (2011)
17. Twitter Inc. FlockDB (2013), https://github.com/twitter/flockdb (accessed August 5, 2013)
18. Yang, S., Yan, X., Zong, B., Khan, A.: Towards effective partition management for large graphs. In: Proceedings of the 2012 International Conference on Management of Data (SIGMOD 2012), Scottsdale, AZ, USA, pp. 517–528 (2012)

An Ambient Casual Game to Promote Socialization and Active Ageing

Raymundo Cornejo, Daniel Hernandez, Monica Tentori, and Jesus Favela

Computer Science Department, CICESE, Ensenada, Mexico
{rcornejo,dahernan}@cicese.edu.mx, {mtentori,favela}@cicese.mx

Abstract. Natural interfaces are facilitating the adoption of videogames by older adults, promoting the development of serious games aimed at encouraging healthy behaviors in this population. In this paper we present the design and evaluation of an ambient game, GuessMyCaption, aimed at enhancing the social networks of older adults, known to have an impact in their wellbeing. GuessMyCaption was deployed during a 5-weeks study in the home of one older adult and twelve relatives. The results demonstrate GuessMyCaption is easy to use and maintains an older adult engaged with exercises while offering new opportunities for online and offline socialization. GuessMyCaption had a positive impact in the perceived wellbeing of the older adult improving her perception on her cognitive skills and physical health, and catalyzing socialization. This research shows that the use of natural interfaces and family memorabilia facilitate the adoption of serious games, improves older adults' perceived wellbeing, and encourage socialization.

Keywords: social networks, exergame, older adults, ambient games.

1 Introduction

The videogame industry has had steady and strong sales over the last years. The average U.S. household owns at least one dedicated console, PC or smartphone with an increased variety of online gaming (e.g. puzzles, card games, strategy) [1].

Online gaming has enabled individuals to perceive playing video games as a social activity. Recent reports have indicated that video gaming triggers online social opportunities for interaction and collocated in-person encounters between players [1].

Although the gamer average age is 30, older adults who do play games are more avid players than their younger peers. Older adults tend to have more free time they could use for playing on a daily basis [2].

The above is particularly interesting since videogames have been shown to become "meeting places" where older adults can initiate social interaction [3]. Paradoxically, as individuals age, their social network is reduced [4], and they tend to concentrate their social ties upon family interactions [5]. Social ties and social integration are important factors to maintain the physical and psychological wellbeing of older adults [6, 7]; therefore the social interactions triggered by videogames might reinforce the desirable social interactions that older adults need.

N. Baloian et al. (Eds.): CRIWG 2014, LNCS 8658, pp. 75–88, 2014.

Along with the social benefits of gaming, researchers have also analyzed the increasing popularity of exergames among the elderly [3, 8]. Exergames have proven to improve physical health and assist during rehabilitation [9]. However, socialization is rarely an issue highlighted in the design of games for older adults.

The present research aims to explore the impact of ubiquitous games with a design revolving around socialization by taking the advantage of the social network of its players. The main contribution is to show how ubiquitous exergames could be designed to encourage socialization that as a consequence have a positive impact in the perceived wellbeing of older adults.

2   Related Work

Serious games could be described as the usage of videogames to help users achieve a specific goal by playing a game [9]. Research with ubiquitous technologies in healthcare has investigated the use of exergames, serious games that encourage players to exercise, to engage players in physical activity in order to play the game. Exergames interaction relies on body movement tracking technology, and activities that may be restricted to a fixed location.

Others have implemented exergames for more specific goals like exercising during therapy [10-13]. For example, Fasola et.al. [11] present a robot as a feedback mechanism for older adults' upper-limb rehabilitation.

Unlike serious games, most casual games can be played in any location (e.g. home, work, transit stop), and require little configuration effort, no previous video game skills or regular time commitment to play [14]. However, researchers have rarely studied the effects of casual games towards specific goals like stress relieve [15] or perceived attitudes on specific themes [16, 17]. For example, OrderUP! [16] is a mobile casual game that requires players to make meals recommendations as healthy and quickly as possible. Its interface is based on tap actions to select the appropriate recommendations.

While serious games serve as a mediator to achieve a specific goal and most casual games are designed for entertaining purposes, their interaction model frequently demands their users to have specific "game skills" to manipulate tangible user interfaces serving as a proxy to control the digital elements of the game.

In contrast, ambient games envision an environment furnished with sensors and actuators [18] to facilitate the use of the game by enabling players to control the game with every day and playful interactions (i.e., not using a dedicated game input device, mouse or keyboard) [19].

The present research goes beyond the combination of ambient games and serious games; exploring the intersection between serious, casual and ambient games facilitating the interaction through the use of everyday and playful interactions while mediating online socialization and face-to-face activities between players. We particularly evaluate the user experience of a casual ambient game and its impact in wellbeing by encouraging socialization.

3   Design Methods

Following a user-centered design process, multiple design methods were used to inform the design of an ubiquitous exergame for older adults. Six participatory design sessions were conducted with potential users, a geriatric nurse, and experts in HCI, Ubicomp and computer vision. The aim of these sessions was to establish appropriate physical and social activities for the older adults along with playful exergames prototypes that incorporate a set of pre-defined activities.

First, two design sessions were conducted with a geriatric nurse and HCI/Ubicomp experts to define appropriate physical activities for older adults; additionally, an 86 year-old woman and two relatives were involved in some design activities (e.g. engagement interviews). Prior the design sessions, the older adult's health was assessed by a geriatric nurse to determine any possible activity restrictions or critical conditions. Several physical activities were discussed during design sessions to establish a set of appropriate alternatives for the targeted older adult yet being general enough to be beneficial for other older adults.

Three additional participatory design sessions were conducted once the set of physical activities was established. These sessions were aimed to supplement our prior understanding on the role of social media in improving social ties [20]. During the first two sessions, HCI/Ubicomp experts, a geriatric nurse and a computer vision expert discussed themes related to the design and integration of the physical activities into a videogame. An additional session involved the older adult's grandson and a granddaughter in-law, with whom we explored different ways in which social media could be used to engage the user. Additionally, a semi-structured interview was conducted with the older adult to gather insights for designing an engaging game experience. The results of these meetings led to the design of two prototypes. Both prototypes were discussed in a final meeting to obtain feedback from all stakeholders. This feedback was used to re-design the prototypes.

Findings from the design sessions suggested three design insights should be integrated into the exergames prototypes: *social media images*, *tangible rewards* and *moderate arm exercises*.

Moderate arm exercises were selected from the first two design sessions as the most appropriate form of moderate activity for older adults. Three physical activities (see Figure 1) were selected: right and left hello (Figure 1a,b), right and left arm raise (Figure 1c,d), and circular arms raise (Figure 1e). According to the geriatric nurse participating in the design team, these exercises help older adults to prevent the loss of arm muscle strength or flexibility, which are common problems among the elderly.

Social media images provide older adults with a continuous update of the relevant events of their social network members. While older adults play a game, opportunities of acquiring sufficient awareness about the whereabouts of their relatives and friends can be triggered by observing iconic representations of their lives in form of photographs. Photographs that are commonly shared by users in Social Network Sites (SNS).

Most basic video game's rewards come in a virtual form (e.g. scores, game boosting abilities, etc.). However, design sessions findings suggest the opportunity of employing *tangible rewards* to engage older adults into videogame tasks. For example, the older adult receives a printout of a photograph of her choice from a SNS (e.g. Facebook) as a reward for completing a video game task.

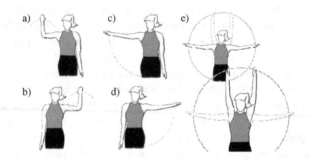

Fig. 1. Upper limb exercises: a) Right hello; b) Left hello; c) Right arm raise; d) Left arm raise; e) Circular arms raise

4   The Ubiquitous Exergame GuessMyCaption

The design findings obtained during our participatory design sessions were used to design two prototypes; we describe one of them called GuessMyCaption.

GuessMyCaption is initiated by the older adult using a "left hello" (see Figure 1a) or a "right hello" (see Figure 1b), and uses the content available in the Facebook's family account to feed the essence of the game. The objective of this game is to guess the photograph that better matches a given description. Three photographs are randomly retrieved from the relatives' accounts on Facebook, and displayed to the user (Figure 2). From this set of photos, one is randomly selected and the description displayed at the bottom is retrieved from the photograph's caption or any given associated comment in Facebook.

A maximum of 45 seconds and two attempts are given to the player to select the correct answer. Three poses can be used to interact with GuessMyCaption; "left arm raise" (see Figure 1c) to select the displayed photograph on the left of the screen, "right arm raise" (see Figure 1d) to select the displayed photograph on the right of the screen and "circular arms raise" (see Figure 1e) to select the photograph displayed at the top of the screen.

Visual and audio feedback is provided to the older adult when a photograph is selected; a glowing green rectangle is displayed around the correct photo and a cheers sound for correct answers. Similarly, a glowing red rectangle is shown and an error sound is played when the answer is incorrect.

A puzzle of a family photograph on the top right corner of the screen is assembled based on the number of attempts that the user required to get the correct answer. If the player deduces the correct photo on the first attempt, then 2 pieces of the puzzle are assembled; if it's the second attempt, then one piece is assembled. Once the puzzle is

Time remaining   Puzzle
Set of 3 photographs   Caption

Fig. 2. Matching game in GuessMyCaption

complete, one virtual photo is given to the older adult. The virtual photo can be exchanged for a printout of a photograph of the older adult's choice. A post is published on the older adult SNS showing the photograph selected to print and the name of the person who shared that photograph (see Figure 3).

4.1   Implementation

The GuessMyCaption prototype interaction model is based on natural user interfaces and was implemented in C# on Windows 7, and runs on an all-in one PC. The Microsoft Kinect sensor was selected to detect the pose of the user. Scene depth information is processed with a decision forest in order to find up to two users interacting with the game. This information is also used to track the user skeleton determined in a three dimensional position of 20 human body joints (i.e., head, neck, shoulders, elbows, wrists, hips, knees, ankles, hands and feet) [21]. Upper limbs of the skeleton are tracked by the game's perception component to identify the five body actions, previously described in Section 3, for manipulating the game.

4.1.1   Body Poses for GuessMyCaption

Left and Right Hello. This body pose corresponds to a natural waving of the hand when saying "hello". Poses are detected when the left or right hand is located near the player's head. Vertical and horizontal thresholds (xThresh and yThresh) are used to delimitate a small rectangular region where the hand is expected to be found. This region is defined by the rules in (1), where α is a small adjustment related to the distance between the user and the sensor; and Xdist and Ydist are the horizontal and vertical distances from the hand to the head respectively. The hand must be placed

Fig. 3. Automatic Post showing the image the player decided to print as award

above the shoulder as another required condition, that is, handYpos – shoulderYpos > 0; next, the hand must be horizontally found between the elbow and the head defined by, handXpos $\in$ [headXpos, elbowXpos]; and finally the elbow is found below the shoulder elbowYpos < shoulderY-pos.

$$\text{If } Xdist \in [30\text{-}xThresh - \alpha, 30 + xThresh - \alpha] \qquad \text{and}$$

$$Ydist \in [15\text{-}yThresh - \alpha, 15 + yThresh - \alpha] \tag{1}$$

Left and Right arm raise. The events for these body poses are triggered when the player's arm is fully extended to the side with the hand at the same level as the shoulder; meaning that the Euclidean distance from the shoulder position and the wrist position should be approximately the distance from the shoulder to the elbow plus the distance from the elbow to the wrist. The pose is detected when the hand is found within the rectangular threshold region defined by (2) and (3):

$$dist(shoulderPos, wristPos) - (dist(shoulderPos, \\ elbowPos) + dist(elbowPos, wristPos)) \\ < distThresh - \alpha \tag{2}$$

$$|handYpos - shoulderYpos| < yThresh - \alpha. \tag{3}$$

Circular arms rise. A circular pose is detected when the player has both arms completely stretched above the head, see (4):

$$handYpos > shoulderYpos,$$
$$elbowYpos > shoulderYpos \qquad (4)$$

The hands and elbows must be directly above the shoulders, |elbowXpos - shoulderXpos| < xThresh − α, |wristXpos - shoulderXpos| < xThresh − α; and finally, hands must be close to each other, |leftwristXpos − rightwristXpos| < xThresh − α. It is through these conditions that the system establishes a rectangular threshold region for pose detection.

Figure 4 shows the detection system in use. The top images show the input captured by the depth perception camera; with this information the system determines the position of the user. The bottom images show the results of the skeletal tracking algorithm used to detect the current pose or gesture.

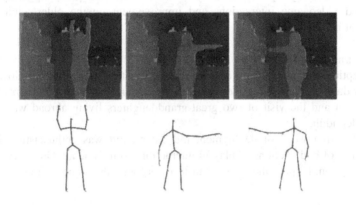

Fig. 4. Images from the detection system

5   In Situ Evaluation

GuessMyCaption was deployed in the home of an independent older adult to confirm prior findings and explore the user experience and possible social interactions in daily family settings. The following section presents the results concerning use and adoption, and perceived wellbeing in terms of socialization, and physical and cognitive health.

5.1   Methods

For a period of 3 weeks, GuessMyCaption was deployed at the home of the older adult who participated in our initial participatory design sessions. Participants (n=12) included a 87-years old woman, 3 of her female-children, 6 of her grandchildren (3 female), 1 grandniece and 1 nephew, all of them were scattered throughout several locations, with almost half of the participants living in the same city as the older adult. Data collection included weekly measurements of arm strength with a dynamometer and weekly semi-structured interviews with participants and logs of computer usage.

Interviews were conducted across the three study phases: pre-deployment (1 week), during the deployment (3 weeks), and post-deployment (1 week). Interviews were individual and semi-structured and were conducted face to face or through instant messaging, video calls, or telephone. During the pre-deployment, the older adult was interviewed by the geriatric nurse to determine her health status, and her self-perceived health. The older adult also received a training session on how to play the game. The interviews were transcribed and we used open coding to reveal evidence in relation to the use and adoption of the game, older adult's perception of the game, health and social support.

5.2   Experiences When Using GuessMyCaption

Data suggest GuessMyCaption was easy to use and well adopted, triggering social traits around videogame interactions and improvement on the older adult's self-perception of her wellbeing.

5.2.1   Use and Adoption

GuessMycaption was played as a casual game, dedicating an average of close to 12 minutes per day (see Figure 10). Nevertheless, days 8 and 9 were two outliers due to a family reunion and the visit of two great-granddaughters living abroad who played with the older adult.

During the first week of deployment the older adult was enthusiastic with the game. The use of body actions as playful interactions to manipulate GuessMyCaption eased its adoption, helping the older adult to incorporate the game as a usual routine in her life.

> OA: "… sometimes around five o'clock [I play GuessMyCaption], it's the time when I am in the bedroom, so I have some time to play"

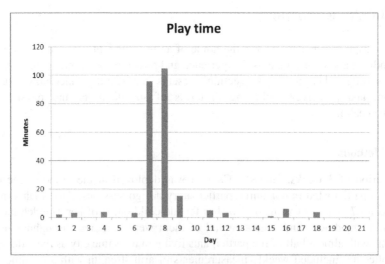

Fig. 5. GuessMyCaption playing time during deployment

The affordances provided by the natural user interface made the system easy to learn. Playing GuessMyCaption with natural and known body actions induced a feeling of certainty on how the game reacts to her arm movements, making clear to her how to perform the actions on the game (see Figure 6), even easier than using current electro-domestics available in her home.

> *OA: "No, no, no, it is not difficult at all. Maybe the challenge is whether or not I remember the context [refereeing to the photograph's caption] of the displayed photograph so I can solve the game. The one I cannot understand [how to use], is my T.V. device [referring to the cable receiver] that is difficult".*

Fig. 6. Older adult playing the matching game

The use of rewards promoted older adult's engagement with GuessMyCaption, and as a consequence catalyzed exercise. The older adult kept playing consistently to earn the printed copy of the digital photographs available in the game.

5.2.2   Perceived Well-Being: Socialization, Physical and Cognitive Health Self-perception

Three major themes on the older adult's self-perception of her wellbeing were observed during the 3 week evaluation: socialization and, physical and cognitive wellbeing.

Socialization. The photographs displayed by GuessMyCaption triggered joyful moments to the older adult while playing the game by herself, and also catalyzed group gatherings to play the game in a "group mode". Indeed, the game was designed for the older adult; however its deployment generated new opportunities for social interaction with her younger relatives. Having the common interest of video games helped the older adult to initiate conversations with younger relatives and invite them to play GuessMyCaption with her.

> *OA: "Just yesterday my great-granddaughters visited me and I invited them to play the game [GuessMyCaption] with me. After a while my granddaughter had to go but the girls did not want to leave. We were having a great time [laughs]"*

GuessMyCaption body based interactions and common themes between the older adult and descendants went beyond the playful interactions and reinforced social traits such as intergenerational conversations.

Cognitive and physical self-perception. During the clinical interview in the pre-deployment phase it was determined that the participant presented minor yet perceptible difficulties raising her arms. Also, she perceived her health to be relatively good for her age.

The regular playing time allowed the older adult to perform the arm exercises in a consistent manner influencing her perception about her health, and that of her relatives. Even though the game presented some false negatives while the older adult performed the circular raise arms gesture, the older adult was aware of the health benefits of repeating the exercise and kept performing the movement until the game reacted properly towards this gesture.

> OA: "The front [referring to the circular arms raise gesture], well sometimes I have to repeat it several times to select the photograph in the middle but that's ok I know it's good for my health. It is like the exercises once my personal doctor prescribed to me a long time ago"

While the measures with the dynamometer do not reflect a significant change in gained strength, the perceived benefit by the older adult was noticeable from the first week of our study. The older adult observed an improvement with her left arm and attributed this change to the exercises performed while playing.

> OA: "I'm feeling really well, my arms feel good. Do you remember that some time ago I told you I had arms-ache? Well something happened, I am sure it was for playing [referring to the use of GuessMyCaption], they do not hurt anymore, especially my left arm".

Furthermore, while the design was intended to require minimum cognitive effort, the older adult observed cognitive challenges in GuessMyCaption. In some occasions the older adult had to be thoughtful while selecting her answer since some of the shared photographs are quite old. Therefore the older adult had to remember the context of the displayed photographs so she could deduct the correct answer. Despite some errors while playing, this thoughtful process alleviated her worries about memory loss and reinforced her certainty of having good memory for her age (see Figure 12).

> OA: "The game has taught me that I still have good memory. I need to have good memory to remember past photos that I have seen and remember who was in the photo or what were they doing so I can select the correct answer."

Fig. 7. Thoughtful playing process

6   Discussion

The primary objective of this work was to explore the adoption and perception of older adults towards ambient casual games that leverage ambient awareness and social capital to encourage exercising and online and offline socialization. Previous research has shown the importance of ambient awareness in family settings [20]; therefore, and given the impact of social networks in the health of older adults health [6, 7, 22], is that we decided the include SNS resources into ambient casual games. GuessMyCaption takes advantage of current gaming technologies (e.g. Kinect sensor and SNSs) to incorporate social resources into ambient casual games and engage players into physical activities that might prevent the loss of arm muscle's strength or flexibility. Leveraging these social resources uncovered different type of uses of ambient casual games; creating private and public spaces for socialization. In a public setting, GuessMyCaption was well adopted and generated feelings of wellbeing and socialization in different manners. It opened opportunities for intergenerational playing, but it was also used as to browse recent photos uploaded to the SNS.

The combination of social resources and natural interfaces into GuessMyCaption facilitated the adoption of the games. GuessMyCaption's natural interaction gestures proved to be enjoyable and easy to perform, and played a crucial role in helping older adults "set up" the environment to start the game. This increased the number of instances the older adult played the game, in contrast with other studies that had emphasized that one of the main barriers when playing games is the time users invest in configuration and personalization. These playful interactions also served to close the gap between younger generations and older adults, as older adults are more open to adopt novel games preferred by younger generations.

Certainly, intergenerational gaming will pose new challenges around adaptability. For example, the gesture recognition algorithm did not contemplate players' physical limitations; limitations that might prevent older adults from playing with younger generations afraid of being outperformed while playing. Thus, games aimed at socialization using shared activities or competition should consider adaptation based on each player's skills and physical capabilities.

Although prior research [23] suggests new technologies for encouraging physical activity should be evaluated in terms of long-term behavior-change, the aim of the present research is not to demonstrate changes in older adults behavior or health, rather we explored perceived well-being. GuessMyCaption results indicate that the social yet simple activity design helped older adults gain increased awareness in self-perception on wellbeing. Open questions remain to explore how these results vary once the novelty effect wears off, however, reflecting from the present research process, we believe ambient casual games could be easily and gradually deployed as new ambient services become available. By doing so, the older adult could try new variations of these games and return at any time to previous games, as gamers usually do. In addition, the fact that these games rely on social media content that is constantly being updated, guarantees that there will continue to be some novelty in the use of the game.

Future longitudinal studies can determine to what extent such long-term behavior changes may be facilitated by this type of games or how this increased perception might be exploited in future designs or other populations –specially, those where socialization plays an important role in health. For example, children, and specially children with autism, can require significant support for both initiating and maintaining social engagements. Using ambient casual games as an assistive technology that takes advantage of social capital could help children to avoid isolation and improve their social cohesion in public spaces such as schools.

Finally, future research is needed to balance "privacy asymmetries" finding an appropriate balance between the privacy that is being lost when moving private information "outside the desktop", and the benefits gained when consulting it [24]. Furthermore, control over the information shared on SNSs might be lost to some degree with this technology; users' privacy preferences might not be enforced properly while the information is transferred to technologically enhanced objects available in a semi-public environment like the home or public spaces like a school. This creates a possible scenario where undesired users can reach information to which they are not entitled.

7   Conclusions and Future Work

The prototype presented here was designed to combine the benefits of ambient games and resources for socialization. Preliminary results suggest positive benefits in older adults' self-perception of wellbeing, evidence of in-person social interactions and the importance of inclusion of social/family components on games. The main contribution of this work is a primary examination of the adoption and perception of older adults towards ambient casual games that encourage online and offline socialization, highlighting that social capital could be used as rewards beyond serving as a support system when exercising. Further analysis is required to investigate the impact of the intervention on the whole family. Also, further research is needed for the design and development of automatic services enabled with Ubicomp sensing that will identify which information is worth displaying and how this rich context would be shared, while preserving privacy concerns.

Acknowledgments. This work was partially funded by CONACYT trough a scholarship provided to the first author.

References

1. Entertainment Software Association: Essential Facts about computer and video game industry. Entertainment Software Association (2012)
2. Lenhart, A., Jones, S., Macgill, A.: Adults and Video Games. Pew Internet & American Life Project (2008)
3. Voida, A., Greenberg, S.: Wii all play: the console game as a computational meeting place. In: Proceedings of the 27th International Conference on Human Factors in Computing Systems, pp. 1559–1568. ACM, Boston (2009)
4. Cornwell, B., Laumann, E.O., Schumm, L.P.: The Social Connectedness of Older Adults: A National Profile. Am. Sociol. Rev. 73, 185–203 (2008)
5. Cagley, M., Lee, M.: Social Support, Networks, and Happiness. Population Reference Bureau (2009)
6. de Belvis, A., Avolio, M., Sicuro, L., Rosano, A., Latini, E., Damiani, G., Ricciardi, W.: Social relationships and HRQL: A cross-sectional survey among older Italian adults. BMC Public Health 8, 348 (2008)
7. Pillai, J.A., Verghese, J.: Social networks and their role in preventing dementia. Indian Journal of Psychiatry 51, 22–28 (2009)
8. Entertainment Software Association: Essential Facts about computer and video game industry. Entertainment Software Association (2011)
9. Garcia-Marin, J., Felix-Navarro, K., Lawrence, E.: Serious Games to Improve the Physical Health of the Elderly: A Categorization Scheme. In: IARIA (ed.) CENTRIC 2011, The Fourth International Conference on Advances in Human-oriented and Personalized Mechanisms, Technologies, and Services, pp. 64–71. IARIA, Barcelona (2011)
10. Doyle, J., Bailey, C., Dromey, B., Scanaill, C.N.: BASE - An interactive technology solution to deliver balance and strength exercises to older adults. In: Pervasive Computing Technologies for Healthcare (PervasiveHealth), 2010 4th International Conference on-NO PERMISSIONS, pp. 1–5 (Year)
11. Fasola, J., Mataric, M.J.: Robot exercise instructor: A socially assistive robot system to monitor and encourage physical exercise for the elderly. In: 2010 IEEE RO-MAN, pp. 416–421 (Year)
12. Gerling, K., Schild, J., Masuch, M.: Exergame Design for Elderly Users: The Case Study of SilverBalance. In: International Conference on Advances in Computer Entertainment Technology, pp. 66–69 (Year)
13. ScienceDaily,
    http://www.sciencedaily.com/releases/2009/06/090611120744.htm
14. Casual Games Association: Casual Games Market Report. Casual Games Association (2007)
15. Russoniello, C.V., O'Brien, K., Parks, J.M.: The effectiveness of casual video games in improving mood and decreasing stress. Journal of Cyber Therapy and Rehabilitation 2, 53–66 (2009)

16. Grimes, A., Kantroo, V., Grinter, R.E.: Let's play!: mobile health games for adults. In: Proceedings of the 12th ACM International Conference on Ubiquitous Computing, pp. 241–250. ACM, Copenhagen (2010)
17. Li, K.A., Counts, S.: Exploring social interactions and attributes of casual multiplayer mobile gaming. In: Proceedings of the 4th International Conference on Mobile Technology, Applications, and Systems and the 1st International Symposium on Computer Human Interaction in Mobile Technology, pp. 696–703. ACM, Singapore (2007)
18. Eyles, M., Eglin, R.: Entering an age of playfulness where persistent, pervasive ambient games create moods and modify behaviour. In: The Third International Conference on Games Research and Development 2007 (Cybergames 2007), Manchester (2007)
19. Eyles, M., Eglin, R.: Ambient games, revealing a route to a world where work is play? Int. J. Comput. Games Technol. 2008, 1–7 (2008)
20. Cornejo, R., Tentori, M., Favela, J.: Ambient Awareness to Strengthen the Family Social Network of Older Adults. Computer Supported Cooperative Work (CSCW) 22, 309–344 (2013)
21. Shotton, J., Fitzgibbon, A., Cook, M., Sharp, T., Finocchio, M., Moore, R., Kipman, A., Blake, A.: Real-time human pose recognition in parts from single depth images. In: Computer Vision and Pattern Recognition, pp. 1297–1304. IEEE (Year)
22. Fratiglioni, L., Wang, H.-X., Ericsson, K., Maytan, M., Winblad, B.: Influence of social network on occurrence of dementia: a community-based longitudinal study. The Lancet 355, 1315–1319 (2000)
23. Klasnja, P., Consolvo, S., Pratt, W.: How to evaluate technologies for health behavior change in HCI research. In: Proceedings of the SIGCHI Conference on Human Factors in Computing Systems, pp. 3063–3072. ACM, Vancouver (2011)
24. Tentori, M., Favela, J., González, V.M.: Quality of Privacy (QoP) for the Design of Ubiquitous HealthcareApplications. Journal of Universal Computer Science 12, 252–269 (2006)

Monitoring Collaboration in Software Processes Using Social Networks

Gabriella C.B. Costa, Francisco Santana, Andréa M. Magdaleno,
and Cláudia M.L. Werner

UFRJ – Federal University of Rio de Janeiro
COPPE – Systems Engineering and Computer Science Department
Zip 21945-970 – Rio de Janeiro – RJ – Brazil – P.O. Box 68511
{gabriellacbc,fwsantana,andrea,werner}@cos.ufrj.br

Abstract. Collaboration monitoring in software process is important to check if the collaboration is indeed happening as planned, but there are few approaches that define how to measure and monitor collaboration. By assessing collaboration during an ongoing process execution, project managers can take corrective actions that might improve the process execution and, consequently, reflect on quality gains of the final product. This research work proposes to evaluate the level of coordination achieved by a running software process through social network analysis metrics.

Keywords: Collaboration, social network analysis, software processes.

1 Introduction

Whitehead *et al.* [1] affirm that "collaboration is pervasive throughout Software Engineering", because most non-trivial software projects require effort and talent of many people working together. However, Software Engineering researchers still discuss which practices, processes, and tools are able to foster and monitor collaboration [2]. We argue that collaboration can be systematically encouraged in software organizations by explicitly considering it as part of the processes.

A software process based approach can be useful to foster collaboration. The assumption that the adopted software process directly influences the quality of the developed product has motivated many organizations to invest in defining processes that are used to produce software products [3]. A software process can be defined as "a coherent set of policies, organizational structures, technologies, procedures, and artifacts required to design, develop, deploy and maintain a software product" [4].

In this paper, we argue that collaboration can be systematically encouraged in software development organizations by explicitly considering collaboration as part of the organization's processes. To manage collaboration in software processes, we created COMPOOTIM [5]. It comprises planning, composing, optimizing, and monitoring software processes with the goal of maximizing collaboration among the members of a team assigned to develop a software project. This paper is dedicated to

N. Baloian et al. (Eds.): CRIWG 2014, LNCS 8658, pp. 89–96, 2014.

planning and **monitoring** stages. We propose to evaluate the level of collaboration (focusing on the coordination aspect) achieved by a running software process, through the analysis of social networks generated with process data execution (actors and tasks) and present a demonstration with data of a toy process to illustrate it.

The remainder of this paper is structured as follows. The next section is dedicated to the collaboration planning stage. Section 3 describes the process execution and data extraction. In Section 4, social networks obtained with the EvolTrack-Process tool are presented and their metrics calculated to monitor its collaboration level. Section 5 discusses some related work. Finally, conclusions are presented in Section 6.

2   Collaboration Planning

To manage collaboration in software processes, we use the COMPOOTIM approach, which aims to provide systematic guidance to plan, compose, optimize, and monitor collaboration in software processes.

In COMPOOTIM, collaboration **planning** should enable software organizations to set goals and define actions to reach higher levels of collaboration. It uses the Collaboration Maturity Model (CollabMM) [6] as a reference to define collaboration levels and the main collaboration characteristics of each level.

CollabMM [6] is divided into four maturity levels: Ad-hoc, Planned, Aware, and Reflexive. Each level contains a set of collaboration practices, considering four aspects to support collaboration: communication, coordination, group memory, and awareness. These practices (PR) were arranged as shown in Table 1.

Table 1. CollabMM Collaboration Practices according to Maturity Levels and Collaboration Aspects

Group Supporting Aspects	Collaboration Practices		
	Level 2 - Planned	Level 3 - Aware	Level 4 - Reflexive
Communication	Communication Planning (PR1)	Distribution of Information (PR5)	Closure (PR9)
Coordination	Group Work Planning (PR2)	Work Monitoring (PR6)	Evaluation (PR10)
Group Memory	Integration of Individual Products (PR3)	Explicit Knowledge Sharing (PR7)	Tacit Knowledge Sharing (PR11)
Awareness	Social Awareness (PR4)	Process Awareness (PR8)	Collaboration Awareness (PR12)

Besides CollabMM, another instrument used in the collaboration planning phase is the Collaboration Measurement Strategy. It was defined and used to estimate the collaboration level of a process and is applied to individual process components, considering that a software process can be seen as a composition of different process components.

Table 2 shows an example of a process component, defined as an encapsulation of data and process behavior at a given level of granularity [14].

Table 2. Process component (Project Responsible Assignment) example

Role: Team Leader	
Collaboration Practices (PR): PR2, PR4	**Collaboration Potential (CP): 4**
Input:	Software Project Information *(Mandatory)*
Output:	Project Open Charter

Using the Collaboration Measurement Strategy, the process engineer can analyze all software process components and determine if it contributes or not for each collaboration practice (PR) of CollabMM. To determine the collaboration potencial (CP), the process component is evaluated in relation to every aspect of supporting collaboration. The component will be scored according to PRs listed in Table 1 to which it complies with. The count of level 2 practices is multiplied by 2 and their level 3 and level 4 counterparts are, respectively, multiplied by 3 and 4. For instance, the process component shown on Table 2, includes PR2 and PR4 practices and its CP is 4 (sum of PR2 and PR4 levels, according to Table 1).

Considering this strategy, the sum of the CPs of all the process components is used to obtain the software process collaboration potential as a whole. The final collaboration level of a complete process is calculated by counting the collaboration practices for each CollabMM level provided by the components comprising the process. A drawback of this strategy is: if the process has some components with high CP, it will raise the process collaboration level, even if the process has most of its components with low values for collaboration potential.

As a third instrument in collaboration planning, we use collaboration characteristics proposed by Santos *et al.*[7] that establishes collaboration features based on SN attributes analysis, using information such as network *density* and *centrality* and collaboration levels defined by CollabMM.

Before starting the process execution and collaboration monitoring, it is necessary to establish the degree of planned coordination. The practices related to the coordination aspect (PR2, PR6 and PR10) and process components in which they appear are listed in Table 3. As shown in this table, most components have practices related to Level 2 (Planned). Level 3 (Aware) was in 'second place', with only one task less. Based on this, we assume the coordination level of this process should achieve Level 2 (Planned). After the collaboration planning stage, in order to monitor collaboration, we need to verify if it was achieved during the process execution as planned.

Table 3. Software process components and associated CollabMM practices

	PR2 Level 2 - Planned	PR6 Level 3 - Aware	PR10 Level 4 – Reflexive
Process components	T2 / T3 / T5 / T9	T7 / T8 / T9	T9
Total	4	3	1

3   Process Execution

As an initial step for process execution, a software process generated using the COMPOOTIM approach [5] was chosen and it was modeled on a Business Process Management System (BPMS) called Bonita BPM[1], as can be seen in Fig. 1.

The software process used in this research came from a real company in Brazil and is based on an agile software development model. The software project was developed in Java by a team with 08 members, and had about 6 months of duration. The specific process was composed by nine process components.

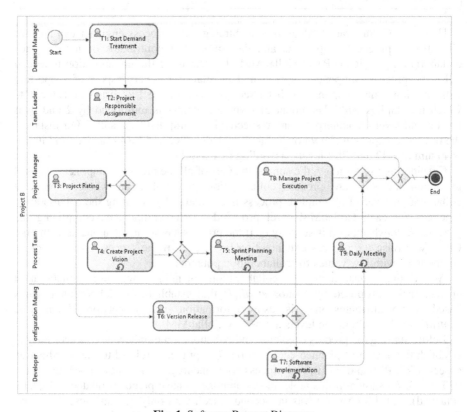

Fig. 1. Software Process Diagram

In order to execute the software process, all components of the chosen process example were modeled as tasks with their roles in Bonita BPM (Fig. 1). Then, 5 distinct process instances were executed. In these executions, 13 actors were used to do all the 9 process tasks. The actors involved in this process execution were divided among the following roles: 8 *developers*, 2 *configuration managers*, 2 *project managers* and 1 *demand manager*. A *project manager* could also act as a *team leader* and all actors (except the *demand manager*) were part of the *process team*.

---

[1] http://www.bonitasoft.com

4   Collaboration Monitoring

In order to monitor collaboration, our approach has four steps:

(i) Process Execution Data Extracting: The first step consists on obtaining data from the software process execution. Bonita BPMS stores process definition and instances data in a database management system and also offers a web REST API to access this information. We developed a data extraction module over the REST API that enables automated extraction of process definition and execution, and transforms this collected data on models, considering *actors* and *tasks* performed data.

(ii) Socio-technical Networks Generation: The collected information needs to be converted into graphs, enabling to evaluate collaboration using SNA metrics. With this aim, we used EvolTrack-Process as an instrument to the visualization and analysis of SN in this step. EvolTrack-Process expands the range of viewpoints offered by EvolTrack [8], adding support to software/business processes. Using EvolTrack-Process, five undirected unweighted graphs have been generated, representing each execution (an example can be seen in Fig. 2).

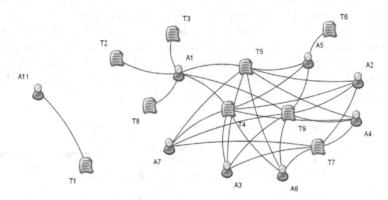

Fig. 2. Socio-technical network showing the a process execution

In generated networks, each node represents either a task (letter 'T' (T1, T2 ... T9)) or an actor letter 'A' (A1, A2, A3 ...) and the edges establish that a specific actor participated in a determined task. In Fig. 2 we can see, for example: task T1 was only performed by actor A11 (or actor A11 performed only task T1); task T5 was performed by many actors (A1, A8 ...); T2, T3 and T8 were executed by only one actor (A1). A sixth SN graph was generated by consolidating data from all the executions, using EvolTrack-Process. This SN can be seen in Fig. 3. The following comments can be made about this figure: (i) task T1 was always performed only by actor A11 in all executed process instances; (ii) most of the actors in SN participate in the execution of tasks T4, T5, T7 and T9, which makes clear the collaborative nature of these tasks; (iii) there is not an actor acting as a 'leader', concentrating the execution of more than one process tasks; (iv) SN members balance their activities and there are no prominent actors in the network.

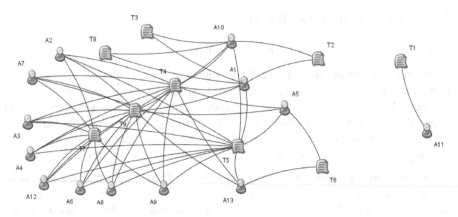

Fig. 3. Socio-technical network showing all executed process instances

(iii) Metrics Evaluation: the selected SN properties were examined according to the method proposed by Santos *et al.*[7], which analyzes collaboration based on social network attributes, such as *network density, network centrality* degree and *network betweenness* centrality. The SN properties values obtained were: *network density*: 22.8%; *network centrality*: 1.8% and *network betweenness*: 13.72%. According to the achieved SN values, one of them (*network density*) characterizes the coordination at process Level 2 (Planned) and the other two values at Level 4 (Reflexive) as can be seen in Table 4. Considering that most of the characteristics values fall within Level 4, we assume that the process achieved Level 4 (Reflexive) of coordination.

Table 4. Achieved coordination levels, based on [7]

	Centralized Coordination Level 2 - Planned	Multiple Coordination Level 3 - Aware	Distributed Coordination Level 4 - Reflexive
Network Density	Low (0,01% a 30,00%)	Medium (30,01% a 70,00%)	High (70,01% a 100,00%)
Network Centrality	High (70,01% a 100,00%)	Medium (30,01% a 70,00%)	Low (0,00% a 30,00%)
Network Betweenness	High (50,01% a 100,00%)	Low (0,00% a 50,00%)	Low (0,00% a 30,00%)

(iv) Analysis of planned and achieved coordination levels: make a comparison between the planned and achieved coordination levels of the software process. Through the results obtained with the example, the planned level of coordination (Level 2 - Planned) does not correspond to the level obtained after the process simulations (Level 4 - Reflexive). It means that a network with a high centralization in a leader was expected, but a distributed coordination was observed. If the absence of this control in tasks and information flow is really not desired for this process, actions can be taken to ensure that the collaborative process occurs as planned. Based on the monitoring of this information, the project manager will be able to make decisions and interfere in the process in execution. For instance, it is possible to plan, using CollabMM collaboration levels and practices, how to change process components to obtain the desired collaboration outcomes for the process.

5   Related Work

Although not focusing specifically on collaboration, Rilling, J. *et al.* [9] present a custom process-centered approach that displays several visualizations of a software maintenance process in an integrated fashion, allowing, for instance, traceability among process requirements and activities to resources (e.g., tools, techniques and users). In another line of work, Suntinger and Obweger [10] present the Event Tunnel, a visualization applied to business processes that assists on the monitoring and optimization of event-based business process by enabling the visualization of these events. Claret [11] defines a set of coordination metrics based on the GQM (Goal, Question, Metric) approach that can be used by managers in monitoring collaboration processes. However, these metrics have not been defined yet.

Erol *et al.* [12] presented new opportunities provided by social software for the design of business processes and discuss the benefits of social software in the business process life cycle. Bruno *et al.* [13] define Agile Business Process Management and advocate that social software allows to satisfy the key requirements for enabling agile BPM. We still do not consider the use of social software in order to foster collaboration in software execution. This issue needs to be analyzed as future work.

6   Conclusion

This work presented how to monitor collaboration, specifying the level of coordination achieved by a running software process, through the analysis of the SN generated by the process data execution. It uses the COMPOOTIM approach to obtain a software process model, CollabMM to define the planned and the achieved collaboration levels and a social network tool, EvolTrack-Process. A real software process was executed and monitored in order to obtain the level of achieved coordination. With this information, the project manager can make decisions / interfere with the process in order to change the collaboration practices, as well as changing process components to obtain the desired collaboration outcomes for the process or creating new strategies aiming to access the expected collaboration level.

More detailed analysis of this example must still be performed with a larger number of instances and other software processes need to be simulated. To complement this work, a real evaluation in industry is planned.

Another future work is to verify how to evaluate other aspects of collaboration (communication, group memory, and awareness). Finally, we can also evolve this approach in order to inform the project manager about which process components can be changed / introduced to obtain the planned collaboration level.

References

[1] Whitehead, J., et al.: Collaborative Software Engineering: Concepts and Techniques. Collaborative Software Engineering 1, 1–30 (2010)

[2] Mistrik, I., et al.: Collaborative Software Engineering. Springer, Heidelberg (2010)

[3] Cugola, G., Ghezzi, C.: Software processes: A retrospective and a path to the future. Software Process Improvement and Practice Journal 4, 101–123 (1998)

[4] Fuggetta, A.: Software process: a roadmap. In: Proceedings of the Conference on The Future of Software Engineering, pp. 25–34. ACM, Limerick (2000)

[5] Magdaleno, A.M., et al.: COMPOOTIM: An Approach to Software Processes Composition and Optimization. In: XV Iberoamerican Conference on Software Engineering (CIbSE), Buenos Aires, Argentina, pp. 42–55 (2012)

[6] Magdaleno, A.M., et al.: A Maturity Model to Promote Collaboration in Business Processes. International Journal of Business Process Integration and Management (IJBPIM) 4, 111–123 (2009)

[7] Santos, T.A.L., et al.: Bringing Out Collaboration in Software Development Social Networks. In: International Conference on Product Focused Software Development and Process Improvement (PROFES) - Short Papers, pp. 18–21. ACM, Torre Canne (2011)

[8] Cepeda, R.S.V., et al.: EvolTrack: Improving Design Evolution Awareness in Software Development. Journal of the Brazilian Computer Society (JBCS) 15(2), 117–131 (2010)

[9] Rilling, J., et al.: Software Visualization - A Process Perspective. In: 4th IEEE International Workshop on Visualizing Software for Understanding and Analysis (VISSOFT), Ontario, Canada, pp. 10–17 (2007)

[10] Suntinger, M., Obweger, H.: The Event Tunnel: Interactive Visualization of Complex Event Streams for Business Process Pattern Analysis. In: IEEE Pacific Visualization Simposium, Kyoto, Japan, pp. 111–118 (2008)

[11] Claret, M.D.: Métricas para Colaboração em Processos de Negócio. In: Workshop de Teses e Dissertações - Simpósio Brasileiro de Sistemas de Informação (SBSI), João Pessoa, PB, Brasil, pp. 31–36 (2013) (in Portuguese)

[12] Erol, S., et al.: Combining BPM and social software: contradiction or chance? Journal of Software Maintenance and Evolution: Research and Practice, 449–476 (2010)

[13] Bruno, G., et al.: Key challenges for enabling agile BPM with social software. Journal of Software Maintenance and Evolution: Research and Practice 23, 297–326 (2011)

[14] Gary, K.A., Lindquist, T.E.: Cooperating process components. In: International Computer Software and Applications Conference (COMPSAC), pp. 218–223. Phoenix, AZ (1999)

Supporting Teleconsulting with Text Mining: Continuing Professional Development in the TelehealthRS Project

Fábio Damasceno[1], Eliseo Reategui[1], Carlos André Aita Schmitz[2],
Erno Harzheim[2], and Daniel Epstein[1]

[1] Federal University of Rio Grande do Sul,
Postgraduate Program in Informatics on Education, Brazil
[2] Federal University of Rio Grande do Sul,
Postgraduate Program in Epidemiology, Brazil

Abstract. In the primary care scenario, telehealth appears as an option for continuing professional development of the professionals involved. Questions submitted by physicians and its respective answers in the Brazilian health ministry telehealth platform were mined using a text mining tool. Graphs about concepts present in database were created using this tool, turning clearer subjects addressed by teleconsultants. A questionnaire, addressing the current answer methodology and possibilities with graphs, was answered by teleconsultants of a telehealth Center, the TelehealthRS Project. Answers obtained showed the importance given by teleconsultants about the current answer methodology and the conviction that graphs can turn the request answering easier and better developed. This is guiding an implementation on the current Brazilian health ministry platform that will impact in the accuracy and speed of response offered to the professional, enhancing their training and helping in the effectiveness of their continuing professional development.

Keywords: Primary Care, Text mining, Continuing professional development.

1 Introduction

The Brazilian scenario of primary health care (PHC) includes a profile of family health teams (FHT) that often lacks qualification in terms of primary care [1]. The evolution of scientific knowledge is turning professional training in an ongoing process to maintain the quality of services [2]. In this context, the telehealth – defined as the use of information and telecommunication activities remotely related to health – appear as an option for the continuing professional development of these professionals. Telehealth becomes a way to develop continuing professional development activities [1].

Many professionals have to deal with clinical questions in every work shift [3]. In this scenario, the practice of referral is quite common, in which the physician refers the patient to another colleague of a given specialty. Telehealth technologies can reduce the number of unnecessary "referrals" and the possibility of iatrogenic problems[1]. However, in telehealth solutions it is often the case that previous knowledge on

---

[1] Any patient harm (lethal or not) resulting from medical intervention.

N. Baloian et al. (Eds.): CRIWG 2014, LNCS 8658, pp. 97–104, 2014.

answered questions is not reused in answering new questions. This gap is addressed in the work proposed in this paper.

The article is divided as follows: section two shows Brazilian epidemiological situation so that one can understand the context of this work; its subsection 2.1 presents family and community medicine as valuable alternative for the Brazilian scenario; subsection 2.2 shows continuing professional development as a policy for training of family health teams; subsection 2.3 presents telehealth as a proper tool for concretize this policy; section three introduces text mining and Sobek Tool, used in proposed work, which is detailed in section four.

2   Brazilian Epidemiological Situation and Primary Care

This project has been developed in Brazil, so it is important to put into context the Brazilian epidemiological situation. Brazil has nowadays an ever growing aging population (demographic transition), accompanied by substantial changes in eating patterns, with increased overweight and obesity problems (nutrition transition) linked to physical inactivity and a reduction in acute conditions, as well as an increasing number of chronic health conditions (epidemiologic transition). In this scenario, health care systems have to be able to deal with such the problems related to modern epidemics of chronic conditions. In the Brazilian context, the country is facing not only an epidemiological transition, but also a rapid demographic transition. The Brazilian population will continue to grow in the coming decades, with growing numbers of ageing citizens and a corresponding increase in chronic conditions [4].

2.1   Family and Community Medicine and Primary Health Care

The analysis of health system indicates that the growth of focal specialization does not bring the expected benefits. Primary care proposed a change to this model [5], being founded on methods, evidence-based practices and socially acceptable technologies available to all, at costs that the community and the country could afford [6].

In this scenario, family and community medicine became valuable alternatives [5]. Although the implementation of these programs has been carried out relatively quickly in Brazil, a problem arose in this context: the availability of trained personnel to compose these teams, as well as the selection of people with a work profile not always suitable for task. This problem compromised the legitimacy of the program, combined with a high number of professionals willing to join the program because of the prospect of immediate employment with salaries above average [5].

2.2   Continuing Professional Development (CPD) as an Alternative in the Training of Health Professionals

The continuing professional development is a necessity for professionals from all areas, seeking the construction of knowledge and the development of new skills and interdisciplinary work [7]. The permanent update of health professionals is quite complex, due to the speed with which knowledge and technological knowledge are renewed in health field, as in addition to the distribution of professionals and their services. Studies on development of new ways to deal with personnel training problems in the health sector propagated throughout Latin America, fostering discussion and production of works in Brazil [8]. In CPD an individual takes control of its own learning and development, engaging in a process of action and reflection. All professionals should have learning opportunities to maintain and enhance their skills [9,10,11].

2.3   Telehealth

Health professionals have to answer clinical questions quite often during their practice. They must have at their disposition a process to express their doubts. Here, telehealth is a viable option. Telehealth is the use of information and telecommunication activities remotely related to health at various levels (primary, secondary and tertiary), enabling the interaction between health professionals or between professionals and patients [1]. In Brazil, telehealth includes the provision of care to support family health teams (FHT) through two modalities: asynchronous (text) and synchronous (teleconsulting video). Both types are characterized as CPD activities. Brazilian health ministry offers a standard telehealth platform, in which the work proposed here takes place. The teleconsulting process itself involves three stages. The initial request is followed by a teleregulation and a subsequent response.

The requesting process is generally conducted in a virtual environment, such as the health ministry platform. It involves a professional of a particular unit and one or more health professionals linked to a telehealth Center.

This process is mediated by a professional who is called teleregulator. He/She mediates the process considering the profession of the requester and the content of the request, choosing the most appropriate teleconsultant. The teleregulator can select from up to five codes to classify the request in the health ministry platform, from International Classification of Diseases [12] and International Classification of primary care [13]. This step aims to contribute in the classification of requests, helping the teleconsultant see its context. These codes used in the requests classification are not search criteria available for the teleconsultant in his answering activities. This shows a gap that will be detailed later.

The answer to a request should be based on the best available evidence for the context of primary care [1] [14]. A search in the current Brazilian health ministry telehealth database is the first step in formulating the response. Checking if a similar issue has already been resolved is an interesting approach, as it can speed up and enhance the whole process. The next step includes the addition of bibliographic references, links and attachments to the answer developed.

In the work of [1], it has been shown that in the experience of TelehealthRS, at every two teleconsultations requested by medical professionals, one patient referral to a higher level of care is avoided. The structure previously presented in this section covers the process of teleconsulting applied by TelehealthRS, approved by the ministry of health [14].

3   Text Mining and Representing Information Extracted from Texts with Sobek Tool

Text mining has its earliest origins around 1960 when Hans Luhn and Lauren Doyle realized that the frequency and distribution of relevant words within text information were notorious for their understanding [15,16]. Text mining can be defined as a process in which a user interacts with a document collection over time using a set of analysis tools. Similarly to data mining, text mining aims for extraction of useful information from specific data sources, by identifying and exploiting patterns. In this case, data sources are collections of documents, in which the interesting patterns are found along unstructured textual data in these documents [17].

The text mining tool Sobek [18] was developed using the n-simple distance graph model, in which nodes represent the main terms found in the text, and the edges used to link nodes represent adjacency information. Therefore, nodes and edges represent how the terms appear together in the text. The method relies on a parameter n to extract the compound concepts with more than one word. According to this parameter, a combination of the current word with the n subsequent words is created, attempting to identify terms represented by the most frequent group of words appearing in the text [19]. A notable work [20] made with Sobek shows the benefits that graphs (which served as base for concept networks) can give when writing about a topic, which can be Telehealth activities in our case.

4   Methodology

In an initial step, Sobek has been used to identify patterns of terms in requests done from October, 2012 to January, 2014, totaling 558 requests. A small Thesaurus with medical terminology was built in order to help the tool identify terms that were expressed in the requests in different ways.

4.1   Teleconsultants' Perception of the Graphs Extracted

With the graphs of the requests and answers, TelehealthRS teleconsultants were questioned about their perceptions on how those graphical representations could support them in their daily work. Twenty-six out of thirty-three teleconsultants participated in the assessment. The questions were the following:

1. In the health ministry's answer protocol the teleconsultants must query the current database before the answer formulation for the request. Do you believe this is important?
2. In the present moment, the database search is based on the 'description' field, where the request is written. Do you believe that the answers to these requests could benefit your work by becoming a search criteria too?
3. In the teleregulator work there is a step in which an ICPC-2/ICD-10 code have to be informed. Do you believe that these codes could benefit your work by becoming another search criteria?

4. Suppose you have to answer the following request: "I would like to know how to address addiction in smokers". The following graph (see Fig. 1) represents related information extracted from the database of previous questions/answers by medical professionals in the telehealth platform.

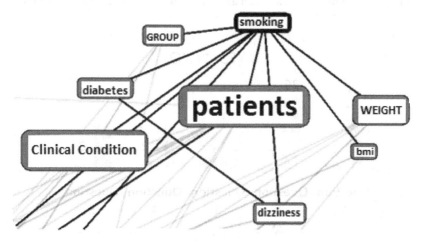

Fig. 1. Graph about concept 'smoking' present in the physician requests

A relationship between the terms 'smoking', 'diabetes', 'group', 'weight' can be seen, as well other relationships. Do you believe that this graphical representation can help your work in searching for previous requests?

5. The following graph (see Fig. 2) shows relationships found for the concept 'smoking' present in the answers made for physician's requests.

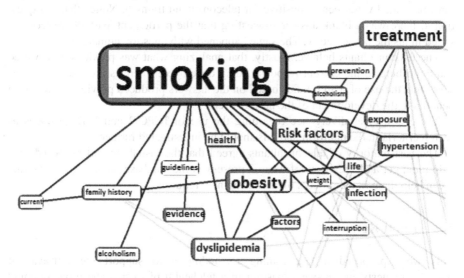

Fig. 2. Graph about concept 'smoking' present in the answers for physician requests

Currently it is possible to search for requests based on their 'description', field written by the health professional. Do you believe that this graphical representation can benefit searching requests?

The teleconsultants could answer each question with 'Yes' or 'No', and they could write an optional commentary for each answer. Figure 3 shows the results obtained:

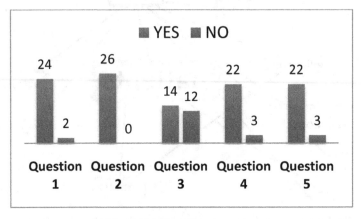

Fig. 3. Teleconsultants' answers

As reported by the teleconsultants, it is difficult to find requests already answered in the past, due to the lack of knowledge about the exact terms that were used in the description field. The relationship among terms can optimize this search according to some of the teleconsultants' comments. It was quoted as a feature that would be very useful to take advantage of the relationships present in the graphs. Besides, the features presented were seen as positive for teleconsultant training. Notice that, for question 4, there was a blank answer suggesting that the participant 'partially agreed', as stated in his the commentary. The same happened with question number 5.

The teleconsultants believe, mostly, that analyzing what was previously answered can contribute to the current response being developed. There is a consensus in the sense that the use of search fields based on the answers produced previously can bring benefits to the development of the new answers.

There is a clear divergence regarding the use of ICPC-2 and ICD-10 codes as search criteria in the platform and this should be investigated further.

The vast majority of teleconsultants agreed that the visual support proposed may help them get a clearer view of what is present in the database requests/answers, and it can help them in the process of answering new requests.

5   Final Remarks

This work proposed a text mining methodology to support the retrieval of previous requests and answers existing in a telehealth platform. The project should be expanded to encompass requests and their respective answers of other professionals – dentists, nurses, community health workers and other health professions present in the health ministry's platform database. With the teleconsultants answers we intend to obtain graphs that may be also relevant for other classes of health professionals mentioned earlier, totaling over 2566 requests and responses.

A new module for the telehealth platform will include these graphs, with the intent to present the graphs to the teleconsultants at the time they receive a particular request.

Another possibility was perceived during the implementation of the Thesaurus: some medical concepts appearing in the graphs had the same meaning or were acronyms. E.g. AIDS and Acquired Immune Deficiency Syndrome, Infarct and heart attack. Since they are considered individually in the mining process, this kind of information should improve the extraction of graphs now and in future steps of the project.

References

[1] Filho, E.D.C., et al.: Telessaúde para Atenção Primária. Secretaria de Atenção à Saúde, Ministério da saúde (2012)

[2] Campos, F., Haddad, A., Wen, C., Alkmin, M., Cury, M.: The National Telehealth Program in Brazil: an instrument of support for primary health care. Latin American Journal of Telehealth (2009)

[3] Gorman, P.N., Helfand, M.: Information seeking in primary care: how physicians choose which clinical questions to pursue and which to leave unanswered. Med Decision Making (1995)

[4] Mendes, E.: As redes de atenção à saúde. Ciência e Saúde Coletiva (2010)

[5] Anderson, M., Gusso, G., Filho, E.: Medicina de Família e Comunidade: especialistas em integralidade. Revista Atenção Primária em Saúde, vol. 8 (January/July 2005)

[6] González, W.: Atención primaria de salud en acción: Editorial Nacional de Salud y Seguridad Social (EDNASSS) (2006)

[7] Oliveira, M.A.: Educação à Distância como estratégia para a educação permanente. Revista Brasileira de Enfermagem - REBEn (2007)

[8] Lopes, S., Piovesan, É., Melo, L., Pereira, M.: Potencialidades da educação permanente para a transformação das práticas de saúde. Comunicação em Ciências da Saúde (2006)

[9] Peck, C., McCall, M., McLaren, B., Rotem, T.: Continuing medical education and continuing professional development: international comparisons. British Medical Journal (February 2000)

[10] Gibbs, T., Bridgen, D., Hellenberg, D.: Continuing Professional Development. South African Family Practice (2005)

[11] Grant, J., Stanton, F.: The effectiveness of continuing professional development Joint Centre for Educacion in Medicine. Reino Unido (1988)

[12] World Health Organization, International Classification of Diseases, manual of the International Statistical Classification of diseases, injuries and causes of death: 10th revision (1993)

[13] Sampaio, M., Almeida, M., Coeli, C., Moreno, A., Camargo Jr., K.: International Classification of Primary Care: A Systematic Review. Meeting of the who collaborating centres for the family of international classifications (October 2009)

[14] Ministério da saúde, Manual da Telessaúde para Atenção Básica/Atenção Primária em Saúde - Protocolo de Resposta (2012a)

[15] Sharp, M.: Text Mining.: Rutgers University (2001)

[16] Soy, S.: (2003),
http://www.gslis.utexas.edu/~ssoy/organizing/l391d2c.htm

[17] Feldman, R., Sanger, J.: The Text Mining Handbook - Advanced Approaches Analyzing Unstructered Data. Cambridge University Press (2007)

[18] Reategui, E., Klemann, M., Epstein, D., Lorenzatti, A.: Sobek: a Text Mining Tool for Educational Applications. In: Int'l Conf. Data Mining, DMIN 2011 (2011)

[19] Schenker, A.: Graph-Theoretic Techniques for Web Content Mining. PhD thesis. University of South Florida, PhD Thesis (2003)

[20] Macedo, A., Behar, P., Terra, B.: Monitoring interaction and collective text production through text mining. ETD - Educação Termática Digital, pp. 67–83 (2014)

Defining a Design Space for Persuasive Cooperative Interactions in Mobile Exertion Applications

Luís Duarte, Paulo Ribeiro, Tiago Guerreiro, and Luís Carriço

Departamento de Informática, Faculdade de Ciências da Universidade de Lisboa,
Edifício C6, Campo-Grande, 1749-016 Lisboa, Portugal
{lduarte,pribeiro}@lasige.di.fc.ul.pt,
{tjvg,lmc}@di.fc.ul.pt

Abstract. This paper presents a design space for persuasive cooperative persuasive interactions for mobile exertion applications. This type of software bridges entertainment with workout activities, providing users with intuitive and fun ways to track their performances. Persuasion and, in particular, cooperative interactions play a pivotal role in user motivation – they are paramount to convince individuals to pursue their goals and overcome obstacles. Applications of this type have started to offer more diverse functionalities, often becoming difficult to label and pinpoint the type of design cues being employed to foster these persuasive and cooperative facets. In this article we propose a design space for this type of features in such software. The design exercise roots itself in existing literature and the results for an online survey we deployed to assess usage habits for this type of applications. Validation is underway and will be achieved via the creation of an application that covers the proposed dimensions.

Keywords: Mobile Exertion Applications, Design Space, Cooperation, Persuasion.

1   Introduction

Sedentary lifestyles are known to originate health conditions that could be prevented or even minimized by exercising a few hours a day [12][9][4]. The increasingly pervasive nature of work in our lives often inhibits individuals from spending a few minutes a day exercising, attempting to improve their physical and cognitive conditions. Unfortunately, many do not have the time to spend exercising, either because they are working late or because there are other matters that take that little time away. Other reasons for this non-existent exercising routine is the physical discomfort that comes associated with these particular kinds of activities [11] and lack of proactive partners to join in on the endeavour. The introduction of specialized mobile exertion applications (MEA) paved the way for users to engage more frequently in this type of activities.

In this paper we propose a design space for the inclusion for cooperative interactions in MEAs. We argue that cooperation is pivotal to engage users in this

N. Baloian et al. (Eds.): CRIWG 2014, LNCS 8658, pp. 105–112, 2014.

type of activities and, ultimately, lead them towards the goal of keeping in shape. The respectable number of products of this type available in different mobile application stores and the continuous interest of researchers in exploring this area testify its importance. Given the importance of cooperative play / interaction in the context of exercising, we considered to be of relevance to address existing caveats regarding the design of MEAs encompassing such features. To our knowledge, no research has attempted at consolidating the existing expertise about the role of cooperation and persuasion in MEAs into a design space. Our goal consists in providing a broad coverage of the dimensions related to persuasion and cooperation featured in MEAs.

2   Related Work

Our review of existing literature focuses on two relevant topics for this research. The first concerns the purpose and characteristics of MEAs, addressing their main features, usage scenarios and benefits. Lastly, we bridge the usage of persuasive technology with the concept cooperation, how they intertwine between each other and which types of cooperation are more commonly seen in the context of MEAs.

Mobile Exertion Applications
Mobile Exertion Applications (MEAs) are tools which translate into a set of effort systems assisting amateur and professional athletes during exercising activities. Endomondo Sports Tracker[1], Nike+[2] or Runtastic[3] are examples of such applications. Typically, these tools are used for hiking, jogging or cycling. By capitalizing on the ubiquitous nature and feature sets offered by modern smartphones such as GPS tracking or health sensor add-ons, developers and researchers envisioned the creation of these applications which aid users in tracking their exertion activities.

MEAs have a strong goal in granting users with physical benefit [6][12][13]. Factors like motivation, physical characteristics and experience contribute to the level of exertion [10]. At the same time, regular physical activity grants cognitive benefits such as improvements in mood, concentration and memory [4], as well as diminishing anxiety or stress [6][12][13]. Most MEAs offer a common set of functionalities: users are able to log their workout sessions for posterior usage as a comparison benchmark with future sessions. These are typically referred to as ghost recordings. Some MEAs also offer back-office tools allowing users to visualize their calories expenditure and performance variation over time.

Role of Persuasion and Cooperation
Maximizing the physical and cognitive gain from these applications requires the user's commitment. Motivation emerges as a pivotal factor to entice users into pursuing these goals. Persuading individuals into engaging in physical activity has been the subject of study of designers and researchers alike [4]. Existing literature

---

[1] Endomondo: http://www.endomondo.com/
[2] Nike+: https://secure-nikeplus.nike.com/plus/
[3] https://www.runtastic.com/

suggests some factors contributing positively towards the adoption of MEAs [3]. In the context of our research, we are focused on one particular aspect: how societal features are fostered as a persuasive factor in MEAs. B. J. Fogg has done extensive research in the domain of persuasion, often tying it with societal factors [5]. Existing research testifies the important of cooperation in user engagement and even performance [8]. Engaging in similar activities with friends or acquaintances has been shown to provide more pleasant experiences than when total strangers are involved. The impact of peer pressure in motivating groups of users is also paramount for the success of certain products and services. MEAs are no exception. Developers typically implement features which allow users to tackle each other's' challenges and compare the results between each other. More recently, the ability to monitor workout sessions live has also spawned new ways for groups of people to interact with each other and exert pressure to motivate each other.

3   Mobile Exertion Applications' Usage Trends

In order to properly create a design space we are required to go beyond the brief review of related literature about MEAs we presented. The first step in envisioning it encompassed deploying an online survey to extract information regarding how individuals utilize MEAs and their preferred setups while working out.

We established a set of goals which capitalized on the related literature we discussed and aimed at bridging the three key components of this research: MEAs, persuasion and cooperation. In the end our objectives consisted in:

- Identifying the most popular MEAs.
- Identifying the most regularly used features.
- Assessing how individuals utilize their MEAs with their peers.
- Assessing how individuals keep track of their progress prior, during and after each workout session.

We distributed an online questionnaire which was disseminated via mailing lists, social networks and direct request. A total of 194 participants responded to the call. Most subjects are young adults between 18 and 35 years old (80% of the subject base), with a gender distribution of 56% males and 44% females.

Technology Usage Characterization
Smartphones were the most used mobile device by our participants (56%). Tablets and basic mobile phones garnered the same percentage (21% each).

In order to run mobile exertion products, these devices need to provide a few set of technologies. Most of these applications heavily use the devices´ integrated GPS sensors and connections to the Internet. As such, we asked our participants if their devices provided these features. 136 participants stated that their devices provided them with GPS functionality (42%). As for Internet connections, 150 participants stated that their devices had 3G telecommunication networks (46%). 34 participants stated to not have access to any of these technologies (10%). 123 participants stating their smartphone provided them with both GPS and 3G communication and 46 participants stating the same for their tablets.

Workout Habits

Most participants led a good relationship with exertion activities, with 51% reporting to at least exercise 2 or 3 times a week. 6% stated to only exercising at least once a month and 14% reported exercising sporadically. 8% stated to not exercise at all. 66% of the population spent between one and two hours in their workout sessions, while only 6% spent more than two hours on their workouts. The favoured sport practiced by the participants was running / jogging (38%). As for the reasons behind their exercising, the indicated motivation was to "keep fit" (31%), followed by "as a hobby" (20%) and to "deal with stress" (19%).

MEA Usage

When asked how familiar they were with MEA´s, most participants stated to have no knowledge of them (38%), while others knew about them but did not use them (32%). Only 59 participants stated that they either have used them (16%) or use them regularly (14%). For these, Endomondo Sports Tracker, Nike+ and RunKeeper[4] were their MEA´s of choice. Some users preferred other non-mainstream applications such as Strava[5], GetRunning[6] or FitBit[7].

Participants with MEA experience use these products because they display personal progress (40%) and because they are able to motivate users to do better (32%). Other reasons chosen by participants ranged from "It makes exercising more fun" (15%), "It allows me to share my results with other people" (8%) and "It allows me to compare my results with other people" (8%), to more specific reasons such as "It allows me to keep a log of my exercises", "Allows me to see info about my tracks" and "Turns it into a game". For MEA power users, the main reason to use the software is the ability to motivate them (36%), followed by the ability to show the users´ progress (34%). We were also interested in finding how often participants used MEA´s. For those that stated to either use them regularly or to have used them, the typical frequency was of at least once a week (15%).

Motivation Catalysts

We tried to assess the impact of motivational strategies which may or may not be present in MEA´s, but which users recur to, prior to or during workout activities. For our purposes, we disregarded participants who stated to either having no knowledge of such strategies or that they have never used them before (effectively reducing the sample to 59 participants). We also looked into persuasive cues usage frequency. The following list covers the most common persuasive approaches:

---

[4] RunKeeper: http://runkeeper.com/
[5] Strava: http://www.strava.com/
[6] GetRunning: http://splendid-things.co.uk/getrunning/
[7] FitBit: http://www.fitbit.com/

- **Performance Messages** – The presence of performance messages containing information about the user's progress made 57 participants feel more motivated (96%). 27 participants reported to use this functionality frequently (45%) and 17 stated to use it rarely (29%).
- **Music Players** – Listening to music while working out is a very common feature used in mobile exertion products. From this study, 22 participants stated that using music player features made them feel more motivated (38%). 63% of the users use persuasive music features frequently.

Role of Peers
As far as social habits are concerned, 42% of the participants stated to exercise by themselves, while 32% stated to doing it in the company of a friend or colleague and 26% within a group.

A second facet of how peers interact between each other concerns the momentum they build before (arranging for the meeting), during (observing each other and cheering for one another) and after (commenting on how good / bad their performance was) the workout session. Since modern MEAs often give support to these features we questioned users about their opinions, reaching the following results:

- **Competing against yourself** – The existence of competition against oneself proved to be the feature with higher positive influence on the participants´ motivation. From the 33 participants that replied to this strategy having some influence on their motivation, 78% stated they felt more motivated. Of these, 49% compete against themselves frequently. Competition against oneself is a natural benchmark for simple performance comparisons. MEA´s often call this feature "ghost" mode since a user is challenging a "ghost" recording of him / herself or of another person in the past, enhancing the importance of cooperation to improve one's condition.
- **Competing against others** – 28 participants responded they use another person as a benchmark for their workout activities and it has some influence on their motivation. 51% of these participants perceive this strategy as being a positive persuasive approach, increasing their motivation. Of these, only 28% recurred to this strategy with some frequency.

4   A Design Space for Cooperative Mobile Exertion Applications

In light of the review of relevant literature in this domain and the results obtained from the online questionnaire we distributed, we envisioned a design space for MEAs with a focus on the cooperative aspect of these applications and related activities. The design space encompasses a set of dimensions and categories which are used to foster cooperative and persuasive interactions between the individual working out and other individuals whether they also participate in the session directly (i.e. other athletes) or not (i.e. audiences).

Activities Overview

As pointed in the questionnaire's results, in the context of exertion activities and, in particular, in the usage of MEAs, persuading via cooperation may be present in events other than the workout session itself, among which:

- **Workout session arrangement & anticipation** – users often pre-emptively arrange for the workout session establishing where it will take place at what time and the duration.
- **Workout session discussion** – after a workout session users also often use social networks or dedicated MEA software to share their accomplishments and their workout performance evolution.

These two activities which take place outside the duration of the workout session possesses a common dimension – time, with one occurring before the session and the latter after it. During the exertion session itself, one can find other relations:

- **Exercising with a human peer** – this is a type of activity which does not necessarily require a MEA to come to fruition (although it can be enhanced by one). One can also state that this type of activity requires the intervenients to perform at the same time and at the same location for it to make sense.
- **Exercising with a virtual peer** – on the other side of the spectrum we can encounter a type of event which is possible due to the usage of MEAs, consisting in working out with virtual peers. The transition from real to virtual peers introduces new possibilities. These ghost recordings can contemplate sessions of users who performed that activity in the same location in which the individual is exercising or a totally different one.

Dimensions and Categories

The previous exercise allowed us to pinpoint a set of activities and scenarios which tie the primary concepts retrieved in our questionnaire. We can now materialize them formally in a design space for the usage of cooperative persuasion in MEAs. We establish the exertion activity (and intrinsically the user) to be at the centre of the design space referential.

Stakeholders. This dimension addresses the individuals involved in the exertion activity as a whole. Other individuals may intervene at different stages of the activity. In light of this frame, we can define the following stakeholders:

- **Spectators** – this entity assumes an indirect influence on the user, since it is not participating directly in the exertion activity itself. Spectators often interact with the user through social networks or, if supported, via the MEA.
- **Partner** – one of the greatest sources of motivation is someone or something one can compare him / herself to. The existence of a partner can be a driving force behind one's motivation during exertion activities. The user may be persuaded by partners exercising in the same instant or be challenged by "ghosts" of those opponents.

- **Self** – the last category addresses scenarios in which users cooperate with themselves to improve their motivation. This can only be carried out if one possesses past data of a workout session – a ghost recording of that user.

Space. The characteristics of the exertion activity are also an important part of the design space. The first we address is tied with the location of the exertion activity. Cooperation plays a role here, especially if we take into account the possibility of adding either real or virtual partners. As such, the space dimension contemplates the following categories:

- **Co-Located** – addresses activities in which the user and his / her partners exercise in the same location.
- **Distributed** – pertains to those activities in which a user may be performing at a determined location, while others are doing it in a totally different place.

Time. Accompanying the previous dimensions we introduce the time dimension. In this overview, we will tie-in most of the already described dimensions and categories along with emphasizing aspects related with the role of cooperation in the persuasive process behind the MEA. Time is divided in three categories:

- **Before** – the user can be persuaded via events which took place before the activity he / she is engaged in: a ghost recording is a log of a session which took place previously but may be in use during the current session; being influenced by partners prior to the session in a discussion.
- **During** – during the session, there are two scenarios in which cooperation plays a role in user motivation: a co-located and a geographically distributed one. In the first, users can keep track of each other's performance visually, while in the latter other communication channels are required to be supported by the MEA (audio for instance).
- **After** – persuasion based on cooperative efforts after the exertion activity can serve two purposes which are linked together. First it can be used as a source of analysis for the user. Secondly, it serves as a build-up catalyst for the next exertion session.

Persuasive Feedback. Here, two sub-dimensions emerge: the first pertains to the scope of the feedback and the second to the instruments used to convey it. Scope can be easily defined: feedback may privately address the user or publicly target all partners partaking in an exertion session.

Our review of existing literature allowed us to categorize feedback according to specific parameters. We found that users often receive feedback in the form of written or audio performance messages (e.g. "you're going faster"). However, some mentioned the usage of music or cheering sounds as a way to motivate themselves. As such, we can define these vehicles according to the following couple of categories:

- **Natural Language** – this type of feedback is capable of being naturally and easily comprehended by human beings without requiring intensive interpretation and with no second meaning involved.
- **Metaphorical Content** – this type of content encompasses messages which are not directly understandable but may be interpreted in a way that delivers a determined message to the user.

5   Conclusions and Future Work

This article presented a design space for cooperative persuasion interactions in mobile exertion applications. Ensuring a sustainable and healthy lifestyle is the aim of several applications available for smart-phones. MEAs attempt to materialize this via persuading their users into exercising more often. Like in other areas, sharing these endeavours with partners and acquaintances can be mutually beneficial, aiding both parties in improving their condition. Validation of this design space in under way, with the development of a MEA which contemplates the presented dimensions.

References

1. Adams, M., et al.: A Theory-based Framework for Evaluating Exergames as Persuasive Technology. In: Proceedings of the 4th International Conference on Persuasive Technology (2009)
2. Chittaro, L., Sioni, R.: Turning the classic Snake mobile game into a location-based exergame that encourages walking. In: Bang, M., Ragnemalm, E.L. (eds.) PERSUASIVE 2012. LNCS, vol. 7284, pp. 43–54. Springer, Heidelberg (2012)
3. Chittaro, L., et al.: Exploring audio storytelling in mobile exergames to affect the perception of physical exercise. In: Proceedings of the 7th International Conference on Pervasive Computing Technologies for Healthcare (PervasiveHealth 2013), pp. 1–8. ICST, Brussels (2013)
4. Gao, Y., et al.: The Acute Cognitive Benefits of Casual Exergame Play. In: Proceedings of the SIGCHI Conference on Human Factors in Computing Systems (CHI 2012). ACM, New York (2012)
5. Fogg, B.J.: Persuasive Technology: Using Computers to Change What We Think and Do. Ubiquity Magazine (2002)
6. Gerling, K., et al.: Full-body Motion-based Game Interaction for Older Adults. In: Proceedings of the SIGCHI Conference on Human Factors in Computing Systems (CHI 2012). ACM, New York (2012)
7. Macvean, A., et al.: iFitQuest: A School Based Study of a Mobile Location-Aware Exergame for Adolescents. In: Proceedings of the 14th International Conference on Human-Computer Interaction with Mobile Devices and Services (MobileHCI 2012) (2012)
8. Mandryk, R., et al.: Physiological indicators for the evaluation of co-located collaborative play. In: Procs of CSCW 2004 (2004)
9. Mueller, F.: Exertion in Networked Games. In: Proceedings of the 4th International Conference on Foundation of Digital Games (FDG 2009). ACM, New York (2009)
10. Mueller, F., et al.: Jogging over a distance between Europe and Australia. In: Proceedings of the 23nd Annual ACM Symposium on User Interface Software and Technology (UIST 2010), pp. 189–198. ACM, New York (2010)
11. Mueller, F., et al.: Designing Sports: A Framework for exertion Games. In: Proceedings of the SIGCHI Conference on Human Factors in Computing Systems (2011)
12. Weinberg, et al.: Foundations of Sport and Exercise Psychology. Human Kinetics (2006)
13. Whitehead, A., et al.: Exergame Effectiveness: What the Numbers Can Tell Us. In: Proceedings of the 5th ACM SIGGRAPH Symposium on Video Games (2010)
14. Calories burned while jogging,
    http://www.myfitnesspal.com/topics/show/134478-accurate-formula-to-determine-calories-burned-jogging

Cooperative Work for Spatial Decision Making:
An Emergencies Management Case

Jonathan Frez[1], Nelson Baloian[2], Jose A. Pino[2], and Gustavo Zurita[3]

[1] Universidad Diego Portales, Santiago, Chile
[2] Universidad de Chile, Department of Computer Science, Santiago, Chile
[3] Department of Information Systems and Management, Business and Economics Faculty,
Universidad de Chile,
Diagonal Paraguay 257, Santiago, Chile
gzurita@fen.uchile.cl

Abstract. Geographical Information systems have been frequently used to support decision processes, especially those involving emergency management. When planning the measures in case of an emergency experts must evaluate and compare many scenarios which arise from different hypotheses about where people may be at the time of the emergency and how will they react. This work presents a tool which can help a group of experts in generating, visualizing and comparing the outcomes of the different hypotheses.

Keywords: Collaborative decision support, GIS, Emergency management.

1   Introduction

There are many decision problems involving spatial issues that might be solved with the help of a GIS. The most recurrent problem can be generally stated in the following way: ***Find a suitable area to "do" something***. For example, Ghayoumian et al. [3] explain how to find locations for constructing artificial water recharge aquifers using floods. In this case, decision makers must be experts on aquifers recharge, but they will need historical information and spatial data in order to design a formula which reflects the correct criteria for selecting the suitable area(s). This formula is used to build a suitability map using a GIS. This map typically shows the suitability level on each point of the map satisfying the requirements. However in ill-structured problems this criterion is complex to build because the goals are not clear and the various decision makers will tend to define different goals according to their own knowledge.

Another difficulty with GIS-based ill-structured problems is that frequently the data available to decision makers is unreliable. In particular, data may be incomplete (not covering the whole space) and/or uncertain (there are doubts on the data accuracy and veracity). A typical case in which this situation occurs is in some Emergency Management response processes [1]. An example of this situation is the people evacuation from coastal areas decisions after a strong earthquake; there is the threat of a possible tsunami and decision makers must decide the evacuation procedure. Various options may be available, but the data to make an easy decision may not be at hand: exactly how many people are now in in each portion of the territory? Do they have

N. Baloian et al. (Eds.): CRIWG 2014, LNCS 8658, pp. 113–120, 2014.
© Springer International Publishing Switzerland 2014

now operating means of transportation (at each location)? Do they have basic supplies (water, electricity, gas…) at each location?, etc.

We have presented a way to deal with incomplete and uncertain spatial data in a previous work [2]. The result is a graphical display of possible values at each location. However, this may not be sufficient. The values are just a consequence of the assumptions made by one expert. Another expert may disagree with them. We deal with a further step in the collaborative decision-making process. Suppose a small number of experts are in a face-to-face meeting and each of them is provided with tools for generating and displaying possible data according to their own assumptions and hypotheses. Of course, each expert's data is different from the others'. How could they somehow get a tradeoff? That is the problem we begin to treat here.

2   Scenario Building

Geographic information is usually represented in a discrete manner, with discrete geometries and deterministic values. However, a variable like temperature over a period of time becomes a random variable. Geographical points can be seen as random variables as well because they are mostly created using sampling procedures (e.g. GPS). Information associated to the geometries of the areas being represented could also have probabilistic and/or epistemic properties, e.g., which is the probability that the boundary runs exactly over a certain line?

The *Dempster-Shafer theory* [5] proposes to use sets of hypotheses regarding a variable, e.g., temperature values in X are always between t1 and t2, associated with a probability of being correct. We explain this theory with an example: Table 1 shows mean number of persons values associated to a certain location. We also have a query Q = [13-23] looking for locations with more than 13 and less than 23 persons. In this case, 3/5 of the locations meet this condition (locations 1 and 3 do not). Now Table 2 contains a "range" of persons registered for each location. In this case, only 2/5 of the locations fully satisfy the condition (positions 4 and 5) and 2/5 more may have a possibility to satisfy it (positions 1 and 2). One location does not fall within any interval of the query range (position 3). The theory defines three types of answers to queries:

— **Plausible:** the probability the random variable takes values within the query range.
— **Certain:** the probability that the whole range of the distribution of the variable is within the range of the query.
— **Uncertain:** no valuable information can be derived from this data.

Table 1. Location vs persons		Table 2. Location vs range	

Location	average #persons	Location	#persons range
1	12	1	[9-21]
2	20	2	[12-23]
3	7	3	[5-10]
4	19	4	[17-20]
5	17	5	[14-22]

Using the *Dempster-Shafer* evaluation, we can compute the hypotheses (Plausibility, Certainty and Uncertainty) for each location of Tables 1 and 2. We see the Certainty level is 40% and Plausibility level is 80% (Table 3). These values are considered as lower and upper bounds of **possibility**, i.e. between 40% and 80% of the locations have some possibility to have a similar number of persons to the queried range. Besides this information, the theory states that a certain weight should be given to each hypothesis. This weight should be assigned by a human expert or a heuristic. Table 4 shows weights assigned by an expert.

Table 3. Location/Persons D-S

Loc.	Persons	Hypothesis
1	[10,20]	Plausible
2	[15-25]	Plausible
3	[10-25]	Uncertain
4	[17-29]	Certain
5	[20-32]	Certain

Table 4. Location/Persons weights

Persons	Weight
[10,20]	20%
[15-25]	15%
[10-25]	35%
[17-29]	20%
[20-32]	10%

In this case, since Q = [13-23], the certainty is 30% (20% from location 4 plus 10% from location 5) and the plausibility is 65% (20% from position 1 plus 15% from position 2 which are plausible, plus 30% from the certain positions 4 and 5).

Using Dempter-Shafer Theory, the expert can define a set of hypotheses expressing his knowledge, e.g., one hypothesis can be *"persons are in shops with a 20% of belief"* or *"persons are in schools or workplaces with a 40% of belief"*. We also define query hypotheses: *"persons are in shops just like in place X,Y"*. The expert can define multiple hypotheses in his/her statement, which are combined using Dempster-Shafer combination rules. Furthermore, these complex scenarios can be designed by experts with no GIS expertise.

The result of combining the hypotheses with real data and fuzzy techniques for spatial representation, gives us a suitability map. A suitability map typically shows the suitability level on each point of the map that satisfies the requirements; in our case, it shows the belief degree of the hypothesis for each evaluated location. This kind of suitability map is what we call a simple scenario

Chile is the country with high seismic activity and 6435 kms. of coast. These characteristics imply complex evacuation conditions under tsunamis. The population has been informed and trained so that in cases of large earthquakes in coastal areas they should evacuate the seaside and seek refuge in higher grounds. However, in some cities in the north of the country a tsunami after an earthquake may occur before people complete their evacuation. This is the case of Iquique (Fig. 1).

In order to make an effective evacuation plan, various experts and stakeholders must collaborate. Each one may have different opinions and hypotheses about which are the best alternatives to elaborate an evacuation plan. In order to exemplify the proposed collaboration method we will assume there are five experts. Suppose two of them believe the best evacuation method is that people must go to higher grounds using any possible way and transport means. The remaining three experts have another hypothesis: most people cannot reach the higher grounds before the tsunami arrival, so they must seek refuge inside high buildings (vertical evacuation).

In order to evaluate both options, organizations responsible for dealing with emergency situations typically relay only on traditional GIS. These systems provide information about the population living in the area, the number of schools, and other stored information.

In case of a tsunami, the evacuation problem can be classified as an ill-problem, because there is no real information about the number of people who must be evacuated, there is not real knowledge about the time available between the earthquake and the arriving of a tsunami to the coast. On top of this, the population in the area varies according to the day and time (an earthquake can occur at any time).

Fig. 1. Left: Iquique from the sea, Right: Evacuation area

We believe the time&belief based scenarios can provide reasonably useful information to decision making when there is uncertain incomplete information and complex modeling.

Using Dempster-Shafer Theory, we can build a set of hypotheses that can tell us where people can be. For example, there is an area at the Iquique coastline where there is a high belief that population will be numerous during daytime, because this area includes a high concentration of universities, restaurants, shopping centers, a popular beach, etc. This area is also far from higher grounds (Fig 2).

For this area, one solution is a vertical evacuation which means people should enter high buildings. This scenario can be described as an evacuation possibility, i.e., if there is a high building nearby then people should enter it. However, an obvious problem occurs if there are too many people and not enough high buildings.

In order to cope with incomplete information and multiples scenarios, we propose to use a Dempster-Shafer based Collaborative Geographical Information System (CGIS). Using the CGIS, the above mentioned five experts can make their own hypotheses evaluation. However, as a result of this process they will have different suitability maps. For example, the first two experts can differ about where people can be at various hours of the day. The other three experts can also differ about how tall the buildings must be or what kind of construction can resist a tsunami.

Fig. 2. The coastline of Iquique. Left: Suitability map with population belief. Right: aerial view of the coastline.

After each expert builds his/her own simple scenario they will have five different suitability maps. The next natural step will group the suitability maps according to both evacuation scenarios. However this does not solve the collaboration problem. Another case can be that each suitability map is based on similar hypotheses, e.g., that during daytime people are in: commercial areas, schools, universities, libraries, banks, bus stations, etc. However, one expert may believe there are many people at commercial areas, and another one may hypothesize these people are in residential areas.

In order to solve the problem, we propose to combine the various suitability maps generated by each expert in a hierarchical order, by combining the various hypotheses according to certain operators, as it has been suggested in [4]. Of course, which suitability maps should be combined using which operation should be collaboratively decided by all experts. At each step the resulting suitability map should be the outcome of the discussion of each scenario possibility including the experts' hypotheses, known information about the area, and relevant factors that must be included. Suitability maps are combined as a result of argumentation and discussion.

Next section describes the proposed combination methods. Some of them are designed to merge data, and others to focus on important factors according to the experts' considerations.

3   Combination Methods

A complex scenario is the combination of various simple scenarios. Building a complex scenario requires cooperative work among various stakeholders like experts in the particular scenario area and decision makers. In order to provide useful tools for collaborative scenario building for a single area we must divide the work in two dimensions: Hypotheses Dimension and Time Dimension (see Fig 3).

The hypotheses scenario dimension is directly related to the collaboration between decision makers and experts who may have different hypotheses about belief function values at a certain time, for example one expert will have a certain hypothesis about the number of people at commercial areas during the morning, noon, evening and night of a working day. When combining suitability maps, experts should consider the same time dimension for stating their hypotheses.

We propose to use five types of operators to collaboratively build a scenario based on combinations of suitability maps. We will use three initial scenarios for the example (Fig 4). Each initial scenario is the result of a specific set of hypotheses.

The **Sum** is probably the simplest operator a decision team should be able to use; it consists of adding the belief value of each scenario for each evaluated location. Graphically, it consists of adding the values of the three bars corresponding to the same cell. Visually the resulting map does not show the sum of the three bars one over the other because the final values for each cell are normalized. This operator can be useful when three independent but related scenarios must be merged. For example police, transit and street maintenance human resources must be combined to evaluate the governmental resources available during an extreme emergency. Using **sum,** the decision maker can easily identify the resources available for each location (Fig 5).

The **Subtract** operator subtracts the belief value of two or more scenarios at each evaluated location. This operator can be useful to evaluate the differences between one scenario and other ones. For instance, if we have possible flood scenarios and possible refuge sites. Using **subtract** the decision maker can easily identify refuge places with low flood belief values (Fig 6).

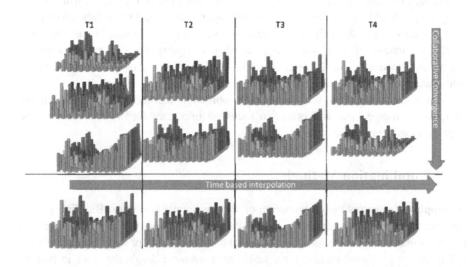

Fig. 3. Hypothesis Dimension and Time Dimension: At T1 experts have three hypotheses

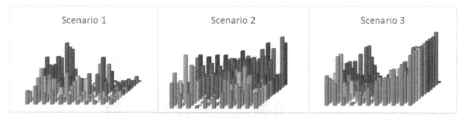

Fig. 4. Initial Scenarios

The **Average** operator is the simple average between the belief values of each location in each scenario. The result of this operator is visually similar to **sum** but can be numerically different. For example, if a cell has value 0 for the belief for two of the experts' maps and 100 for another one, the sum will be 100, but the average 33.3. This operator can be used in order to find places where to deploy scarce resources. For example, places to deploy police forces according to criminality (Fig 7).

An **OWA** operator is the weighted average given an ascending or descending order. The OWA operator has been already used to combine data using the Dempster-Shafer Theory [4]. Assuming the values are ordered, only two results can be obtained (values and weights are ordered ascendingly or descendingly). Given this, we define two operators: OWA-ASC and OWA-DESC.

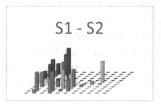

Fig. 5. Sum of the three scenarios

Fig. 6. Subtract of scenarios

Fig. 7. Average of scenarios

When using the OWA-DESC operator, values and weights are ordered both in descending order. This combination can emphasize the largest belief values of each scenario, avoiding that a certain important fact known by one of the experts could be ignored because of simple averaging. For example, if a criminality scenario has a large belief degree at a certain location, using **average** this information can be mixed with low degree values from other scenarios (Fig 8).

When using the OWA-ASC operator, values are ordered in a ascending sequence and the weights are also ordered in an ascending sequence. This combination emphasizes the belief when the values are constantly high in all scenarios. This operator is similar to average, but it is not susceptible to isolated big values. It can be applied to allocate specific and limited resources that can support multiple scenarios. It can also be used to identify critical areas.

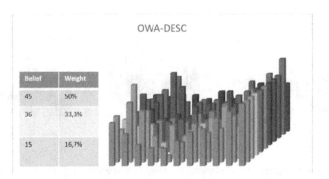

Fig. 8. OWA-DESC applied to 3 scenarios

4   Discussion

The proposed approach for cooperative work in spatial decision making scenarios may be used to collaboratively analyze spatial information using decision making tools. This platform could be used in several disaster scenarios. For example: floods, volcano eruptions, air or water pollution, tsunami, and in general any kind of disaster that could be modeled using GIS technologies.

The main drawbacks of the proposed technology are strongly related to Dempster-Shafer Theory. Currently, a suitability map is the graphical representation of belief values, not magnitude values. The relevance of this issue is that computing the number of people is not possible given a belief value or suitability map. Furthermore, after applying an operator between two or more suitability maps, the relation between the belief values and the original hypotheses set is not clear. However, the resulting combination could still be useful as a visualization and decision making tool.

References

[1] Chen, R., Sharman, R., Rao, R., Upadhyaya, S.J.: Coordination in emergency response management. Communications of the ACM 51(5), 66–73 (2008)
[2] Frez, J., Baloian, N., Zurita, G.: Getting Serious about Integrating Decision Support Mechanisms into Geographic Information Systems. In: Proc. 9th Int. Conf. on Computer Sc. and Inf. Technologies (CSIT), pp. 1–11. IEEE Computer Society Press, Yerevan (2013)
[3] Ghayoumian, J., Ghermezcheshme, B., Feiznia, S., Noroozi, A.A.: Integrating GIS and DSS for identification of suitable areas for artificial recharge, case study Meimeh Basin, Isfahan, Iran. Environmental Geology 47(4), 493–500 (2005)
[4] Merigó, J.M., Casanovas, M.: Induced aggregation operators in decision making with the Dempster–Shafer belief structure. International J. of Intelligent Systems 24, 934–954 (2009)
[5] Shafer, G.: A mathematical theory of evidence, vol. 1. Princeton University Press, Princeton (1976)

Architecture of Mobile Crowdsourcing Systems

Frank Fuchs-Kittowski[1,2] and Daniel Faust[2]

[1] Hochschule für Technik und Wirtschaft (HTW), Berlin, Germany
`frank.fuchs-kittowski@htw-berlin.de`
[2] Fraunhofer Institute for Open Communication Systems (FOKUS), Berlin, Germany
`{frank.fuchs-kittowski,daniel.faust}@fokus.fraunhofer.de`

Abstract. This paper proposes a general architecture and a classification scheme for mobile crowdsourcing systems, which are illustrated by two example applications. The aim is to gain a better understanding of typical functionalities and design aspects to be considered during development and evaluation of such collaborative systems.

Keywords: crowdsourcing, crowdsourcing system, mobile crowdsourcing, crowdsourcing application, architecture, classification scheme.

1   Introduction

Many organizations are increasingly using crowdsourcing as a new model for value creation, where new web technologies are used to outsource tasks, which are traditionally performed by a specialist or a small group of experts, to an undefined large group of people [1]. Meanwhile, mobile devices (phones, smartphones, tablets, and in the near future glasses, watches, and so on) have become ubiquitous and a tool for crowdsourcing [2]. In mobile crowdsourcing mobile devices are used for data-collection tasks delegated to a larger number of people as well as for the coordination among the people involved.

In the recent years numerous mobile crowdsourcing applications have been realized and have shown the potential for business and society [3]. As the popularity of these applications increases, our understanding of how to design and deploy successful mobile crowdsourcing systems must improve [4]. Many systems described in the scientific literature were individual, task-specific, ad-hoc implementations [5]. Without a profound theoretical foundation, the development of such mobile crowdsourcing applications is still a difficult task and costs as well as the time needed for each development can be high.

The objective of this paper is to gain a better understanding of typical functionalities and design aspects to be considered during development and evaluation of such collaborative systems. Thus, the way how mobile crowdsourcing applications are developed will shift from an ad-hoc manner to a planned routine. Based upon an extensive literature review, a categorization of existing applications of mobile crowdsourcing systems, and an overview of typical design aspects of mobile crowdsourcing

N. Baloian et al. (Eds.): CRIWG 2014, LNCS 8658, pp. 121–136, 2014.

systems, a classification scheme and a general architecture for mobile crowdsourcing systems are described.

The remainder of the paper is structured as follows: The second section gives an overview of related conceptual work in the domain of mobile crowdsourcing. A classification scheme for mobile crowdsourcing applications is presented in the subsequent section. In section four, a general architecture for mobile crowdsourcing systems is proposed. Based on this architecture and classification scheme two example applications are illustrated in section five. Finally, main conclusions are drawn and further research tasks in this field are identified.

2   Related Work

Our understanding of the notion 'mobile crowdsourcing' is that a group of people voluntarily collects and shares data and information using widely available mobile devices (smartphones etc.), where this data is processed and provided via a data-sharing infrastructure to third parties interested in integrating and remixing this data.

Similar approaches are Volunteered Geographic Information (VGI [6]), Public Participatory GIS (PPGIS [7]), Participatory Sensing (PS [8]). But while all these terms are close, they emphasize different aspects. VGI focuses on presenting the captured geospatial data on maps. PS emphasizes the use of the built-in sensors of mobile devices for data capturing. PPGIS focuses on storage and processing of data in GIS. All approaches have in common the volunteer and participatory nature of the data capturing and sharing process.

Over the last several years crowdsourcing has been a vital research field [9].There has been much research regarding (i) the *definition* of crowdsourcing (e.g. [10], [11]), (ii) the characterization of the crowdsourcing *process* (e.g. [12]), (iii) the development of a crowdsourcing *taxonomy* (e.g. [13, 14, 15, 16, 17]), and (iv) the introduction of a *conceptual framework* that supports the designing of crowdsourcing systems (e.g. [18, 19]). However, only little research was conducted to define a crowd sourcing system and its *technical design* precisely [20].

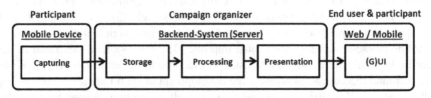

Fig. 1. General architecture for Crowdsourcing

In the scientific literature mobile crowdsourcing applications have two essential subsystems/components (see Figure 1): data capturing and data processing [21]. These main components are generally organized as client-server architectures with a mobile client (mobile device) for ubiquitous data capturing, and a server (backend system) for data storage, processing, and visualization [22].

Existing approaches to general architectures focus either on more refined functions of certain subsystems (data capture, data processing, campaign management etc.) or non-functional aspects (privacy etc.). [21], [23] present common architectural components with a special focus on data capturing and leveraged processing. [5] emphasize the functional components regarding campaign management. The architecture presented by [24] pays special attention on recruitment of participants. Several architectures of applications emphasize task management, e.g. the distribution of data capturing tasks and software to the participants resp. their mobile devices [25, 26, 27, 28]. Enhancements of general architectures to solve privacy issues are presented in [29, 30]. Medusa [31], MoCoMapps [32], PRISM [27], and AnonySense [33] are frameworks for cost-efficient development of mobile crowdsourcing applications and with focus on generality, security, scalability, and privacy. A detailed view on the common functional components (and their interfaces) of mobile crowdsourcing applications still does not exist.

3 Research Method

A systematic literature review (SLR) was conducted to improve the understanding on functional and technical aspects of mobile crowdsourcing systems. A SLR provides a well-structured and repeatable procedure to identify, evaluate and interpret existing literature relevant to a specific research question [34]. The main goal of a SLR is not only to methodically aggregate scientific studies in a certain research domain but also to support the development of evidence-based guidelines for practitioners [35].

After formulating the research question (What type of components and functions of mobile crowdsourcing systems can be conceptualized?) an appropriate search strategy was derived and applied: As search resources web search engines (e.g. google search) as well as scientific literature databases such as ACM Digital Library, IEEE Xplore Digital Library, ScienceDirect, SpringerLink and Wiley were used. Any publication type (from application web site to peer-reviewed journal paper etc.) was considered with a restriction to German or English language publications. The literature review includes contributions that (i) describe a certain mobile crowdsourcing application, (ii) address design issues, or (iii) classify or give an overview of applications.

The authors identified 124 publications, which were carefully read. Keywords were collected which either addressed a component or a function of a mobile crowdsourcing application. Iteratively, specific keywords were aggregated to more generic terms. Finally, a concept map was created that maps all relevant literature to one or more of the derived generic components and function terms.

4 Characteristics of Mobile Crowdsourcing Applications

Regarding business cases and application domains, the scientific literature shows a large variety of possible mobile crowdsourcing applications. Based upon the observed object, these applications can be categorized as either people-centric or environment-centric [29]. People-centric applications collect data about the user. They mainly focus on observation of the user's health (e.g., physical effort), documenting activities (e.g., sport experiences), and understanding the behavior of individuals (e.g., eating disorders). Based on this data, the users can be provided real-time support, e.g. in case of an asthma or heart attack. In contrast, environment-centric apps capture information about the surroundings of the user, i.e. environmental parameters (e.g., air quality, air pressure, noise pollution, thermal, road condition, and traffic in cities) or interesting events (e.g., accidents, damages, disasters). Typical application domains are maintenance of man-mad infrastructures (e.g., report of damages in urban areas or observation of depots to reduce costs of maintenance and inspections), environment and nature protection (e.g., mapping of species to take targeted protective measures, monitoring specific geographic areas, and mapping of damages), disaster management (e.g. mapping of damages after natural disaster to use limited resources more efficiently, collecting of real-time-data about a certain area to distribute relief units).

Table 1. Morphological analysis for classification of mobile crowdsourcing applications

	Characteristic	Feature		
Device	Type	Special device		Standard device
	Sensor technology	Manual input	Embedded sensors	External Sensors
Data	Capturing	Automatic	Manual	Context-related
	Spatial data	Point	Line	Polygon
	Transmission	Real-time		Delayed
	Anonymization	None	Anonymized	Authenticated
	Processing	None (raw data)	Analysis	External systems
Participation	Admission	Own initiative	Non-binding enquiry	Binding request
	Selection	A priori		Dynamically
	Knowledge/Skills	Low/none	Medium	High
	Selection Criteria	Person	Role	Location
	Registration	Anonymous	Known	Formal relation
	Assessment	Automatically	Participant	Organizer
Involvement	Degree	Active	Limited	Passive
	Task type	Initial collection	Update	Verification
	Creation of result	(Pure) Data Collection	Data processing	Knowledge generation
	Use of result	Publically available	Collective use	Personal use
Campaign	Duration	Fixed	Implicit limited	Unlimited
	Location	Limited		Unlimited
	Type of group	Single person	Closed	Open
	Monitoring	Periodic		Continuous
	Application field	Environment	People	Hybrid

Specific mobile crowdsourcing-based applications differ with regard to a number of various dimensions. This section mainly focuses on differences of higher level systems that enable general propositions about the design aspects. Due to significant differences between application domains, their specific methodological aspects are not covered in this paper.

4.1 Mobile Devices (Type, Sensors)

Crowdsourcing campaigns usually use mobile devices to collect user data („Bring your own device!" philosophy). These are typically widely available Internet-capable and GPS-enabled devices such as smartphones or tablets. Cases in which task-specific hardware is needed to gather user data are rare. Mobile crowdsourcing applications differ with regard to the sensors used for data collection. Users typically enter all data manually (free text, forms etc.). But also sensors installed on mobile devices are often used to collect data. There are several sensor technologies for data collection: position sensors (GPS, magnetic field/compass, proximity sensors etc.), motion sensors (accelerometer, gyroscope etc.), and ambient sensors (light, air pressure, temperature, and humidity sensors). Furthermore, data can be collected with cameras, microphones and by using video, audio and image recording. Prospectively, further sensors will be fitted as standard in mobile devices. In special cases, additional sensors are connected to the device, for example to monitor vital functions of the human body (pulse, skin resistance etc.) or to measure air pollution (pollen, exhaust fumes etc.). It is often possible to find combinations of manual entered and collected sensor data (e.g. rare picture and manually-entered textual description of the place where it was found).

4.2 Data (Collection, Location, Transmission, Anonymization, Processing)

Data capturing can be done manually, automatically, or context-aware [21]: In the manual mode, the participants are personally involved and trigger the collection of data themselves when they detect relevant events (e.g., a severe weather situation). In the automatic mode, the participants are not directly involved and the data is collected automatically by the embedded sensors of the mobile devices (e.g., noise and air pollution). In the context-aware mode, the embedded sensors monitor their environment and activate the manual or automatic capturing function when previously set thresholds are exceeded. Data captured are often location-based [36]. The location coordinates, mostly simple point coordinates, are determined by an embedded location sensor (GPS etc.) that, if necessary, can be validated. Some applications are also capable of capturing line geometries (e.g. distance walked), surface geometries or polygons (e.g. disaster area). Data captured have to be transmitted from mobile devices to a central collection server. Data can be transmitted in real time (e.g., via mobile phone networks) or with a delay (e.g., when WLAN is available). Usually, data that are used for communication purposes (SMS, mobile phone network) are transmitted in real time. In offline situations, delayed transmission may be essential to meet security requirements. In such a case, data transmission would start manually (or, if necessary, automatically) when a connection to the Internet is available (e.g., via WLAN) or other means of communication are in use (LAN, flash drive etc.). Data collected are often transferred in anonymous form. However, there are applications that require user and data authentication (e.g., medical applications for monitoring vital functions of the human body). Some applications produce raw unprocessed data (e.g. noise measurement) or forward data to downstream external systems (e.g., weather or flood forecast). These applications often carry out complex data processing and data analysis operations (e.g., aggregation, recognition of duplicates, visualization), and provide users with post-processed data.

4.3   Participation (Admission, Selection, Criteria, Registration, Evaluation)

The admission of a participant for inclusion into a data collection campaign can be based on the participant's own initiative or on a request by the campaign organizer to a certain person. The latter can be a binding order or a non-binding enquiry. The required a priori knowledge of the user can be low/none (e.g., automatic sensing of traffic), medium (e.g., capturing of damages in urban areas) or high (e.g., observation of natural processes, air pollution reports, wild animal classification). The selection of the people to be requested can be role-based, person-based or location-based or can be carried out by combination of these different factors. Thereby, the participants of the campaign can be fixed at the beginning of the campaign or participants can also be included during the course of the campaign (or selected on demand). The participants can participate anonymously or can be registered users, i.e. known to the organizer. Furthermore, known participants can be in a formal relation to the organizer, i.e. employee, member, partner, customer etc. of an organization. Often the participants are members of the general public ("motivated crowd"). The assessment of the contributions of the participants can be carried out automatically by the system, by other participants or by the organizer (or other people in charge).

4.4   Involvement (Degree, Task Type, Creation of Result, Use of Result)

The degree of participation can have a significant influence on the willingness of the participants to contribute to a campaign – with their time and their own device. It can be distinguished between active and passive participation: In case of an active participation the participant is actively involved in the process. The task of the participant then is not limited to the data capturing, but can contain more tasks ranging from development and formulation of goals to the generation of information and reports based on the data collected during the campaign. In case of a passive participation the participant does not have to become active, because the data capturing application runs on his mobile device in the background and collects and transmits data automatically. The data collection tasks can be initial or non-recurring data collections of new data. They also can concern updates, changes or verifications of existing data. The tasks of the participants are not limited to (pure) data collection. They can involve other tasks of the knowledge and result generation process, like further processing of the collected (raw) data (e.g., model calculation and simulation) and evaluation of the results (reports etc.). With regard to the use of the results public, collective or personal projects can be distinguished. In case of a public project, the results are available to the general public, e.g., collection of damages and shortcomings in a city. In case of a collective project a group of people with a shared interest selects data with regard to a common goal, e.g., proof of pollution caused by a factory. In case of a personal project the participants are focused on self-awareness and self-improvement. They exclusively use the results for their personal purposes, e.g. monitoring and optimization of eating habits, sleep, and sport performance. Possibly these data are shared with friends (e.g., via social networks) or (in anonymized form) made available publically, to learn one from each other.

4.5   Campaign (Duration, Location, Group Type, Monitoring, Application Field)

Temporally, the data collection campaign can be limited (fixed duration) or unlimited (open-end). The temporal frame of a campaign may also be implicitly defined, e.g. a collection is carried out until a certain amount of data has been gathered. Locally, especially environment-centric campaigns take place within a spatially delimited area, region or place, e.g. capturing of environmental damages or road conditions. In contrast, people-centric applications often are not limited to a certain, spatially delimited area or place. Different types of tasks may require different numbers of participants and forms of groups. Sometimes, only a single person is necessary to accomplish a certain data collection task. In other cases, a closed group with specific knowledge for a problem solution is required. And in other cases, an open group is formed, because the general public is invited to participate in the data collection task [37]. The monitoring of a campaign (activity of the participants, quantity and quality of the collected data etc.) can be carried out continuously or periodically. In the former case, campaign management is a permanent task, like it is required for large campaigns. The latter requires monitoring of the campaign only ad-hoc, on-demand or at certain moments in time. In general, a mobile crowdsourcing campaign and application can be assigned to one of the aforementioned application fields in the area of people-centric and environment-centric applications. But some applications can be assigned to several application fields, e.g. personal health applications (like monitoring of asthma patients), that also consider environmental conditions (e.g. air pollution).

5   Architecture of Mobile Crowdsourcing Systems

From the analysis of existing mobile crowdsourcing applications and architectures the following architecture of a general crowdsourcing solution (see Figure 2) is derived. This architecture provides a more detailed view on the common functional components (and their interfaces) of mobile crowdsourcing applications.

Typical roles within this architecture are: **a) Campaign organizer:** initiates and monitors the targeted data collection effort (crowdsourcing campaign), including the definition of the campaign as well as recruitment, control and coordination of well-suited participants. **b) Participant:** contributes to the geo-crowdsourcing campaign by (voluntary) capturing and sharing geospatial data using their own mobile device.

c) End user: accesses and processes the data captured by the participants according to their interests and needs, e.g. integrating, analyzing and remixing this data.

The general system architecture system is divided into two independent runtime systems: **1) Backend system** (server) with the main components *Campaign Management* (campaign planning, participants management, recruiting, tasking, and campaign monitoring) and *Data Management* (pre-processing, storage, processing and provisioning), and **2) Mobile device** (mobile client) with the data capturing and data transfer components.

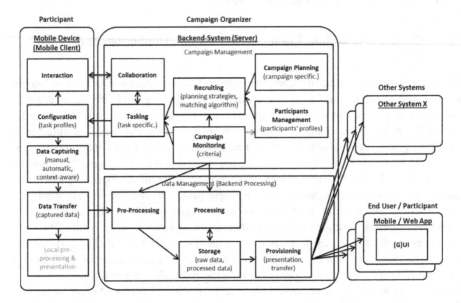

Fig. 2. Architectural components and roles of mobile crowdsourcing applications

Typical functional tasks of these architectural components are:

5.1   Campaign Management

Campaign Planning: For each data-collection campaign the relevant parameters must be defined in advance. These campaign specifications may involve a number of factors including overall budget, kind of data to be collected (what), geographic coverage (where), timespan, interval or frequency of data collection (when), participants' device capabilities, demographic diversity, and social network affiliation (who).

Participant Management: For recruitment, tasking and monitoring of participants certain information about the participants are necessary, that are managed in this component in a participant profile. For recruitment, it is necessary to determine the expertise or skill level of a participant [38]. Relevant characteristics could be participants' device capabilities, reputations as data collectors, motivational structure, skills and availability in terms of geographic and temporal coverage. For authentication registration information is necessary. For tasking, participants' movement pattern could be helpful. The monitoring of participants considers acceptance and rejection decisions of the user's contributions [39]. Additionally, ranking scores can be provided, that present the skill level, the reputation or the quality of the participant [40].

Recruiting: The recruitment component takes campaign specifications and participants' profiles as input and recommends participants for involvement in data collections. For finding a fit between diverse participants (profiles) and the crowdsourcing campaign (specification) certain planning strategies and matching algorithms can be used. In general, participants for campaigns must meet minimum requirements. Advanced approaches identify which subset of individuals is best suited (according to the campaign specification, e.g. budget, maximal coverage etc.).

Tasking: This component distributes the assigned data-collection task to the participants' mobile devices. Tasking can be proactive (the data-collection task is pushed to the participant' device) or passive (user chooses the data collection task from a set of tasks provides by the organizer). The task specifies the data capture modalities based on the campaigns requirements including criteria when, where, what data to capture (e.g., take a 2 megapixel picture at location x in time frame y twice a day).

Campaign Monitoring: This component monitors the progress of the campaign according to predefined criteria in the campaign definition such as quantity and quality of the collected (raw and processed) data. The check of the criteria can occur continuously or periodically. If necessary (e.g., a criterion is below a threshold) campaign organizers may be alerted and measures are taken like recruiting additional participants (e.g., if the coverage of a specific area has to be improved or not enough data have been collected) and create additional tasks (e.g., to verify outlier data or to get more data on detected hot spots). In automatic data capturing mode, the tasking action can be completely automated such as changing sampling rates (adaptive sampling) or turning sensors on/off (actuation). In the manual mode, a trigger message can be sent asking participant to collect data on certain location or change route.

Collaboration: This component facilitates collaboration and coordination. Participants need to interact to solve a task collaboratively. Additionally, interaction between participant and organizer may be needed, e.g. to provide feedback to the results of the participant (from organizer to participant), and to ask for more details regarding the task specification of the organizer (from participant to organizer). This component can also be used to assign tasks newly.

5.2   Mobile Device

Configuration: This component manages the assigned tasks on the mobile device of the participant. For each task a profile with the task specification is maintained. The participant can be informed about tasks to fulfill, navigated to locations, where data should be captured, or the sensors of the participant's mobile device can be controlled for data capture manually and automatically. In the configuration component, participants can select different data transfer strategies. Participants also can manage parts of their participant profile (role, place, time etc.) to update and specify the kind of tasks they are willing to fulfill (light grey line to "participant management" in fig. 2).

Interaction: This component provides functionalities to facilitate interaction among participants included in the task profile (in case of a collaborative task) and between participant and organizer. These are functionalities to communicate and share information, manage personal identities, maintain relationships, re-assign tasks among the participants or collaboratively document knowledge using the servers' collaboration component.

Data Capturing: This component collects different kinds of data requested in the campaign definition and task profile respectively using the mobile device's data-input mechanisms and built-in or plugged-in sensors - prevalently camera (picture, video), microphone (audio), clock (time), GPS (location), gyroscope, accelerometer etc. Capturing can be done manually, automatically, or context-aware (see 4.2).

Data Transfer: This component ensures the transmission of the data collected by the capturing component to the backend system (server). The data can be transferred immediately or delayed (see 4.2). This component also ensures short-time storage of the data to be transmitted to the backend system (server).

5.3   Data Management

Pre-processing: This component analyzes the transmitted data and prepares them for storage. Usually, several application-specific processing steps are necessary. But in many cases it may also be very simple. Examples of those steps are: a) Data extraction and transformation of the data into the internal data structure, for instance by applying audio analysis (speech recognition to extract words, sound classification to match sounds, for instance for detecting birds voices etc.) or image analysis (e.g., determining the water level with a picture of a gauge board etc. or optical character recognition to scan text, object recognition to find objects). b) Data assessment and analysis, for instance data cleaning, quality assessment with data filtering (eliminating duplicates etc.), data fusion and aggregation of multiple or similar data.

Storage: This component ensures the long-term storage of the (raw and processed) data. The data is stored in relational databases or databases specially adapted the management of spatial data or sensor data. In some cases a history and a comparison of current data with historical trends are necessary, so that a robust long-term storage is a central requirement.

Processing: This component processes the stored raw data to extract features of interest und get insights about the observed phenomenon. This step is typically application specific and potentially involves an immense number of (spatial) data processing methods ranging from numeric modeling to descriptive statistics to image processing and sophisticated machine learning algorithms.

Provisioning: This component prepares the data for presentation or for transfer to other systems. In case of a standalone system this component provides presentation and visualization services to present the results obtained by the processing component to the end user. The results can be visualizations of the raw data or the processed data and are usually presented through web-portals or on mobile devices (often on maps). In case of external systems interested in using the data, this component provides services to access the data including necessary transformations.

6   Example Applications

The architecture proposed in section 5 was successfully applied in several projects. Two examples from the field of disaster management are described in this section.

6.1   Flood Risk App "Hochwasserrisiko"

Floods belong to the natural disasters which regularly take human life and cause high damages to property [41]. In order to take precautionary measures for protection, the authorities responsible for and the citizens endangered by flood need up-to-date information on the water levels and the actual state of risk [42]. In the first example, a community of volunteer water level and gauge observers transmits via a smartphone app the geo-referenced measurements and a photo of the scale of non-automatic gauges to the authority.

With the aid of the app "Hochwasserrisiko" [43] the current water and gauge levels at rivers can be collected and entered by interested citizens themselves with a photo and metadata (1 & 2 in Fig.3). This information can be displayed on a map or in augmented reality in the mobile devices of the users (3 & 4 in Fig. 3). In addition, this information is further processed by the flood authority's warning center for more precise flood prognosis and warning. These flood warnings again can be displayed on the mobile devices of the users (red icon in 3 & 4 in Fig. 3). In addition, with this data flood risk maps [44] can be processed, which inform about the current risk potential of floods (5 & 6 in Fig. 3). They can be viewed on a map (bird's eye view) or through the smartphone's live camera (augmented reality view). Using augmented reality, the virtual water level of a flood scenario can be visualized in reality. This allows a realistic display and improves the perception and analysis of the scenario's hazard.

Fig. 3. Screens of the flood risk app "Hochwasserrisiko"

This scenario works with a fixed set of a priori selected citizens, which receive an order with detailed instructions, where to collect water and gauge levels. The users can see the places where the information is needed on a map (see 1 in Fig. 3) and select and accept the places (tasks). The users then move to the places selected and collect the information manually by entering in a form and taking some pictures (see 2 in Fig. 3). This data is then transferred to a central server of the flood authority's warning center, where this information is controlled by an operator who compares the photo with the measurements, thus ensuring the same quality of measurements as with automatic gauges. In this case the data filtering and fusion is performed by an operator. This is necessary because the crowdsourcing data is used as an input for further model processing. This is only applicable when the data characteristics and the data quality are adequate for being used as a basis for further algorithmic hazard prediction methods, in other words when it can be assured that the mobile crowdsourcing data can substitute the data of a physical sensor. The processed data (flood warnings and flood risk maps) is than available on the mobile devices of the users and the general public (see 3-6 in Fig. 3).

6.2   Disaster Management App "Emergency Help"

Modern communication technologies such as mobile applications, that inform people about natural disasters occurring in their vicinity, are increasingly gaining in importance. For example, the KATWARN app offers location-based disaster warnings for small areas. KATWARN warnings are issued by the local emergency management agencies such as fire departments [45]. Mobile crowdsourcing offers possibilities for sustained improvement of danger prevention processes.

The app „Emergency help", which is an expansion of the KATWARN app, helps to manage disaster relief volunteers, which are skilled to provide support and assistance, for example trained in first aid. The app informs people in the affected area about an upcoming hazard and how they should behave to stay safe, for example to leave the danger zone. Moreover, it allows preregistered disaster relief volunteers to provide certain information about their skills (first aid, languages etc.), to signal their willingness to give assistance in case of a hazard, and to inform about the completion of assigned assistance tasks. Another user group of the app is people in need of help that are not able to follow all security procedures, e.g. elderly people with mobility impairments who cannot leave the danger zone on their own. The „Emergency help" app offers such people the possibility to provide information on their physical and health problems, and to signal their need for aid in case of a hazard.

Fig. 4. Screens of disaster management app "Emergency help"

After a warning is sent, an additional button is displayed in the app. People in need of help can use this button to request help. The person's disability information and location data are sent to the app's central service. There, based on the request received, the service searches for voluntary helpers in the vicinity. The helper selected receives a notification that his assistance is needed. The helper can now confirm his willingness to assist. Then, he receives the (anonymized) data of the person in need, e.g. his or her location. Based on the exchanged data the app supports the helper, e.g. information sharing with the person in need (name or address if needed), navigation to the location of the person in need or evacuation routes. The app shows the person in need that the request was accepted, and who and how far the helper is. The successful completion of a rescue operation is reported to the central service. , so the responsible emergency management agencies have an overview about all registered people in a case of a crisis.

In this scenario, a dynamically forming group of skilled and unskilled participants is participating in the campaign ad-hoc, voluntary and possibly anonymous, based on the participants own initiative. Thereby, point coordinates (location of the user, which is provided by the mobile standard device's GPS localization function) are automatically captured in real-time, anonymized, and transmitted with other manually captured, structured data (profile, incl. role of user). These raw data are not further processed on the server's side, but only used for the coordination functionality provided by the system. Temporally, the campaign is implicitly limited to the duration of the dangerous situation. Locally, the campaign is spatially delimited to the affected area.

In this scenario, the campaign management subsystem plays a crucial role. The configuration component is used to specify the role of the user (helper or in need) and updates the participant profile in the participant management component. The participant management and the recruiting component are important for managing, selecting and matching the disaster relief volunteers. The tasking component then provides the necessary information for coordination among the people involved, which interact using the interaction and collaboration component. The data capturing component only collects data about location of user and completeness of assistance tasks. Campaign monitoring provides information about the progress and monitors people still in need of help.

7   Summary and Outlook

In this paper we investigated different dimensions of and typical functionalities of mobile crowdsourcing systems. The main goal of this paper was to gain a better understanding about design aspects to be considered in the development of such systems. Therefore, the paper presented a classification scheme and a general architecture with the typical roles, components and functionalities of mobile crowdsourcing systems. The aim was to generalize and study mobile crowdsourcing applications, and to facilitate the understanding of current systems and the design of new ones.

On the surface these systems appear deceptively simple: participants are asked to collect data, the data is collected and then analyzed for a variety of purposes. But the current diversity of mobile crowdsourcing applications, which is represented within the classification scheme, shows the difficulties and the complexity, when designing and implementing such an application. This requires not only profound knowledge how mobile crowdsourcing works and when it can be applied [46], but also how it is technically designed and implemented. Therefore, the general architecture may be used as a blueprint and to guide decisions during the development process.

However, the design and deployment of successful systems require meeting simultaneously a diverse set of technical and people-centric challenges. Even though (mobile) crowdsourcing has been a vital research area over the last several years, there are still several research questions in the field, e.g. data quality (potential for accidental submission of bad data or malicious submissions) [47], [48], privacy concerns of the users (risk of loss or theft of the device with personal data, profiling, central data archives) [30], and the motivation issues that participants may have within the system [49]. Other issues are related to the scalability of the technical infrastructure and the campaign organization [50].

Acknowledgement. This paper was created as part of the project "KoPmAn". The project is funded by the European Union (European Social Fund).

References

1. Greengard, S.: Following the crowd. Communications of the ACM 54, 20–22 (2011)
2. Mea, V.D., Maddalena, E., Mizzaro, S.: Crowdsourcing to Mobile Users - A Study of the Role of Platforms and Tasks. In: Cheng, R., Sarma, A.D., Maniu, S., Senellart, P. (eds.) Proceedings of First VLDB Workshop on Databases and Crowdsouring (DBCrowd 2013). CEUR, vol. 1025, pp. 14–19 (2013)
3. Chatzimilioudis, G., Konstantinidis, A., Laoudias, C., Zeinalipour-Yazti, D.: Crowdsourcing with Smartphones. IEEE Internet Computing 16(5), 36–44 (2012)
4. Chon, Y., Lane, N.D., Kim, Y., Zhao, F., Cha, H.: A Large-scale Study of Mobile Crowdsourcing with Smartphones for Urban Sensing Applications. In: UbiComp 2013 (2013)
5. Abecker, A., Braun, S., Kazakos, W., Zacharias, V.: Participatory Sensing for Nature Conservation and Environment Protection. In: 26th International Conference on Informatics for Environmental Protection (EnviroInfo 2012), Shaker, Aachen, pp. 393–401 (2012)
6. Goodchild, M.F.: Citizens as voluntary sensors: spatial data infrastructure in the world of Web 2.0. International Journal of Spatial Data Infrastructures Research 2, 24–32 (2007)
7. Sieber, R.: Public participation geographic information systems: A literature review and framework. Annals of the American Association of Geography 96(3), 491–507 (2006)
8. Burke, J., Estrin, D., Hansen, M., Parker, A., Ramanathan, N., Reddy, S., Srivastava, B.: Participatory sensing. In: Workshop on World-Sensor-Web (WSW 2006): Mobile Device Centric Sensor Networks and Applications, pp. 117–134 (October 2006)
9. Zhao, Y., Zhu, Q.: Evaluation on crowdsourcing research: Current status and future direction. Information Systems Frontiers, 1–18 (2012)
10. Brabham, D.C.: Crowdsourcing as a Model for Problem Solving - An Introduction and Cases. Convergence 14(1), 75–90 (2008)
11. Estellés-Arolas, E., González-Ladrón-de-Guevara, F.: Towards an integrated crowdsourcing definition. Journal of Information Science 38(2), 189–200 (2012)
12. Geiger, D., Seedorf, S., Schulze, T., Nickerson, R., Schader, M.: Managing the Crowd: Towards a Taxonomy of Crowdsourcing Processes. In: 17th Americas Conference on Information Systems, pp. 1–11 (2011)

13. Rouse, A.C.: A preliminary taxonomy of crowdsourcing. In: ACIS 2010, paper 76 (2010)
14. Schenk, E., Guittard, C.: Towards a characterization of crowdsourcing practices. Journal of Innovation Economics 1(7), 93–107 (2010)
15. Erickson, T.: Some Thoughts on a Framework for Crowdsourcing. In: Workshop on Crowdsourcing and Human Computation, CHI 2011, pp. 1–4 (2011)
16. Yuen, M.-C., King, I., Leung, K.-S.: A Survey of Crowdsourcing Systems. Privacy, security, risk and trust. In: IEEE 3rd Int. Conf. on Social Computing, pp. 766–773 (2011)
17. Doan, A., Ramakrishnan, R., Halevy, A.Y.: Crowdsourcing systems on the World-Wide Web. Communications of the ACM 54, 86–96 (2011)
18. Kazman, R., Chen, H.-M.: The metropolis model a new logic for development of crowdsourced systems. Communications of the ACM 52, 76–84 (2009)
19. Malone, T.W., Laubacher, R., Dellarocas, C.: The collective intelligence genome. IEEE Engineering Management Review 38, 38–52 (2010)
20. Hetmank, L.: Components and Functions of Crowdsourcing Systems – A Systematic Literature Review. In: 11th Int. Conf. on Wirtschaftsinformatik, Leipzig, pp. 55–69 (2013)
21. Estrin, D.: Participatory Sensing: Applications and Architecture. In: 8th ACM International Conference on Mobile Systems, Applications, and Services (MobiSys), pp. 3–4 (2010)
22. Tilak, S.: Real-World Deployments of Participatory Sensing Applications: Current Trends and Future Directions. ISRN Sensor Networks 2013, Article ID 583165 (2013)
23. Khorashadi, B., Das, S.M., Gupta, R.: Flexible architecture for location based crowdsourcing of contextual data. Patent US 8472980 B2 (2011)
24. Reddy, S., Estrin, D., Srivastava, M.: Recruitment Framework for Participatory Sensing Data Collections. In: Floréen, P., Krüger, A., Spasojevic, M. (eds.) Pervasive 2010. LNCS, vol. 6030, pp. 138–155. Springer, Heidelberg (2010)
25. Yan, T., Marzilli, M., Holmes, R., Ganesan, D., Corner, M.: mCrowd - A Platform for Mobile Crowdsourcing. In: SenSys 2009, Berkeley, USA, pp. 347–348 (2009)
26. Lasnia, D., Bröring, A., Jirka, S., Remke, A.: Crowdsourcing Sensor Tasks to a Socio-Geographic Network. In: 13th AGILE Conf. on Geographic Information Science (2010)
27. Das, T., Mohan, P., Padmanabhan, V., Ramjee, R., Sharma, A.: PRISM: Platform for Remote Sensing Using Smartphones. In: 8th Int. ACM Conference on Mobile Systems, Applications, and Services (MobiSys 2010), San Francisco, California, USA (2010)
28. Luqman, F., Griss, M.: Overseer: A Mobile Context-Aware Collaboration and Task Management System for Disaster Response. In: 8th Int. Conf. on Creating, Connecting and Collaborating through Computing (2010)
29. Christin, D., Reinhardt, A., Kanhere, S., Hollick, M.: A survey on privacy in mobile participatory sensing applications. J. of Systems and Software 84(11), 1928–1946 (2011)
30. de Cristofaro, E., Soriente, C.: PEPSI - Privacy-Enhanced Participatory Sensing Infrastructure. In: ACM Conf. on Wireless Network Security (WiSec 2011), pp. 23–28 (2011)
31. Ra, M., Liu, B., La Porta, T., Govindan, R.: Medusa: A Programming Framework for Crowd-Sensing Applications. In: 10th International Conference on Mobile Systems, Applications, and Services (MobiSys 2012), pp. 337–350 (2012)
32. Hupfer, S., Muller, M., Levy, S., Gruen, G., Sempere, A., Ross, S., Priedhorsky, R.: MoCoMapps: mobile collaborative map-based applications. In: Conf. on Computer Supported Cooperative Work (CSCW 2012), pp. 43–44. ACM, New York (2012)
33. Cornelius, C., Kapadia, A., Kotz, D., Peebles, D., Shin, M., Triandopoulos, N.: Anonysense: Privacyaware people-centric sensing. In: ACM MOBISYS, pp. 211–224 (2008)
34. Kitchenham, B.: Guidelines for performing Systematic Literature Reviews in Software Engineering. EBSE Technical Report, Keele University, Keel, UK (2007)

35. Kitchenham, B., Pearl Brereton, O., Budgen, D., Turner, M., Bailey, J., Linkman, S.: Systematic literature reviews in software engineering – A systematic literature review. Information and Software Technology 51, 7–15 (2009)

36. Gonzalez, A.L., Izidoro, D., Willrich, R., Santos, C.A.S.: OurMap: Representing Crowdsourced Annotations on Geospatial Coordinates as Linked Open Data. In: Antunes, P., Gerosa, M.A., Sylvester, A., Vassileva, J., de Vreede, G.-J. (eds.) CRIWG 2013. LNCS, vol. 8224, pp. 77–93. Springer, Heidelberg (2013)

37. Fraternali, P., Castelletti, A., Soncini-Sessa, R., Ruiz, C.V., Rizzoli, A.E.: Putting humans in the loop: Social computing for Water Resources Management. Environmental Modelling; Software 37, 68–77 (2012)

38. Corney, J.R., Torres-Sánchez, C., Jagadeesan, A.P., Yan, X.T., Regli, W.C., Medellin, H.: Putting the crowd to work in a knowledge-based factory. Advanced Engineering Informatics 24, 243–250 (2010)

39. Mashhadi, A.J., Capra, L.: Quality control for real-time ubiquitous crowdsourcing. In: 2nd International Workshop on Ubiquitous Crowdsouring, pp. 5–8. ACM, New York (2011)

40. Schall, D.: Expertise ranking using activity and contextual link measures. Data & Knowledge Engineering 71(1), 92–113 (2012)

41. Blöschl, G., Montanari, A.: Climate change impacts - throwing the dice? Hydrological Processes 24(3), 374–381 (2010)

42. Hornemann, C., Rechenberg, J.: Was Sie über den vorsorgenden Hochwasserschutz wissen sollten. Umweltbundesamt, Dessau (2006)

43. Fuchs-Kittowski, F., Simroth, S., Himberger, S., Fischer, F.: A content platform for smartphone-based mobile augmented reality. In: Arndt, H.-K., Knetsch, G., Pillmann, W. (Hrsg.) 26th International Conference on Informatics for Environmental Protection (EnviroInfo 2012) – Part 1: Core Application Areas, pp. 403–412. Shaker, Aachen (2012)

44. Europäische Gemeinschaft: Richtlinie 2007/60/EG des Europäischen Parlaments und des Rates vom 23. Oktober 2007 über die Bewertung und das Management von Hochwasserrisiken. ABl. L 288 vom 06.11.2007 (2007)

45. Meissen, U., Faust, F., Fuchs-Kittowski, F.: WIND - A meteorological early warning system and its extensions towards mobile devices. In: Page, B., et al. (Hrsg.) 27th Int. Conf. on Environmental Informatics (EnviroInfo 2013), pp. 612–619. Shaker, Aachen (2013)

46. Thuan, N.H., Antunes, P., Johnstone, D.: Factors Influencing the Decision to Crowdsource. In: Antunes, P., Gerosa, M.A., Sylvester, A., Vassileva, J., de Vreede, G.-J. (eds.) CRIWG 2013. LNCS, vol. 8224, pp. 110–125. Springer, Heidelberg (2013)

47. Jordan, L., Stallins, A., Stokes, S., Johnson, E., Gragg, R.: Citizen Mapping and Environmental Justice: Internet Applications for Research and Advocacy. Environmental Justice 4(3), 155–162 (2011)

48. Thogersen, R.: Data Quality in an Output-Agreement Game: A Comparison between Game-Generated Tags and Professional Descriptors. In: Antunes, P., Gerosa, M.A., Sylvester, A., Vassileva, J., de Vreede, G.-J. (eds.) CRIWG 2013. LNCS, vol. 8224, pp. 126–142. Springer, Heidelberg (2013)

49. de Vreede, T., Nguyen, C., de Vreede, G.-J., Boughzala, I., Oh, O., Reiter-Palmon, R.: A Theoretical Model of User Engagement in Crowdsourcing. In: Antunes, P., Gerosa, M.A., Sylvester, A., Vassileva, J., de Vreede, G.-J. (eds.) CRIWG 2013. LNCS, vol. 8224, pp. 94–109. Springer, Heidelberg (2013)

50. Liu, C.H., Hui, P., Branch, J.W., Bisdikian, C., Yang, B.: Efficient network management for context-aware participatory sensing. In: 8th Annual IEEE Communications Society Conference on Sensor, Mesh and Ad Hoc Communications and Networks (SECON), pp. 116–124 (2011)

A Semantic Approach
to Shared Resource Discovery

Kimberly García[1], Salma Velasco[2],
Sonia Mendoza[2], and Dominique Decouchant[1]

[1] Department of Information Technologies, UAM-Cuajimalpa, México
[2] Department of Computer Science, CINVESTAV-IPN, México DF., México

Abstract. The current available technologies have not been fully exploited to assist collaborators to perform activities that are considered as time/effort wasting and tedious, but which cannot be omitted. A person working in an organization can be in need of a resource they do not own at any moment, but in a place consisting of multiple buildings that are full of resources (e.g., computer devices, files, software and even people) it can get really hard for such a person to locate or even be sure whether there is a resource or a set of them able to fulfill their request. Some works have been already proposed intended to discover services, but their focus is mainly on applications looking for other applications. Therefore, those proposals do not consider information characterizing real life environment that involves human users, whose conditions are constantly changing (e.g., a person's availability). In this paper, we propose a matchmaking service, which takes a semantic approach for resource discovery in collaborative environments by paying special attention to the effects of human interaction over the availability of resources. The proposed matchmaker has been implemented as a key service of the RAMS Architecture (Resource Availability Management Service) which is able to provide users with a pervasive experience for resource discovery.

Keywords: ubiquitous collaborative environments, resource discovery, semantic matchmaking.

1 Introduction

Pervasive technologies [1] have become really popular. Their aim is to make every day environments much more comfortable by providing users with information and services in a transparent way. Thus, users do not need to worry about complicated configuration tasks or, in an ideal case, they would not need to even ask for services, since they would get the effect of those services according to their identity, role, characteristics, habits, location, etc. Several industries, such as healthcare, automotive and appliances [2] [3], have turned their heads to the idea of the smart version of their products. Those products offer single user applications (e.g., a driver tutoring system [3]) which provide responses based on information retrieved by sensors about the user and their environment. However,

N. Baloian et al. (Eds.): CRIWG 2014, LNCS 8658, pp. 137–152, 2014.
© Springer International Publishing Switzerland 2014

pervasive computing support for collaborative environments arises different challenges, as information from many more sources (i.e., users) should be correctly handled. Moreover, social relationships and their effects should be considered, as they might affect the availability of a resource (i.e., access rights and usage restrictions). In order to bring together Pervasive Computing and Computer-Supported Cooperative Work, we have proposed the RAMS Architecture [4], which is integrated by several components that provide relevant information for managing resource sharing (e.g., computational devices, files, applications, and even people's schedule).

In this paper, we explore in detail the very core of our architecture, which is the matchmaking service. This service is in charge of evaluating information captured by multiple components of the RAMS Architecture. Such information can be either static, concerning technical characteristics or capabilities that hardly change about the entities involved in the resource sharing process, or dynamic, which refers to the data that helps determining the current status of such entities or the environment. Thus, the best resources available and suitable would be chosen to satisfy a consumer's request.

Concerning the two research areas we want to cover, throughout this paper we find out that the RAMS Architecture takes a step towards the creation of a pervasive environment, since it automatically gets current information about the environment and the entities interacting within it and then evaluates such information, in order to bring resources closer to users. As for the collaborative field, the goal of our work is to consider as many human interaction variables as possible to provide controlled resource sharing among people in an organization.

Following, in section 2 the related work is presented, section 3 summarizes the RAMS Architecture. Then, section 4 details the proposed matchmaker, followed by its actual implementation and validation tests presented in section 5. Finally, in section 6 we present our conclusions and future work.

2   Related Work

Several works were developed to discover network services, such as Service Discovery Protocols (SDP) [5], which target users are applications requiring services. These works represent the first attempt to ease the process of searching resources that might be available, but they would remain underexploited if users do not know about their existence. However, these SDPs lack accuracy, since most of them perform syntactic matches to find services by their type. These results could be valid for client applications, but not for humans. Thus, some frameworks and architectures were proposed to deal with service discovery in a more complex way, incorporating contextual variables and offering semantic matches.

DAIDALOS

This framework has been developed to perform service discovery considering context [6]. It adopts a publish/subscribe approach to work over a traditional Service Discovery Protocol [5]. To publish a resource or initiate a search, a service provider or a user requesting a service supplies: 1) basic and semantic information about the resource being published or looked for, 2) a pointer to their context source, which handles a set of sensors providing current information about the service or the requester, and 3) their requirements, which are expressed either in terms of the user and the environment for a service being published, or in terms of the environment and the service, in case of a request. The Service Discovery Server (SDS) component is in charge of storing the published services and answering requests. When the SDS receives a request, it is processed by three filters: 1) considering the user basic query and the basic records of the services stored in the SDS, the basic filter evaluates the published resources by using the filter of the traditional SDP (usually a syntactic match) in which DAIDALOS is deployed, then 2) the semantic filter executes the semantic query of the requester, which consists of additional characteristics expressed in an ontological way, and finally 3) the context filtering process compares the current state of the environment, the service and the user, in order to find matches among the published resources that have passed the previous two filters and the requirements of both the own service and the user. The resulting list of these three filtering processes is then sent to the requester.

AIDAS

This framework allows to describe the properties and capabilities of the services and users in a semantic way, by creating a profile for each one of these entities [7]. A service profile includes a dynamic and a static part. The static part is integrated by the identification data of the service, and its capabilities, requirements (i.e., access restrictions), and service interface, which describes the required input and output parameters. The dynamic part corresponds to the service state. A user is also described by a profile that contains information about their location and current state, as well as information about the characteristics and requirements of the requested resource. The AIDAS discovery management service performs a semantic match by comparing the characteristics offered by a service to the ones a user has requested. Each characteristic is treated separately, since a user might have determined a characteristic as more important than another. When comparing an offered characteristic to a requested one, a matching degree is assigned to it, which can be: exact, subsumption, or plug in. Once the evaluation of all the capabilities has finished, a general score for such a service is obtained. This score helps creating an ordered list of all the services that match a user request in a higher or lower degree.

UDDI with Matchmaker

UDDI is a standard registry for Web Services that has been enabled working as a search engine capable of performing service capability matching [8]. Before a new service is registered into the UDDI Business Registry, a tool on the provider side extracts, from the WSDL service description, the service semantic information, which is stored by the Semantic Service Matchmaker. The rest of the service description is sent to the UDDI registry. To search for a resource, a user specifies the type of service they are looking for in UDDI augmented, i.e., a user adds a semantic specification containing their needs, in terms of classes in an ontology, I/O parameters, and restrictions. The semantic matchmaker considers each characteristic individually and gives it a match degree, which can be of the type: exact, plug-in, or relaxed. A second matchmaking process is performed over the whole service description. Thus, a user looking for a resource can select one or several of the following filters: a) namespace, which ensures that the offered and requested services have a common namespace, b) text, in case of existing human readable information, this filter detects matches by using knowledge retrieval technics, c) I/O type, which verifies whether the ontologies input and output match or not, and d) constraint, which confirms the offered service is less constrained than the requested one, through polynomial subsumption. Finally, from the score of the selected filters, the matchmaker computes the service degree in order to answer the user request.

CASSD

This architecture combines the traditional characteristics of service discovery with semantic and contextual discovery [9]. A service description and a service request are stored in the CASSD registry service by specifying a profile that includes the offered or requested characteristics, and if desired, the context preferences in form of rules. As this architecture provides semantic and contextual matchings, the former match is performed in order to pair both the requested and offered service descriptions, since they might use different names for the same concept. From this pairing, a degree of similarity is obtained. Following, the offered and requested service descriptions are evaluated in terms of context. Thus, an inference engine executes the rules specified in the preference section of the service description. Such an inference engine is aided by a context manager responsible for handling and storing information coming from sensors that denote the current state of the offered service and the requester. Finally, the requester receives a list of the service discovery results.

Discussion

The presented related works increase the attention to semantic and context information in service discovery. DAIDALOS and AIDAS incorporate location as their contextual variable and handle some information about services in a semantic way. CASSD is an architecture that exploits ontologies to pair service descriptions that might be expressed in concepts named differently, but which mean the same. Finally, the UDDI matchmaker adds semantic information to traditional WSDL Web services descriptions. Although these works consider some information about the user and the environment, they still present several deficiencies, as they do not take into account social relationships that affect the service availability. Moreover, the dynamism of a real environment is not fully acknowledged, since the contextual variables and constantly changing data that these works evaluate are minimal.

3   The RAMS Architecture

The Matchmaking Service is one of the most important components of the RAMS Architecture, since it is in charge of discovering shared resources by evaluating all the information automatically and explicitly gathered by the rest of the architecture components. However, before explaining in detail our matchmaker proposal, we present a summary of the whole architecture, in order to provide the reader with a vision of the significance of our work.

The RAMS Architecture is based on an asynchronous publish/subscribe model [10], which has been customized to fulfill the needs of a collaborative environment, in which the social relationships among users exchanging messages should be acknowledged. Thus, despite the traditional publish/subscribe model, in which the provider and consumer of messages do not know each other, for our requirements, it is really important to have this information. A user interacting with a RAMS-based application could be a consumer or a producer at any moment. A producer is in charge of sharing a resource with their colleagues and generating events about the current state of the resource they are sharing. In turn, a consumer subscribes themselves to RAMS by expressing their interest in receiving information about a specific resource or a type of them. Since the RAMS Architecture is meant to be a computational support for resource sharing in real collaborative environments, it is crucial to consider as much information regarding the preferences of both types of users as possible, so it would be feasible for a RAMS-based application to deliver a personalized, accurate and responsible answer about available resources for a user request. Thus, in order to guaranty a producer that the resource(s) they share would be safe, a producer is able to grant access rights, and restrictions, as well as specify usage policies. Moreover, to guaranty that a consumer would receive updated and tailored information, they are able to determine the technical resource features and their own contextual preferences (e.g., location and current status of the resource). All the information regarding the description of either a consumer, a provider or even the environment, is expressed in a semantic manner, by following the structure of a set of ontologies designed to represent all of the entities involved in a resource sharing process [11]. This set of ontologies was designed in a generic fashion, so they can be used and customized in any type of organizations as the RAMS Architecture pretends.

Figure 1 shows the RAMS Architecture, which components are grouped according to the service they offer. The Human Interaction set (see Fig. 1-A) contains the Publication and Subscription Services. The former service (see Fig. 1 step #1) allows developers to implement applications that receive information from producers wanting to share a new resource. A producer is able to specify the technical characteristics of such a resource, to define usage policies and to give access rights to colleagues they trust. The Publication Service sends

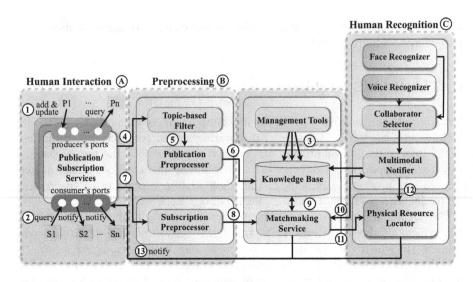

Fig. 1. RAMS Architecture

that resource related information to the Topic-based Filter (see Fig. 1 step #4), which classifies it into the right ontology according to the type of resources that is being published. The Publication Preprocessor structures the classified information received from the Topic-based Filter (see Fig. 1 step #5) to make it comprehensible for the RAMS Ontologies and stores it into the Knowledge Base (see Fig. 1 step #6).

The Subscription Service (see Fig. 1 step #2) is in charge of enabling the interaction between a consumer collaborator and a RAMS-based application. This service allows a consumer to describe the resource(s) they need, in terms of the technical capabilities and their own personal preferences (e.g. location and current status of the resource). Once this service has received a resource request, it is sent to the Subscription Preprocessor (see Fig. 1 step #7) which transforms the consumer request into queries that can be understood by the Matchmaking Service (see Fig. 1 step #8). This Service selects authorized resources, from the RAMS Knowledge Base (see Fig. 1 step #9), which attributes correspond to the technical consumer requirements. The selected resources are then evaluated by this same Matchmaking Service, in terms of the dynamic information that might affect their availability status. Such dynamic information is provided by the Multimodal Notifier (see Fig. 1 step #10) and by the Management Tools (see Fig. 1 step #3). The Multimodal Notifier delivers the decision of the Collaborator Selector, which is in charge of determining a human resource's presence and location by considering information coming from the Face and Voice Recognizers. The Management Tools allow producers to modify their availability or the one of their published resources at anytime.

In case a consumer makes a physical resource request, the set of suitable and available resources selected by the Matchmaking Service is transmitted to the

Physical Resource Locator (see Fig. 1 step #11), which asks the Multimodal Notifier for the consumer's current location (see Fig. 1 step #12) in order to compute the closest resource and the path they should follow to reach it. The results produced by either the Physical Resource Locator [4] (when looking for a physical resource) or the Matchmaking Service (when searching for a human or virtual resource) are finally delivered to the consumer (see Fig. 1 step #13).

4 The RAMS Matchmaking Service

The Matchmaking Service of the RAMS Architecture is in charge of processing information about resources, in order to find the resource(s) that best satisfies a consumer request. However, which could be the most suitable resource for one collaborator, it could not be the best resource for another. We claim that each consumer request should be evaluated in a personal manner by considering the current conditions of the consumer, provider and environment. Moreover, the ideal resource, which is just awaiting for a consumer to use it, might not always be found. Thus, our matchmaker proposal considers the consumer needs and preferences allowing them to decide the availability state(s) a resource should be in to be considered as a good response for such a consumer request. Following, in section 4.1 we present the availability states that the RAMS Matchmaking Service considers. Section 4.2 describes the type of contextual preferences that can be expressed by a consumer, in order to enhance their resource search. Then, in section 4.3, we present the matchmaking process by detailing the three phases in which our proposal consists. Later in section 4.4 a B method-based formalization of the proposed resource availability states is presented.

4.1 Availability States

When a consumer makes a request, they can select as many availability degrees as they want. However, the *Allowed* state is a default state, so for a resource to be considered available in some degree or another, it should be at least in this state. The ideal available resource would be in all the following six states:

* ★ *Allowed*. A physical, virtual or human resource will be in this state, if the provider of the resource or the own resource, in case of a human resource request, agree to share such a resource or their information with a specific consumer.
* ★ *Free*. A physical or a virtual resource will be free if it is not used by anybody.
* ★ *Satisfy restrictions*. A physical, virtual or human resource will be at this state, if a consumer fulfills the restrictions defined by the provider of the resource. In case of human resources, this state denotes that the schedule they have defined is being followed.
* ★ *Reachable*. A physical or a virtual resource (i.e., stored in a physical device) located inside a restricted place will be in this state, if someone allowed to give access to the resource is inside the same place as the resource.

⋆ *Available provider*. Human, some physical and virtual resources can be in this state. A human resource will be considered to be open to interact with someone else, if they are available. A physical resource or a virtual resource contained in a physical one, which is located at a restricted place will be in this state if the provider collocated with the resource agrees to be disturbed at a specific moment, in order to give access to it.

⋆ *Satisfy context restrictions*. A human, physical or virtual resource will be in this state, if it fulfills the contextual preferences a consumer defines.

4.2   Contextual Preferences

In order to provide a consumer with a pervasive experience, in which they receive personalized information according to their current state, a consumer can express their contextual preferences, which can be of the organizational or physical type. An organizational contextual preference can be expressed in terms of:

- *Process:* For a physical, human or virtual resource, a consumer can determine the preferred action in which the resource is involved, e.g., the consumer asks for a room where a professors reunion is taking place.
- *Role:* A consumer looking for a human resource is able to determine the role this resource should be performing, in order to fulfill the consumer expectations, e.g., the consumer is interested in a debates coordinator.
- *Group:* Similar to the role, a consumer looking for a human resource can determine the group they prefer the resource belongs to, e.g., the consumer is interested in a team working in a cryptography project.

A physical contextual preference can be expressed in terms of location. A consumer can decide whether they are interested in getting information about physical, virtual, or human resources situated in a physical space or equipment.

4.3   Matchmaking Process

The Matchmaking Service of the RAMS Architecture performs syntactic and semantic matching processes, in order to find the best resource (or a set of them) that satisfies a consumer request. The syntactic match is performed in the Characteristic Search Phase, since the set of resources that fulfill the technical requirements and capabilities of a user's request is retrieve by comparing strings representing data. On the other hand, the semantic match is carried out within the State Analysis and Query Execution Phases, in which relationships among ontological representations of the resources and collaborators are looked for. Such relationships refer to the availability states in which the resources are involved. These two types of matching processes retrieve information from the RAMS Knowledge Base, which has been developed as a set of ontologies expressed in the OWL language.

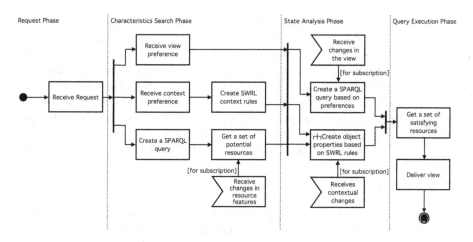

Fig. 2. Matchmaker Activity Diagram

Request Phase. In this phase, the Matchmaking Service receives a resource request from a consumer who specifies the features they are looking for (see Figure 2). It is important to highlight that a consumer can be either: a) a human resource, who is a collaborator asking for a resource, b) an activity, in which a group of collaborators asking for a resource can be involved, and c) a role, which is played by a human resource at a specific moment (e.g., a student can be a debates coordinator in a forum). A request can be of one of two types: one-response or subscription. In the former case, information about the resources managed by the RAMS Architecture is evaluated just once, and the resources that best satisfy the consumer request are presented. In the latter case, a consumer subscribes to the types of resources they are interested in. When the RAMS Architecture receives changes in the resources, the consumer context, or the environment, it updates the set of resources that best fulfills the consumer request.

Characteristics Phase. Three actions are executed in parallel in this phase: a) receiving the consumer preferences about the resources fulfilling a given availability degree; b) receiving the consumer preferences about the context conditions of the resources they are looking for, and c) creating a SPARQL query based on the consumer request (see Figure 2). For the first action, a consumer selects one or more of the states described in section 4.1, to express the degree of availability that the resources suggested by the Matchmaking Service should fulfill, in order to be of the consumer interest. The second action performed in this phase corresponds to the reception of the consumer preferences regarding the contextual conditions of the requested resource, which can be expressed as described in section 4.2. For the last action, a SPARQL query is created by considering the request made by the consumer. This query is then executed, performing the syntactic match, which gets a set of potential resources that fulfill the technical characteristics a consumer is asking for. Even though, not just the static features of a resource (e.g., type of consumables used or display resolution) are considered in this action, but also its current context determined by its characteristics (e.g., available storage space or free memory). This set of potential resources cannot be considered as the best match. These resources need to be also evaluated in terms of their dependencies with other resources as well as the consumer context. It is important to notice that in the case of a subscription, the execution of the SPARQL query corresponding to the set of potential resources is performed every time a change in the current characteristics of a resource occurs, e.g., current free memory in a computer. Thus, updated information is always evaluated.

State Analysis Phase. In this second phase (see Figure 2), two actions are performed in parallel. One is to create a new SPARQL query according to a consumer viewing preferences. These preferences can come from the *Characteristic Search* phase or in case of a subscription, a change in the viewing preferences of a consumer is likely to appear at any moment from a signal denoting this change. The other action performed in this phase is the creation of relationships, called object properties, between the representation of the consumer and the one of a resource from the set of potential matches retrieved in the *Characteristic Search* phase. These object properties denote the availability states.

To create such properties, the semantic match process is accomplished by evaluating SWRL rules, which are constructed at different moments. The rules determining that a resource satisfies a consumer context preferences are created by the Matchmaking Service when they make a request. For the rest of the states, SWRL rules were defined at design time and they are triggered when the matchmaking process takes place. However, all these rules are applied considering the current state of the resource, consumer, producer and environment. In case of subscriptions, the rules are activated when a contextual change is detected.

Query Execution Phase. Once the SPARQL query and the object properties are created, in the *Query Execution* phase, the query is executed and resources in the right state(s), according to the consumer preferences are retrieved, in order to create a view (see Figure 2). This is also a semantic match, since relationships among ontological representations of resources and consumers are looked for.

4.4   Availability States Formalization

The proposed matchmaker algorithm works over the RAMS Knowledge Base, which is integrated by a set of ontologies built in OWL language [12], [13]. This set describes resources as individuals of a class representing a type of resources (e.g., computer and file). These individuals are related by object properties, which are in charge of giving meaning to the information stored in the ontologies. Thus, the availability states described in section 4.1 are translated into object properties, which relate individuals representing the resources that fulfill the features a consumer is looking for and an individual representing such a consumer. According to the B formalizing method, we define the following sets:

- C is the set of individuals from the `HumanResource`, `Activity`, `Role`, and `Group` classes.
- R represents the set of individuals from the `HumanResource`, `VirtualResource`, and `PhysicalResource` classes.
- Y is the set of individuals from the `VirtualResource` and `PhysicalResource` classes.
- S denotes the set of individuals from the `UsageStatus` class.
- T denotes the set of individuals from the `Task` class.

We also use the following variables:

- *canPerform* denotes the relationship between individuals from the `HumanResource`, `Activity`, `Role`, and `Group` classes and individuals from the `Task` class.
- *isPerformedOn* presents the association between individuals from the `Task` class and individuals from the `HumanResource`, `VirtualResource`, and `PhysicalResource` classes.
- *isAllowedFor* defines the relationship between individuals from the `HumanResource`, `VirtualResource`, and `PhysicalResource` classes and individuals from the `HumanResource`, `Activity`, `Role`, and `Group` classes.
- the *hasUsageStatus* variable denotes the association between individuals from the `VirtualResource` and `PhysicalResource` classes and individuals from the `UsageStatus` class.
- the *isFreeFor* variable presents the relationship between individuals from the `VirtualResource` and `PhysicalResource` classes and individuals from the `HumanResource`, `Activity`, `Role`, and `Group`.

Following, we present the rules designed to establish the relationships that determine a resource is at the *Allowed* and *Free* states. It is important to remember that, according to the type of such a resource, it can be at one or many states, which means that the resource can be involved in one or several relationships with other individuals.

Allowed State

The association *canPerform* is a partial function from the set C of individuals belonging to the `HumanResource`, `Activity`, `Role`, and `Group` classes to the power set $P(T)$ of the set T of individuals from the `Task` class:

$$canPerform \in C \nrightarrow P(T) \tag{1}$$

The partial function *canPerform* relates a consumer to a possibly empty set of tasks. More precisely, for each consumer $c \in dom(canPerform)$, the set *canPerform(c)* contains the tasks it can carry out.

The association *isPerformedOn* is a partial function from the set T of individuals belonging to the `Task` class to the power set $P(R)$ of the set R of individuals from the `HumanResource`, `VirtualResource`, and `PhysicalResource` classes:

$$isPerformedOn \in T \nrightarrow P(R) \tag{2}$$

The partial function *isPerformedOn* relates a task to a possibly empty set of resources. More specifically, for each task $t \in dom(isPerformedOn)$, the set *isPerformedOn(t)* contains the resources over which the task can be performed.

The association *isAllowedFor* is a partial function from the set R of individuals from the `HumanResource`, `VirtualResource`, and `PhysicalResource` classes to the power set $P(C)$ of the set C of individuals from the `HumanResource`, `Activity`, `Role`, and `Group` classes:

$$isAllowedFor \in R \nrightarrow P(C) \tag{3}$$

The partial function *isAllowedFor* relates a resource to a possibly empty set of consumers. More precisely, for each resource $r \in dom(isAllowedFor)$, the set *isAllowedFor(r)* has the consumers allowed to interact with the resource.

Expressions *canPerform* (1), *isPerformedOn* (2), and *isAllowedFor* (3) fulfill the following invariant:

$$\forall c,t,r \cdot (c \in C \wedge t \in T \wedge r \in R \wedge t \in canPerform(c)$$
$$\wedge\, r \in isPerformedOn(t) \implies c \in isAllowedFor(r) \tag{4}$$

Invariant (4) denotes that for every consumer $c \in dom(canPerform)$ that is able to perform a task $t \in (canPerform(c) \cap dom(isPerformedOn))$ on a resource $r \in (isPerformedOn(t) \cap dom(isAllowedFor))$, such a consumer has rights on that resource.

Free State

The association *hasUsageStatus* is a total surjection from the set Y of individuals from the `VirtualResource` and `PhysicalResource` classes to the set S of individuals from the `UsageStatus` class:

$$hasUsageStatus \in Y \twoheadrightarrow S \tag{5}$$

For every resource $y \in Y$, *hasUsageStatus(y)* denotes its corresponding usage status. More precisely, every resource y is just related to a usage status $s \in S$ by means of the total surjection *hasUsageStatus*, while every usage status s is at least associated with a resource y.

The association *isFreeFor* is a partial function from the set Y of individuals from the `VirtualResource` and `PhysicalResource` classes to the power set $P(C)$ of the set C of individuals belonging to the `HumanResource`, `Activity`, `Role`, and `Group` classes:

$$isFreeFor \in Y \nrightarrow P(C) \tag{6}$$

The partial function *isFreeFor* associates a resource with a possibly empty set of consumers. More precisely, for each resource $y \in dom(isFreeFor)$, the set *isFreeFor(y)* contains the consumers that can see the resource as not being used.

Expressions *isAllowedFor* (3), *hasUsageStatus* (5), and *isFreeFor* (6) fulfill the following invariant:

$$\forall\, c, y, s \cdot (\, c\, \in C\, \wedge y \in Y \wedge s \in S \wedge$$
$$c \in isAllowedFor(y) \wedge hasUsageStatus(y) =$$
$$s \,\wedge\, s = free \Longrightarrow c \in isFreeFor(y) \tag{7}$$

Invariant (7) denotes that for every resource $y \in$ *(dom(isAllowedFor)* $\cap$ *dom(hasUsageStatus)* $\cap$ *dom(isFreeFor))* that has a usage status s, such that *hasUsageStatus(y)* $= s$ and $s = $ *free*, and y is accessible to a consumer $c \in$ *(isAllowedFor(y)* $\cap$ *isFreeFor(y))*, it can be said that such a resource is free for that consumer.

5   Usage Scenario

In order to validate our proposal, let us consider that a Computer Vision team (CV) in a multidisciplinary research center is working with a team from a Neuroscience Hospital (NS). The CV team is developing a tool for detecting scars in the cerebral cortex, by analyzing Magnetic Resonance Imaging (MRI) samples provided by the NS team. Let us suppose that Matt, the NS team leader received an invitation to give a talk next Friday in the Chemistry Department of the research center in which the CV team is. As Matt's agenda is always really busy he contacts Finn, the CV team leader to arrange a team meeting the same day, just after the talk in the Chemistry Department, in order to discuss the recent advances on the development of the tool. Since the meeting is really soon, just in few days, Finn immediately starts setting up the reunion. Finn is not only the team leader of this project, but also a professor and he has to coordinate many activities in the department, so he does not have time to be running around asking for the resources that are needed for the meeting. Luckily, the RAMS Architecture is deployed in the research center, thus Finn has at his disposal several applications to ask for the resources he needs and to notify his team about the meeting. Such resources are:

- A meeting room available on friday from 3pm to 6pm
- A high definition 3D projector
- Access to the horizon imager during the meeting schedule. This device can print high quality medical images, but due to its cost and the expensiveness of its consumables it is restricted. However, research team leaders involved in medical projects are allowed to use it, by just respecting the established schedule and the maximum amount of printings allowed per month.

In order to better illustrate the matchmaking process, we will briefly describe how Finn's request is processed in each phase of the matchmaker:

Receive Request. In this phase, the Matchmaking Service receives Finn's request from the Subscription Service. Such a request states that Finn is looking for a meeting room for 15 people, a 3D projector, and the horizon imager, which are going to be used on Friday from 3pm to 6pm.

Characteristic Phase. At this phase, Finn expresses his viewing preferences in terms of the availability states that the resources he is asking for should fulfill. Let us suppose that today is Wednesday, so he decides to see resources that are currently in the *Satisfy Restrictions* and *Satisfy Context Restrictions* states. That way, he will be able to see the options and decide whether he wants to reserve any of them. Finn also determines that he prefers a meeting room that is located in the same floor as the horizon imager because the department owns just one of such devices and it cannot be moved. From this context preference, a SWRL rule will be created by the Matchmaking Service. It will also create a SPARQL query to look for a set of resources that satisfies the technical description that have been specified by Finn. This is a syntactic match, since the characteristics of the resource are expressed as data properties, which can be evaluated in terms of their value. Following, the SPARQL query is executed to retrieve such a set of resources. From this syntactic evaluation, the Matchmaking Service retrieves the following technically suitable resources:

- The capacity of the meeting room 2301 is of 25 people
- The meeting room 4001 is able to hold up to 20 people
- The capacity of the meeting room 1203 is of 18 people
- Projector Sony 4K-VPL is a full HD 3D Projector
- Projector Sony VPL-H250 is a full HD 3D Projector
- There is a horizon imager

State Analysis Phase. Two main actions are performed in parallel in this phase, the former one refers to the creation of a SPARQL query based on Finn's viewing preferences, which determine that he is interested in the *Satisfy Restrictions* and *Satisfy Context Restrictions* states. Here, the query is just constructed, but it is not executed yet. The latter action executes the SWRL rules, which have being just created in the previous phase, about Finn's context preferences, and the predefined SWRL. The execution of these rules determines whether any of the resource selected as technically adequate for Finn's request is also available and contextually suitable for him. Thus, for each positive evaluation regarding the SWRL rule being executed, an object property that determines the resource availability state is created between the evaluated resource and the individual representing Finn. Hence, considering the obtained resources in the previous phase, the following relations are created:

- The meeting room 4001 is associated to Finn by the *satisfiesCxtRestFor* object property, since it is located in the fourth floor as the horizon imager. In fact, this resource is across the hallway from the room 4001.
- Finn cannot fulfill a restriction associated to meeting room 2301. This restriction states that he can use the room from Tuesday to Friday before noon.

– Finn does not have to satisfy any schedule restrictions associated to the meeting rooms 4001 and 1203, but he is allowed to use these resources 6 hours per week. As he has not consumed this time, the restriction associated to the meeting rooms 4001 and 1203 are related to Finn by the *isSatisfiedBy* object property.

– Finn also satisfies the restriction of the number of hours allowed to use the Sony VPL-H250 projector per month. Thus, this restriction is related to Finn. However, he does not fulfill the restrictions related to the projector Sony 4K-VPL, so this resource will not be in the *Satisfy Restrictions* state for Finn's request.

– Finally, the restrictions linked to the horizon imager are also related to Finn by the *isSatisfiedBy* object property. As Finn has not used the resource this month, he satisfies the usage restrictions.

Query Execution Phase. The SPARQL query created in the previous phase is now executed to retrieve the resources that best satisfies Finn's request, which are the meeting room 4001, since it is at the *Satisfy Context Restrictions* and *Satisfy Restrictions* states. Finn's response also includes the Sony VPL-H250 projector and the horizon imager, as both of them are at the *Satisfy Restrictions* state. Once Finn is notified about these resources, he can reserve them.

6   Conclusion and Future Work

In this paper, we proposed the Matchmaking Service of the RAMS Architecture, which is in charge of determining the best resources available for a consumer request by performing a four phase matching process, which evaluates information in a syntactic and semantic way. This matching process is possible thanks to the set of ontologies that conforms the RAMS Knowledge Base. Such ontologies have been designed to provide a generic but close to reality representation of a collaborative environment in which resources of different kinds are shared. We have customized the proposed ontological model to suit our research center in order to validate our Matchmaking Service proposal. Unlike works already done concerning resource discovery, our proposal focuses in providing support for human users, who represent a bigger challenge than application clients, since to offer collaborators information about available resources for a specific request, it is necessary to evaluate not just the own resources descriptions, but also the current conditions of such resources, updated information about the collaborators involved in the resource sharing process, and the conditions of the environment in which the collaborators are interacting. Such information is gathered and managed by several components of the RAMS Architecture. Moreover, the Matchmaking Service we propose allows collaborators to express their resource availability states preferences when making a request. These availability states represent the fulfillment of a set of conditions in a specific moment; in this way a collaborator would always receive really accurate and personalized information.

The current state of our research work lead us to perform several tests in complex real environments, in order to detect flaws in its operation. We would also like to take advantages of some standard measures for matchmaking evaluation, such as precision and recall. The results obtained through such tests and measures will helps us to place our current proposal among similar works and will highlight the improvements that can be made to our proposal.

References

1. Weiser, M.: The Computer for the 21st Century. In: SIGMOBILE Mobile Computing and Communications Review, pp. 3–11. ACM, New York (1999)
2. Tentori, M., Favela, J., Rodriguez, M.D.: Privacy-Aware Autonomous Agents for Pervasive Healthcare. IEEE Intelligent Systems 21(6), 55–62 (2006)
3. Karvonen, H., Kujala, T., Saariluoma, P.: In-Car Ubiquitous Computing: Driver Tutoring Messages Presented on a Head-Up Display. In: Intelligent Transportation Systems Conference, ITSC 2006, pp. 560–565. IEEE (September 2006)
4. Garcia, K., Mendoza, S., Decouchant, D., Rodriguez, J., Perez, T.: Determining and locating the closest available resources to mobile collaborators. Expert Systems with Applications 40(7), 2511–2529 (2013)
5. Zhao, W., Schulzrinne, H.: Enhancing Service Location Protocol for Efficiency, Scalability and Advanced Discovery. Journal of Systems and Software 75(1-2), 193–204 (2005)
6. Suraci, V., Mignanti, S., Aiuto, A.: A Context-aware Semantic Service Discovery. In: Proceedings of the 16th IST Mobile and Wireless Communication Summit, pp. 1–5. IEEE Computer Society, Budapest (2007)
7. Toninelli, A., Corradi, A., Montanari, R.: Semanitc-based Discovery to Support Mobile Context-aware Service Access. Computer Communications 31(5), 935–949 (2008)
8. Paolucci, M., Sycara, K., Kawamura, T., Hasegawa, T., Ohsuga, A.: Web services Lookup: A Matchmaker Experiment. IT Professional Journal 7(2), 36–41 (2005)
9. Patel, P., Chaudhary, S.: Context aware semantic service discovery. In: The World Conference on Services, pp. 1–8. IEEE Computer Society, Bangalore (2009)
10. Muehl, G., Fiege, L., Pietzuch, P.: Distributed Event-Based Systems, pp. 258–263. Springer, Heidelberg (2010)
11. Garcia, K., Kirsch-Pinheiro, M., Mendoza, S., Decouchant, D.: An Ontological Model for Resource Sharing in Pervasive Environments. In: The 2013 IEEE/WIC/ACM International Conference on Web Intelligence (WI 2013), pp. 179–184. IEEE Computer Society, Atlanta (2013)
12. Horridge, M., et al.: A Practical Guide To Building OWL Ontologies Using Protege 4 and CO-ODE Tools, Edition 1.3, pp. 7–106. University of Manchester (March 2011)
13. Dean, A., Hendler, J.: Semantic Web for the Working Ontologist: Effective Modeling in RDFS and OWL, pp. 62–112. Elsevier (2008)

Performance Effects of Positive and Negative Affective States in a Collaborative Information Seeking Task[*]

Roberto González-Ibáñez[1] and Chirag Shah[2]

[1] Departamento de Ingeniería Informática
Universidad de Santiago de Chile
Avenida Ecuador #3659, Estación Central, Santiago, Chile
roberto.gonzalez.i@usach.cl
[2] School of Communication and Information (SC&I)
Rutgers, The State University of New Jersey
4 Huntington St., New Brunswick, NJ, 08901-1071 USA
chirags@rutgers.edu

Abstract. Collaborative information seeking (CIS) is a common process carried out by groups in a wide variety of situations and contexts. From family activities to business tasks, people typically engage in collaborative search practices while working toward a common goal. In collaborative settings, various aspects of human behavior influence the way people interact with each other and make decisions. One of these aspects corresponds to emotions and related affective processes such as mood and feelings. Studies in social psychology have suggested that group dynamics and their performance may be affected by the interaction of affective processes, in particular positive and negative ones. Although such findings have been derived in different group situations, to the best of our knowledge none of them refer to the particular case of CIS. Based on previous studies, we investigate to what extent positive and negative affective states relate to group performance in CIS. To carry out this study, we designed an experiment with 45 dyads distributed in three configurations based on initial affective states: (1) positive-positive, (2) positive-negative, and (3) negative-negative. To achieve these initial conditions, members of each dyad were individually exposed to affective stimuli. Following, each dyad worked on a precision-oriented search task. Our results suggest that the three interactions of affective states have different implications on the performance of dyads. In particular, the negative-negative configuration performed significantly better than the other two configurations. Conversely, performance of the positive-negative condition was found to be significantly lower than the other two conditions. Findings from this work have practical implications for applications such as team design in tasks involving CIS.

Keywords: collaboration, information seeking, affective states, performance.

---

[*] The work reported here is supported by the US Institute of Museum and Library Services (IMLS) Early Career Development Grant #RE-04-12-0105-12.

N. Baloian et al. (Eds.): CRIWG 2014, LNCS 8658, pp. 153–168, 2014.

1   Introduction

Studies in psychology and neuroscience have pointed out that affective states (e.g. emotions, mood, and feelings) can shape the way people behave [7], make decisions [2], and interact with others [6]. Indeed, researchers have shown that emotions as well as other affective processes have the potential to define the performance of people in particular tasks. For instance, in business contexts it has been shown that positivity and negativity in the right proportions may distinguish high performance teams from low performance ones [9,18]. Although research on affective processes has been carried out in several contexts, their role in particular processes such as collaborative information seeking (CIS) has been barely explored. From family activities to complex business tasks, CIS is a rather common process carried out in group contexts [23]. Consider for example a couple planning their wedding, a group of students working on a term project, or a business team making strategic decisions; in all these scenarios, information needed to accomplish common goals may be sought as part of a collaborative process.

In this context, if initial affective processes - in particular positive and negative ones - have the potential to shape team performance in collaborative projects, to what extent, if any, this level of influence applies to CIS as a particular sub-process of such projects? To address this research question we conducted an experiment with 90 participants in 45 dyads (pairs). Dyads were distributed in three configurations that define all possible combinations of initial affective states of their members, that is to say: (1) positive-positive, (2) positive-negative, and (3) negative-negative. To ensure participants' initial affective processes, participants were individually exposed to stimuli designed to elicit positive and negative affective states. Then, each dyad worked on a precision-oriented search task consisting of a set of *A Google a Day*[1] questions. To evaluate the performance in this task we focused on two particular set of measures: (1) search performance, which is expressed in terms of search effectiveness and search efficiency; and (2) task performance, which relates to the overall answer precision in the search task.

Results from this study suggest that the three interactions of affective states have different implications on the performance of dyads. In particular, the negative-negative configuration performed significantly better than the other two configurations. Conversely, performance of the positive-negative condition was found to be significantly lower than the other two conditions. These findings have implications in practical applications such as the design of high performance teams in tasks involving CIS.

2   Background

Studies of the affective dimension in information seeking have been limited. Not surprisingly, as presented below, studies of affective processes in CIS are even scarcer. Even though relevant works about affective processes and individual

---

[1] http://www.agoogleaday.com/

information seeking have been produced over the years, their findings do not necessarily apply to information-related situations in which collaboration takes place. The participation of the social dimension in CIS makes research about affective processes more complex than in individual settings. Such heightened complexity is due to the fact that affective states in such scenarios may derive from and/or may influence not only information-related processes, but also group dynamics [10]. As noted by Wilson et al. [29], "Most commonly, in CIS, we model relationships between collaborating searchers in terms of roles. We expect there to be searchers who differ in terms of search expertise and knowledge [...] Further, we can consider them to be taking different tactical approaches to divide up the tasks [...] While these role-focused models account for behavioral changes, they have so far not modeled affective changes. How do people feel being watched, when they have different roles or abilities? How do people perform if they feel anxious or judged by collaborators?" [p. 1]

When looking at relations between collaboration, information search, and affective dimension, multiple questions arise. For example, based on findings on affect infusion [6,7] and positive psychology [18,9] one could ask what role, if any, do positive or negative affective states play in the selection of relevant information? How do being happy or unhappy change the way people formulate queries and assess information? In collaborative contexts, can team dynamics and the affective processes of team members change how they deal with information? Or could prior affective states serve as predictors to anticipate search performance, success, or failure? This list of questions can be easily expanded, which make research in this domain quite complex.

While some research in CIS acknowledges the participation of affective processes as part of collaborative practices carried out by team members when searching, collecting, evaluating/assessing, and using information, the majority of such works refer tangentially to the affective dimension [14]; that is, the authors suggest possible links between affective states and information-related practices of team members but do not study nor develop these ideas. In general terms, affective studies in CIS have investigated the participation of affective processes within the search process of teams [14,25]. Others have also referred to affective states as subproducts of the collaboration process, which is typically expressed in terms of affective load [26]. To the best of our knowledge, no studies have been carried out to investigate affective states as initial conditions in CIS and their implications in performance.

Perhaps, two studies closely related to what we present in this article are described in [25,10]. In [25] the authors conducted an exploratory user study with dyads performing exploratory search tasks. The authors attempted to describe group members' search behaviors using Kuhlthau's ISP [17], a well established model for individual searchers. With regard to the affective dimension, the authors found mixed feelings (i.e. positive and negative) during the initiation and selection stages, and also during the transitions between stages. In the remaining stages, however, positive feelings were found to be predominant. Another interesting aspect involving affective processes was identified in the selection of relevant information. According to the authors, "[i]t was [.] observed that the selection of relevant information was first done by an individual and then subjected to the group's judgment and reflection" [p. 7]. This phenomenon was later referred in [10] as Group's Affective Relevance (GAR), where the authors claim that relevance judgments are socially constructed through both objective and emotional discourse, meaning that team members share their opinions (e.g. "This page contains useful information"), reactions (e.g. "I loved this page"), and objective comments (e.g. "This information came from the president of the company") with respect to the information they find and share.

With regard to performance evaluation, in a preliminary evaluation of GAR [25], the authors "found that the closer the distance between the number of positive, negative, and neutral information judgments, the higher the performance of teams in terms of precision [15]. This indicates an interesting correlation between expressed emotions and performance of a group" [25, p. 318]. Furthermore, it was argued that social interactions carried out when selecting relevant material may dynamically shape feelings, engagement, and the confidence of team members in their actions within the group. The evaluation approach of the authors focused on particular products of CIS, in particular modified versions of traditional measures in information retrieval, namely: precision and recall [15]. Note that product-based performance evaluations have been previously used in CSCW studies to formally evaluate collaborative work [1].

One of the major problems of affective research is the ability to distinguish different levels of the affective dimension. As Palmero et al. [20] suggest, four categories or levels of affective processes can be distinguished: feelings (subjective component), emotions (objective, expressive, and categorical component with short duration), mood as well as affect (objective and internal processes with medium to long term durations described in terms of dimensions such as valence/tone and arousal/intensity). Note that affective processes, affective states, and affective dimension are used in this article to refer in a broad sense to these four categories. Such distinctions have practical implications in the way studies are designed and also in the way affective processes are measured.

Different methods are available to conduct affective research. For example emotion elicitation techniques to manipulate people's affective states [19,5] are commonly used in psychology to investigate effects of particular affective processes in areas such as perception, decision making, and information processing. To measure affective processes, different instruments have been developed and validated. For example, questionnaires such as the Self-Assessment Manikin (SAM) [3] and PANAS [28] are used to measure self-reported affective experiences (e.g. feelings). This study uses emotion elicitation techniques under the assumption that they can effectively change paople's internal affective states, this in spite of their subjective experience (feelings) as reported through questionnaires like SAM.

The following section presents the methodological approach designed to conduct the experimental study presented in this article.

3 Method

To address the research question introduced earlier, we conducted a controlled experiment. This section provides a detailed description of the study including: experimental design, sample, task description, session workflow, laboratory setup, and instrumentation used to collect data.

3.1 Experimental Design

Taking into account findings that indicate that affective states may shape the way information is processed, we designed a multiple-group design involving three experimental groups. These conditions correspond to dyads configuration in terms of initial affective states: positive-positive ($C1^{++}$), positive-negative ($C2^{+-}$), and negative-negative ($C3^{--}$).

As depicted in Table 1, the overall experimental design consisted of two stages: (1) Affective induction and (2) Evaluation of prolonged effects (P.E.). In the first stage, the participants in the three experimental groups were individually treated with affective stimuli (X) in order to elicit positive (+) and negative affective states (−). Stimuli were presented to participants while working alone (collaboration was not enabled on this stage) on a precision-oriented search task consisting of multiple search challenges. More details about the affective stimuli stage and the search task are provided below. In the second stage (evaluation of prolonged effects) dyads in the three groups performed a search task similar to that of the first stage, but in the absence of affective stimuli. It is important to note that in the second stage collaboration was enabled by allowing participants to communicate via text chat, share information, and make decisions with their teammates. It is also important to consider that while the participants had to complete multiple search challenges, the overall study is not a repeated-measure design because participants were not exposed to multiple experimental conditions.

Table 1. Experimental design summary. (R): Random placement, (PreS): Pre-Stimuli, (PostS): Post-Stimuli, (O_i): observations, (X): treatment/stimuli, and (MT): main task.

	Stage 1: Affective Induction	Stage 2: Evaluation P.E.	
$C1^{++}$	R O1 PreS O2 X^{++} On PostS	On+1 MT	On+m
$C2^{+-}$	R O1 PreS O2 X^{+-} On PostS	On+1 MT	On+m
$C3^{--}$	R O1 PreS O2 X−− On PostS	On+1 MT	On+m

Note that the affective induction stage depicted in Table 1 is divided into three parts: (1) Pre-stimuli evaluation (PreS), in which the participants performed a search task before being exposed to affective stimuli; (2) stimuli exposure, in which the participants addressed a search task for 10 minutes while receiving affective stimuli; and (3) post-stimuli evaluation (PostS), which aimed to evaluate the efficacy of affective stimuli. Before and after each search challenge, observations were made through different instruments (details are provided below). Overall, the affective induction stage lasted 20 minutes, whereas the following stage lasted 25 minutes.

3.2   Task Description

To properly investigate affective processes in short span of time, we designed a precision-oriented search task based on multiple-step fact finding. The task comprised a set of questions that were presented sequentially to the participants. Questions are independent from each other. In each question, dyads were given a maximum of five minutes to find the responses.

The questions used in the study were obtained from *A Google a Day*, which is a puzzle-based game implemented by Google to train and evaluate search skills. The puzzle consists of questions that can be answered by searching information online. As stated in the main website, "there is no right way to solve it, but there's only one right answer". One interesting aspect of *A Google a Day* is that for each question an ideal search path to find the answer as well as the answer itself are provided for past questions.

The collection of questions used in this study corresponds to those posted on *A Google a Day* between April 7th and August 31st of 2011. Questions were collected along with answers and their corresponding ideal search paths. After the collection process, the search path of each question was objectively rated based on the number of steps or queries suggested to find the answer. As a result, the number of steps - also referred to as complexity level - for the entire set of questions ranged between 2 and 5. Results from the pilot study showed that level-2 questions were more adequate in terms of perceived difficulty, response precision, topic familiarity, and response time [12]. For this study, a random set of level-2 questions were used in the affective induction stage and in the evaluation of prolonged effects. Note that questions that required non-textual information (e.g. videos, maps, audio, or images) to find the answer were removed from the set of questions used in this study. Following, an example of a level-2 question that requires textual information is presented:

Question: You're a detective at a crime scene with no visible evidence. On a hunch, you spray the carpet with a light-emitting solution and it glows, revealing blood. What component of hemoglobin catalyzed the reaction?

How to find the answer: Search [light-emitting solution crime scene] to find that luminol is the active chemical of choice for blood detection. Search [luminol blood] to learn that it reacts with iron to emit a slightly bluish light for about 30 seconds.

Answer: Iron.

3.3   Sample

The study was carried out with a sample of convenience that consisted of 45 pairs of undergraduate students (15 per condition) from Rutgers, the State University of New Jersey. Recruitment was conducted through open calls (e.g. announcements posted on campuses' bus stops, facilities, and email lists). Participants were compensated with $10 in cash for one-hour session. As an extra motivation, they were offered with the possibility to win cash prices based on performance ranking at the end of the study (i.e. $50 first place, $25 second place, or $15 third place). To ensure common ground [4] within dyads, candidates were required to sign up with someone with whom they had previous experience collaborating.

Ages among the participants ranged between 18 and 24 years (M=20.29; SD=1.32). In this sample, 57.78% of the participants were women. Regarding participants' search skills, these were reported as intermediate to high. Participants were all English native speakers. Note that English was a restriction imposed since the recruitment stage in order to avoid effects on cognitive and affective load associated to information processing in a non-native language [16], which is beyond the scope of this study.

We decided to work with dyads as the minimal group size in order to avoid effects of particular intervening variables that are present in larger groups. We made this decision based on previous work that has shown that as group size increases, interactions among group members become more complex, which makes misinterpretation and misunderstanding more plausible [27].

3.4   Elicitation of Positive and Negative Affective States

As indicated above, the first stage of the experimental design aimed to set the initial affective conditions of the participants for the evaluation of prolonged effects. To elicit positive and negative affective states, different techniques, like those described in [5,19], were investigated and evaluated. As a result of this process, it was found in a pilot study [12] that game feedback [19], also referred to as false-feedback [5] was effective in eliciting affective states in the context of the search task used in the experimental evaluation. This technique consists on providing either positive (e.g. "You are doing great!") or negative (e.g. "Wrong. That was disappointing") feedback to participants regardless of their actual performance when working on a given task. Side effects of this approach include frustration, disinterest on performing the task, and overconfidence, which were overcome by balancing the number of positive and negative feedback provided to participants.

Stimuli consisted of a box containing a text message and a blinking emoticon (smiley, frowning, or neutral face) on top of the box. First, text messages were composed by words from the positive and negative categories listed in the Linguistic Inquiry and Word Count (LIWC) [21]. Second, the blinking effect was implemented to grab the attention of participants while working in the task. Third, the size and position of the color box were adjusted based on eye fixations captured during the pilot study using an eye tracker. Fourth, boxes and emoticons were presented in three different colors, namely, green for positive stimuli, red for negative stimuli, and yellow for negative-neutral and positive-neutral stimuli. The latter types of stimuli were used to achieve balanced feedback as indicated above by expressing slightly negative messages (e.g. "So so. You can do it") for participants receiving positive feedback in $C1^{++}$ and $C2^{+-}$ and slightly positive messages (e.g. "Just a little better this time") for participants exposed to negative feedback in $C2^{+-}$ and $C3^{--}$. This was implemented in order to avoid canceling the expected affective response specified by each experimental condition. Finally, regarding stimuli duration, color boxes with the corresponding

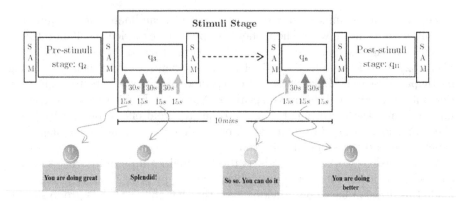

Fig. 1. A closer look to the stimuli stage of positive condition (C1$^+$)

messages and blinking emoticons were displayed to participants for 15 seconds in intervals of 30 seconds. Stimuli were also presented at the moment of submitting the answers to each question. A closer look to the stimuli stage for participants receiving positive stimuli is depicted in Figure 1.

Note that the efficacy of stimuli relied on the assumption that these were able to elicit internal affective changes regardless of the subjective experience of the participants (i.e. feelings).

3.5   Laboratory Setup

The study was conducted in an interaction laboratory with two rooms for the participants (one in each room) and one for the study supervisor. Rooms were isolated so that visual and auditory contact between participants and also with the supervisor were not possible. Each participant worked on a desktop computer equipped with a 19" display, full size keyboard, mouse, and headset. The 64-bit version of Windows 7 Enterprise (Service Pack 1) was used as the operating system. Note that the supervisor of the study located in the other room monitored the study, performed non-participant observations, and assisted participants in case they had questions or technical difficulties.

3.6   Session Workflow

Each session was conducted following a research protocol devised to ensure consistency in all the sessions and provide better documentation for possible replications of the study. As part of this protocol, first, participants were briefly introduced to the study. Second, participants signed consent forms. Third, participants filled out a demographic questionnaire. Fourth, participants watched a brief tutorial that explains how to use the system. Fifth, participants familiarized with the system while working on a practice question. Sixth, affective induction stage was carried out. Seventh, evaluation of prolonged effect was performed.

Finally, end-session questionnaires were responded. Overall sessions lasted approximately 60 minutes. Note that the different stages described in the session workflow were automatically guided by the system described below.

3.7 System

In order to conduct the study and collect data from participants' search processes, a system based on Coagmento [24,11] was implemented. This system, named Coagmento Collaboratory[2] [12], was designed to support experiment protocols, experimental designs involving affective treatments, timed tasks, multiple-stages, multiple sessions, and enhanced logging capabilities. The system comprises three main components: (1) a toolbar, (2) a sidebar, and (3) a server-side Web application. First, the toolbar provided two buttons: search and snip. The search button directed the participants to Google home page (the search engine enabled during the experiment) with parameters to retrieve pages indexed before April 7th of 2011. Other search engines were blocked by the toolbar. These decisions were made to avoid participants having access to pages that could contain *A Google a Day* questions and their answers, which are usually available online shortly after questions are posted on the main site. The snip button, on the other hand, allowed participants to save snippets of text from Web pages along with their sources. Participants were required to save snippets that helped them to find the answers to the questions. In addition, the toolbar captured all browsing activity and different users' actions within the browser, which were submitted to the server-side application through Web services. Data were logged along with local timestamps in order to facilitate synchronization with data captured with other tools (e.g. screen capture software, keystroke logger, webcam, and other sensors).

Second, the sidebar contained different elements depending upon the stage of the task. Specifically, on top of the sidebar the remaining time for each question and for the corresponding stages were displayed. Then, the current question and a button to jump to the answer form were provided. Below, saved snippets for the corresponding question were listed. During the second stage (i.e. evaluation of prolonged effects), where collaboration was enabled, a chat system was provided. Finally, and only during the affective induction stage, stimuli were displayed on the bottom part of the sidebar. Content in the sidebar was automatically updated with AJAX calls to the served-side Web application. A snapshot of the system and Firefox 11 during an actual session is presented in Figure 2. Note that both the toolbar and the sidebar container were implemented as a single Firefox add-on. Additionally Firefox was customized to only display toolbar, sidebar, tabs, and navigation buttons. Other components such as address bar were hidden to restrict participants' attempts to visit search engines other than Google with the specific parameters described above.

Third, the server-side application provided a set of web services that were used by both the toolbar and the sidebar. A snapshot of the system during an actual session is presented in Figure 2.

[2] http://www.coagmento.org/collaboratory.php

Toolbar   Main container   Sidebar

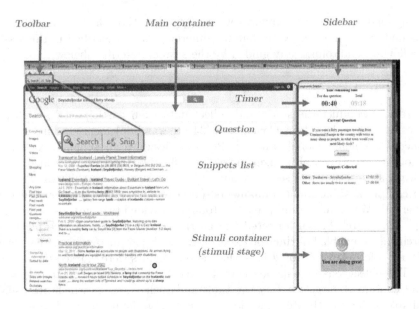

Fig. 2. Snapshot of the system during an actual session

3.8   Performance Measures

To evaluate the performance of dyads in the completion of the search task, we used two set of measures. The first focused on search performance, whereas the second focused on task performance. Search performance was measured in terms of search effectiveness and search efficiency, which are formulated as follow:

$$SearchEffectiveness(p,q) = \frac{UsefulCoverage(p,q)}{TotalCoverage(p,q)} \tag{1}$$

$$SearchEfficiency(p,q) = \frac{SearchEffectiveness(p,q)}{n(Queries(p,q))} \tag{2}$$

First, search effectiveness (Equation 1) of a pair p in search challenge q (each search challenge corresponds to *A Google a Day* question) is expressed in terms of *UsefulCoverage* and *TotalCoverage*. On the one hand, *UsefulCoverage* correspond to the total number of pages in which participants spent 30 seconds or more [8,30]. On the other hand, *TotalCoverage* consists of the total number of pages visited by pair p.

Second, search efficienty (Equation 2) of a pair p in search challenge q corresponds to the ratio between Search Effectiveness of p in q and the number of queries used by participants in p to complete q.

Both measures have been proposed in [13] to measure performance in information search task from the searcher's perspective. This as an alternative to traditional measures such as precision and recall that are widely used in information retrieval studies.

Regarding task performance measures, this was measured in terms of the overall response precision (Equation 3) in the search task. That is to say, the ratio between the number of correct answers provided by pair p and the total number of answers provided during the entire task. This measure is formulated as follow:

$$ResponsePrecision(p) = \frac{CorrectAnswers(p)}{TotalNumberOfAnswers(p)} \tag{3}$$

3.9 Instruments

In order to collect data and perform observations during the experimental sessions, multiple instruments were incorporated in the study. These resources are organized into three groups, namely, hardware, questionnaires, and software. Following, brief descriptions of the instruments and systems under each group are presented.

3.10 Hardware

In order to monitor room conditions, an Acurite digital humidity and temperature monitor (model 00325) was installed in the participants' rooms. This instrument allowed the researcher to monitor and regulate the temperature and humidity of the room so that all the participants worked in similar conditions.

3.11 Questionnaires

To monitor participants subjective experience after being exposed to affective stimuli, we used the SAM questionnaire [3]. Using SAM, the participants self-reported how they felt immediately before and after working in each question. SAM is a non-verbal scale that allows individuals to indicate their affective states or reactions in three dimensions, namely, pleasure (happy-unhappy), arousal (excited-calm), and dominance (controlled-in control). There are different variations of the scale; however, in this study the 9-points version of the scale was used.

3.12 Software

The final group of data collection resources consists of software that helped the researcher to keep track of browsing activity, record responses to questionnaires, and perform observations. The list of software include the system Coagmento Collaboratory introduced in the previous section, Morae Recorder to capture desktop activity along with keystrokes and mouse actions, and Morae Observer to perform non-participant observations.

4 Results and Discusion

This section provides a description of the quantitative analyses carried out to investigate the effects, if any, of initial affective states (positive and negative) in dyads performance in a CIS task. First, a brief overview of the results of affective induction from the participants' perspective is provided. Then we focus on the evaluation of prolonged effects (PE) with special emphasis on performance.

4.1   Affective Induction

As explained in the previous section, participants in $C1^{++}$, $C2^{+-}$, and $C3^{--}$ were individually exposed to affective stimuli aiming to elicit positive and negative affective states. More specifically, prior starting the task two major groups of participants were defined: (1) participants exposed to stimuli designed to induce positive affective states (C^+) and (2) participants exposed to stimuli designed to induce negative affective states C^-. As part of the pretest-posttest (PreS and PostS in Table 1) evaluation, it was found that the participants exposed to positive stimuli (C^+) presented significant variations ($p<.05$) of search behaviors in PostS with respect to PreS as reported by Friedman and the Wilcoxon signed-rank tests. Variations on search behaviors were expressed in a number of measures such as information coverage, number of queries, exploration of search result pages (SERPs), precision, recall, and time spent (dwell time) in content pages as well as in SERPs. Among the significant changes reported by the statistical tests, the following are highlighted: increased coverage of relevant webpages (pages from where snippets were collected) in C^+ (on average 0.37 more relevant pages than those covered in PreS, $p<.05$), decreased dwell time on content webpages in C^+ (on average 4.62 seconds less than in PreS, $p<.05$), and increased dwell time on SERPs in C^- (on average 1.18 seconds more than in PreS, $p<.05$).

Additionally, from the participants' perspective (as reported through SAM), positive affective stimuli were 71.67% effective in the elicitation of positive affective states in C^+. On the other hand, negative affective stimuli were 65% effective in the elicitation of negative affective states in C^-. Note that these results are only referential to illustrate the participants' subjective experiences after the affective induction stage. Nevertheless, based on the literature on emotion elicitation [19,5] it was assumed that affective stimuli were able to elicit the desired internal affective states (objective) regardless of the participants being aware of such variations or that they were not able to express this accurately through the SAM questionnaire.

4.2   Performance Results

To evaluated performance we measured (1) search performance and (2) task performance in accordance to the equations presented in the previous section.

Search Performance. is computed for each search challenge q completed by a dyad p as described above. Since all dyads were given a maximum of five minutes to complete each search challenge, there were cases of pairs that spent less time working in some questions. This expected behavior resulted on dyads with different number of *A Google a Day* questions completed. Overall, we found that all dyads were able to at least respond eight of these questions. Note that questions were presented to all dyads in the same order. This allowed us to perform comparisons with each question as the unit of analyses.

In terms of search effectiveness (Equation 1), Kruskall-Wallis test reported significant differences only in the eight question. In this question search effectiveness was found to be significantly lower at $p < .05$ in $C1^{++}$ (Mdn=0.67) than in the other two conditions. On the other hand, search effectiveness was found to be significantly higher $p < .05$ in $C2^{+-}$ (Mdn=0.75) than in the other two conditions.

With regard to search efficiency (Equation 2), our results showed consistent differences between $C2^{+-}$ and $C3^{--}$ in the first three questions. According to Kruskal-Wallis test search efficiency was significantly higher at $p < .05$ in $C2^{+-}$ than in $C3^{--}$ in all three questions with medians ranging between 0.13 and 0.17. Search efficiency of $C1^{++}$ was found to be significantly higher (Mdn=0.17) at $p < .05$ than in the other two conditions only in the first question. In the second question search efficiency became significantly higher at $p < .05$ in $C2^{+-}$ (Mdn=0.24) than in the other two conditions. This difference disappeared in the fourth question and they only showed up again in the eight question with search efficiency being significantly higher at $p < .05$ in $C2^{+-}$ (Mdn=0.22) than in the other two conditions. Contrary to what happened in the first question, in this question search efficiency was significantly higher at $p < .05$ in $C3^{--}$ (Mdn=0.22) than in $C1^{++}$ (Mdn=0.20).

The fact that search effectiveness was no significantly different in the majority of the search challenges suggests that pairs behaved similarly in terms of the exploration of useful information. This can be explained by two specific aspects. First, due to the particular characteristics of the task, it was expected that participants were able to complete search challenge with a combination of specific facts obtained from two different Web pages. Second, generally speaking dyads tried to complete search task as quick as possible. The average time in each question ranged between 155.07 and 185.16 seconds (approximately three minutes) in the three groups. This particular aspect indicates that dyads employed similar amount of time to find and evaluate information prior completing search challenges.

With regard to search efficiency, our results suggest that dyads whose members initiate the task in negative affective states ($C3^{--}$) tend to formulate more queries than those in a condition with mixed affective states ($C2^{+-}$). Note that these particular results are not necessarily related to task performance as we discuss in the following section. Indeed, neither search effectiveness nor search efficiency consider whether answers are right or wrong.

Task Performance. focused on the quality of work performed by dyads. To conduct this analysis we focused on response precision (Equation 3), which as described in the method section corresponds to the ratio between correct answers and the total number of answers provided by the participants.

Based on a comparative analysis with the Kruskal-Wallis test it was found that response precision was significantly higher at $p < .05$ in $C3^{--}$ (Mdn=0.85) than in $C1^{++}$ (Mdn=0.76) and $C2^{+-}$ (Mdn=0.73). In turn, response precision was found to be significantly lower at $p < .05$ in $C2^{+-}$ than in the other two conditions.

Results suggest that a mismatch in affective states, which is the case of $C2^{+-}$ may produce negative effects in the quality of results in a CIS task. Contrary to what theory in positive psychology suggest, a combination of negative affective states in $C3^{--}$ seems to produce better results in this particular CIS task. This could be explained by information processing strategies as suggested in [6,7]. According to studies around affect infusion, negative affective states are related to more systematic information processing strategies than those used by people in positive affective states [22]. It is also possible that these results are a direct effect of the type stimuli applied. That is to say, participants who received positive stimuli might have become more confident about their performance, whereas those who received negative stimuli during the first stage of the study might have become more careful when evaluating information.

5   Conclusion

In this article we studied the effects of initial affective states in the performance of dyads in a CIS task. Specifically, we investigated their influence in (1) search performance and (2) task performance. To achieve this research objective we designed an experiment to contrast three configurations of dyads based on initial affective states, namely: positive-positive ($C1^{++}$), positive-negative ($C2^{+-}$), and negative-negative ($C3^{--}$).

Results from this study suggest that different combinations of initial affective states in the formation of dyads have different effects expressed in terms of search performance and task performance. On the one hand, a positive-negative configuration ($C2^{+-}$) seems to lead dyads to a more efficient search process than homogeneous configurations, specially when the initial affective states of both members are negative ($C3^{--}$). However, these results are not related to the quality of the work achieved by dyads in each configuration. Indeed, our results for task performance showed the opposite. This is, negative-negative configuration ($C3^{--}$) seems to be more precise in solving fact-finding challenges than other dyads configurations.

We attribute this result to two possible factors: (1) affect infusion and/or (2) affective stimuli. First, regarding affect infusion [6,7], participants who initiated the search task in negative affective states could be more systematic in terms of their information processing strategies than those who initiated the task in positive affective states. This conclusion is consistent with previous findings on affect infusion [22]. This may be a decisive factor in the successful completion of fact-finding tasks like the one used in this study design. Second, with respect to affective stimuli, it may be possible that the type of stimuli used in this study had direct effects on participants' stand with respect to the task. For instance, it may be possible that participants who received positive stimuli in the affective induction became overconfident, whereas those exposed to negative stimuli became more cautious in their decision making process about what information is relevant and what is not.

While the results of this study suggest that initial affective states would influence search performance and task performance, it is necessary to consider underlying limitations such as sample, group size, stimuli, search task, communication channel, collaboration format, and assumptions. Although these aspects contribute to internal validity, they limit the generalization (external validity) of our results.

At the theoretical level the major implication of the results presented in this article is that initial affective states may be determining factors for the way search processes are carried out. This extends what traditional models in information science have suggested in the past, in which affective processes are typically investigated as intrinsic factors to the search process.

The results and theoretical implications of this study could have different practical implications. By way of example, our findings could be applied in the design of dyads that perform precision-oriented search tasks. According to our results, dyads in which both members initiate a search task in negative affective states could achieve better results than other configurations. It may be possible that critical feedback to dyad members in a particular training process has direct implications in how systematic they will become when performing a CIS task.

Future work in this research line should explore variations of our study design in order to address the limitations listed above. By doing this, it could be possible to validate our results in different settings toward the generalization of our findings.

References

1. Baeza-Yates, R., Pino, J.A.: A first step to formally evaluate collaborative work. In: ACM Group Conference, pp. 56–60. Phoenix, AR (1997)
2. Bechara, A., Damasio, A.R.: The somatic marker hypothesis: A Neural Theory of Economic Decision. Games and Economic Behavior 52, 336–372 (2005)
3. Bradley, M.M., Lang, P.J.: Measuring emotion: The Self-Assessment Manikin and the semantic differential. Journal of Behavior Therapy and Experimental Psychiatry 25(1), 49–59 (1994)
4. Clark, H.H., Brennan, S.E.: Grounding in communication. In: Resnick, L.B., Levine, R.M., Teasley, S.D. (eds.) Perspectives on Socially Shared Cognition, pp. 127–149. The American Psychological Association (1991)
5. Coan, J.A., Allen, J.J.B. (eds.): Handbook of emotion elicitation and assessment. Oxford University Press (2007)
6. Forgas, J.P.: Affective influences on partner choice: Role of mood in social decisions. Journal of Personality and Social Psychology 61, 708–720 (1991)
7. Forgas, J.P.: The Affect Infusion Model (AIM): An integrative theory of mood effects on cognition and judgments. In: Martin, L.L., Clore, G.L. (eds.) Theories of Mood and Cognition: A User's Guidebook, pp. 101–136. Lawrence Erlbaum Associates Publishers, Mahwah (2009)
8. Fox, S., Karnawat, K., Mydland, M., Dumais, S., White, T.: Evaluating implicit measures to improve Web search. ACM Transactions on Information Systems (TOIS) 23(2), 147–168 (2005)
9. Fredrickson, B.L., Losada, M.F.: The positive affect and the complex dynamics of human flourishing. American Psychologist 60(7), 678–686 (2005)
10. González-Ibáñez, R., Shah, C.: Group's Affective Relevance: A proposal for studying affective relevance in collaborative information seeking. In: ACM Group Conference, Sanibel Island, FL, USA, November 6-10, pp. 317–318 (2010)
11. González-Ibáñez, R., Shah, C.: Coagmento: A system for supporting collaborative information seeking. In: ASIS&T 2011, New Orleans, LA, USA, October 9-13 (2011)

12. González-Ibáñez, R., Shah, C.: Investigating positive and negative affects in collaborative information seeking: A pilot study report. In: ASIS&T 2012, Baltimore, MD, USA, October 26-30 (2012)
13. González-Ibáñez, R., Shah, C., White, R.: Pseudo-collaboration as a method to perform selective algorithmic mediation in collaborative IR systems. In: ASIS&T 2012, Baltimore, MD, USA, October 26-30 (2012)
14. Hyldegard, J.: Collaborative information behaviour - Exploring Kuhlthau's information search process model in a group-based educational setting. In: Information Processing and Management, vol. 42(1), pp. 276–298. Pergamon Press, Inc., Tarrytown (2006)
15. Kent, A., Berry, M.M., Luehrs Jr., F.U., Perry, W.: Operational criteria for designing information retrieval systems. American Documentation 6(2) (1955)
16. Kirkland, M.R., Saunders, M.A.P.: Maximizing student performance in summary writing: Managing cognitive load. TESOL Quarterly 25(1), 105–121 (1991)
17. Kuhlthau, C.: Inside the search process: Information seeking from the user's perspective. Journal of the American Society for Information Science 42(5), 361–371 (1991)
18. Losada, M., Heaphy, E.: The role of positivity and connectivity in the performance of business teams: A nonlinear dynamics model. American Behavioral Scientist 47(6), 740–765 (2004)
19. Martin, M.: On the induction of mood. Clinical Psychology Review 10, 669–697 (1990)
20. Palmero, F., Guerrero, C., Gómez, C., Carpi, A.: Certezas y controversia en el estudio de la emoción. Revista electrónica de motivación y emoción (R.E.M.E) 9, 23–24 (2006)
21. Pennebaker, J.W., Francis, M.E., Booth, R.J.: Linguistic inquiry and word count (LIWC): LIWC2001. Erlbaum Publishers, Mahwah (2001)
22. Sinclair, R.C., Mark, M.M.: The effects of mood state on judgemental accuracy: Processing strategy as a mechanism. Cognition and Emotion 9(5), 417–438 (1995)
23. Shah, C.: Collaborative information seeking: A literature review. Advances in Librarianship 32, 3–33 (2010)
24. Shah, C.: Coagmento - A collaborative information seeking, synthesis and sensemaking framework. Integrated demo at CSCW 2010 (2010)
25. Shah, C., González-Ibáñez, R.: Exploring information seeking processes in collaborative search tasks. In: American Society of Information Science and Technology (ASIST), Pittsburgh, PA, October 22-27 (2010)
26. Shah, C., González-Ibáñez, R.: Evaluating the synergic effect of collaboration in information seeking. In: Annual ACM Conference on Research and Development in Information Retrieval (SIGIR 2011), Beijing, China, pp. 913–922 (2011)
27. Tang, A., Pahud, M., Inkpen, K., Benko, H., Tang, J.C., Buxton, V.: Three's company: Understanding communication channels in three-way distributed collaboration. In: Proceedings of CSCW 2010 (2010)
28. Watson, D., Clark, L.A., Tellegen, A.: Development and validation of brief measures of positive and negative affect: The PANAS scales. Journal of Personality and Social Psychology 54(6), 1063–1070 (1988)
29. Wilson, M.L., Wilson, M.: Social anxieties and collaborative information seeking. In: Collaborative Information Seeking Workshop at GROUP 2010, Sanibel Island, FL, USA (November 7, 2010)
30. White, R.W., Huang, J.: Assessing the scenic route: Measuring the value of search trails in Web logs. In: Annual ACM Conference on Research and Development in Information Retrieval (SIGIR), Geneva, Switzerland (2010)

Promoting Elderly-Children Interaction
in Digital Games: A Preliminary Set of Design Guidelines

Ana I. Grimaldo, Alberto L. Morán, Eduardo Calvillo Gamez, Paul Cairns,
Ramón R. Palacio, and Victoria Meza-Kubo

{ana.grimaldo,alberto.moran,mmeza}@uabc.edu.mx,
eduardo.calvillo@gmail.com,
paul.cairns@york.ac.uk,
ramon.palacio@itson.edu.mx

Abstract. In this paper, we propose a set of guidelines to facilitate the design of digital games to support elderly-children interaction. We conducted a literature review to identify preliminary elements in the elderly-children interaction process and assessed elder-child interaction during their using two digital games to complement the initial findings. Based on these results, we proposed a set of guidelines to aid in the design of digital games. To validate the guidelines, we conducted an evaluation on the design of digital games with a group of 12 postgraduate students from two local Computer Science programs. Results are promising, indicating a high perception of usefulness, ease of use and intention of use of the proposed guidelines by the group of developers.

Keywords: Digital games, collaboration, design guidelines, elderly, children.

1 Introduction

The demographic change seen as an aging population tsunami is a worrying situation worldwide [1]. Thus, one of today's main challenges is to promote the quality of life of the elderly [2]. In order to prevent some of the negative effects of aging, researchers propose increasing socialization as a strategy to support wellness in elders [3-5]. The use of technology such as digital games provides the opportunity to facilitate socialization [6, 7]. Nevertheless, sometimes when elderly people are exposed to the use of technology, they face difficulties [8]. As a consequence, frequently they prefer to be technologically isolated, which could gradually affect their health [5, 9].

Research suggests that the elderly need a subtle process when they face changes in their environment such as the interaction with technology [2]. For that reason, we propose to include elderly people in the social digital context by means of playing digital games with children. In order to design digital games that are suitable for the elderly, children and their interaction, it is necessary to consider characteristics to meet the needs and skills of both kinds of users [10]. Therefore, in this research we developed a set of guidelines to design digital games that promote the interaction between the elderly and the children.

N. Baloian et al. (Eds.): CRIWG 2014, LNCS 8658, pp. 169–176, 2014.

The paper is organized as follows. Section 2 briefly introduces the use of digital games by children and the elderly. Section 3 and 4 present the methodology followed and a set of partial results obtained during the development of the guidelines. Section 5 describes the guidelines proposed to aid in the design process of digital games that foster elderly-children interaction, while Section 6 presents the results of a study focused at validating the use of the guidelines in the design of digital games to promote the interaction between children and the elderly. Finally, section 7 presents a general discussion of our findings, our conclusions and directions of future work.

2   Use of Digital Games

In order to understand the interaction between the children and the elderly in the use of digital games, we firstly conducted a literature review. Based on its results, we developed a taxonomy of factors implicated in elderly-children interaction while using digital games (see Figure 1). The factors are grouped into three categories: usability, user experience and social interaction [11]. Usability is a factor that may hamper the adoption and use of digital games. User experience is related to the use of digital games to provide pleasurable experiences to the users. Finally, social interaction is focused on elderly-children interaction during the game.

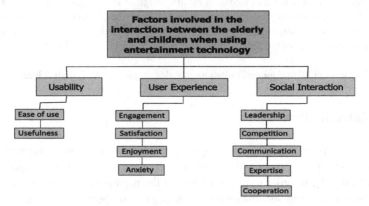

Fig. 1. Taxonomy of factors implicated in elderly-children interaction while using entertainment technology. (Image adapted from [11]).

3   Assessing Elderly-Children Interaction with Digital Games

In order to understand elderly-children interaction with digital games, we conducted two studies. We evaluated the use of digital games while elder-child pairs interacted together. In both experiments, we used natural interfaces [12, 13, 14] to facilitate the interaction, particularly for the elderly.

In the first study we used a touch-based game for learning to add Mayan numbers. The participants were 13 elder-child pairs (avg. age 60 years old and avg. age 7.07

years old, respectively). In the second study, we used a Kinect-based leisure and exercise game to catch a dragon and fly a kite. We recruited 9 elder-child pairs (avg. age 69.7 years old and avg. age 8 years old respectively).

During both studies, participants interacted together with the game. In the first study, participants were taking turns to interact with the touch screen computer. In contrast, in the experiment with the Kinect both participants were interacting with the game at the same time.

4   Defining Elderly-Children Interaction with Digital Games

We used Activity Theory [15] to analyze the videos obtained from the evaluations conducted. In addition, we applied Grounded Theory [16] to analyze the interviews carried out. As a result, we identified three interaction levels while elders and children played together, including: a) Interaction between participants, b) Interaction with the device and c) Interaction through the game.

Interaction between participants: It includes verbal and non-verbal interaction such as: verbal interaction (e.g. conversation), gestural interaction (e.g. nodding) and corporal interaction (e.g. giving a hug). The interaction between participants can be seen as an alternative to involve both participants in the interaction with the game. This can be illustrated with the next conversation that occurred during the interaction with Mayabaco. Elder: "How many items do you need to represent number 3?" Child:"I need 3 dots". Elder: Well done! (While giving a hug to the child).

Interaction with the device: It describes the actions taken by the user to use the devices in order to interact with the game (e.g. click a mouse button, push a UI button or select an item in a touch screen). In addition, this interaction may occur in parallel (e.g. both participants interacting with the Kinect) or sequentially (e.g. as when participants take turns to use a touch-screen computer) in collaborative activities.

Interaction through the game: It is composed by the meaning of the actions that participants perform to achieve something in the game while manipulating the devices (e.g. both players press a button that make their characters jump in the game). The main goal in the interaction with the game is focused on achieving an objective.

5   Design Guidelines to Foster Elder-Children Interaction

From the information presented in the previous section, we integrated the findings into a non-exhaustive set of 6 guidelines. The aim is to facilitate the identification of important factors in the design of elderly-children interaction with digital games. A brief description, the development and underpinnings for each guideline follow.

DG1: Promote De Use of Natural Devices and Interfaces

The guideline is focused on understanding the difficulties that participants face when they interact with complex devices and interfaces. As a result, users have difficulties in the interaction with complex devices. The lack of experience of the elderly in the use of technology increases the difficulties for them while they interact with digital games accompanied by children. As a solution, it is proposed to provide natural interfaces and devices to ease the use of the game, especially for the elderly.

DG2: Allow Users to Perceive the Usefulness of the Application

There are times when users do not perceive the usefulness of the interaction with technology; as a consequence, they do not feel motivated to use it. In addition, having fun is the main motivation of children to use technology. In contrast, the elderly need to perceive that children can benefit from the interaction with digital games [17, 18]. As a solution, it is important to motivate both participants during the interaction process. On the one hand, the elderly must perceive their own participation as necessary during the use of digital games accompanied by children. On the other hand, the children must perceive that they will have fun during the interaction with the game.

DG3: Provide Different Difficulty Levels in the Game

Sometimes people are exposed to activities that are inadequate to their skill level in the use of digital games. In order to facilitate the interaction with the game, it is important to provide different difficulty levels within the activities of the game while maintaining the consistency on the dynamic of the interaction and theme. A solution is based on activities that correspond to the "Zone of proximal development". Therefore, it is necessary to provide small increments to meet the needs and desires of both the elderly and children during the execution of increasingly difficult tasks [19].

DG4: Allow for Short Game Sessions to Promote Repetition

When participants have long game sessions, opportunities to recover and start a new session are reduced. Furthermore, sometimes the use of technology implies the execution of unfamiliar activities for the user. In addition, interaction with technology is usually a novel activity that is learned over time. Children need a training process to increase confidence in them regarding their development in the game [19]. Meanwhile, the elderly need to get familiar with using the devices and understanding the dynamics of the interaction with the game. As a solution, it is proposed to provide short sessions in the interaction with the game to allow for repetitions and to facilitate the learning of the dynamics of the game.

DG5: Provide Diversity of Activities in the Game

For a digital game to be successful, it is necessary that participants perceive it not only as useful or easy to use, but also as providing a range of entertainment activities which keeps them interested and persuaded to use it in the future. The lack of activities may affect the motivation of participants to be interested in using the application to conduct the proposed activities. If this is not the case, it would be difficult to keep the user motivated. This may result in rejection of the application or, if used, in an unpleasant user experience. As a solution, it is proposed to provide diversity in the activities during the interaction with the game but maintaining consistency in the theme and performance.

DG6: Promote Active and Balanced Participation

During pair-based interaction in digital games, it is difficult to balance the participation of both users. In the case when turn-taking interaction is not required, there are times when just one of the participants takes and active role in the game. According to the interaction levels, the elderly and children can interact through different levels: i) interaction between participants, ii) interaction with the device and iii) interaction through the game. To mediate this, the theme of the game is an important element. When the elder has more experience than the child in the theme, the interaction between them could be increased, as the elder could help the child to solve the challenges in the game. In addition, the interaction with devices provides an opportunity to use devices, either in parallel or by taking turns, which provides an opportunity to manage the interaction for both participants. For that reason, it is important to facilitate the active and balanced interaction for the elderly and children by facilitating the interaction through the interaction levels.

6   Validating the Design Guidelines

In order to analyze the impact of using the guidelines in the design of digital games to promote elderly-children interaction, we conducted an exploratory validation study. We used an approximation of the expert-based evaluation described in [20], in which participants firstly identified usability issues from the interfaces provided by using only their expertise. Then, participants used ergonomic criteria to complement their previous result.

Our approach is also similar to that proposed in [21] to conduct a validation of a set of design guidelines for the development of usable cognitive stimulation applications. In our study, participants firstly proposed an initial design for a digital game prototype, which supports elderly-children interaction, based on their own expertise. Then participants used the proposed guidelines to enhance their previous designs. Finally, the initial and final designs were compared to determine the impact of using the proposed design guidelines.

In order to conduct the study, 12 participants were invited to participate in the evaluation. They were 11 graduate students from two local computer science programs and 1 part-time junior researcher. 6 students declared to have "beginner" experience in HCI while 5 students and the part-time researcher declared to have "medium / advanced" experience in HCI.

At the beginning of the study, participants were asked to analyse 5 online digital games. Then, they were asked to identify requirements and to design an initial prototype of a digital game that supports elderly-children interaction. Subsequently, they were asked to complete a TAM-based questionnaire and to determine the perceived usefulness of each guideline using a 5-Likert scale questionnaire. Participants were asked to "Think aloud" during the whole process, which were recorded for an ulterior analysis.

Quantitative Results

We analyzed the results obtained from the TAM-based questionnaire. According to the results, participants in all teams perceived a high usefulness (6.3/7) and ease of use (6.3/7) of the guidelines. They also reported to have perceived a high intention to use them (6.6/7). Although the results show little differences between the perception of beginners (higher scores) and advanced (lower scores) teams, the results are still high for all of them. A possible explanation could be that participants with more experience in HCI demanded a higher level of detail in the contents of guidelines.

Qualitative Results

In addition to the quantitative analysis reported in the previous section, we conducted a qualitative analysis based on the features considered in the initial and final prototype designs of each team. Differences were identified about the elements included in the initial prototypes compared to the re-designed prototypes of two teams (1 Beginner team, and 1 advanced team).

Begginer team: According to the analysis, the main changes in the re-designed prototype included: i) An expert level to increase the challenge (DG3: Difficulty level); ii) Additional mini-games appearing randomly after solving some rounds (DG4: Short duration of the sessions and DG5: Diversity in the game); iii) Replacing the game when the participants have problems with the actual challenge (DG3: Difficulty level); and iv) Using touch-screen interaction to play the game (DG1: Devices and interfaces).

Advanced team: The elements changed during the re-design prototype included: i) the use of multi-touch screen interaction (DG1: Devices and interfaces); ii) New elements in the scenario (DG3: The difficulty level); iii) A new challenge to foster learning for the children (DG2: Usefulness); iv) New elements in the game to provide different scenarios (DG5: Diversity in the game); and v) Participation of both (elder and child) at the same time (DG6: Active and balanced participation).

7   Discussion and Conclusions

The guidelines presented in this work are based on a literature review and on the results of the evaluations conducted. The proposed taxonomy defines three categories of factors in the interaction between the elderly and the children with digital games. Then, the interaction level model derived from the assessments with digital games helped us identifying ways to include the elderly and the children during the interaction with digital games. Subsequently, the development of guidelines provides different elements to help us understanding them and facilitating its use.

According to the assessments in the use of the proposed guidelines, we obtained evidence about their impact on the design of digital games. Quantitative results indicated high scores on the perceived usefulness, ease of use and intended use for all participants. In addition, with the qualitative results we are able to appreciate some differences between the use of the guidelines by beginner and advanced teams.

Regarding the guidelines: DG1 (Devices and interfaces) was considered until the re-design phase by the beginner teams. In contrast, advanced teams considered it as indispensable during the first design phase. DG2 (Usefulness), in relation to the importance in the context of elderly-children interaction, it was considered as less important for the beginner team while for the advanced team it was considered a contribution to their expertise with the target audience. In addition, guidelines DG3 (The difficulty level), DG4 (Duration of the sessions) and DG5 (Diversity of activities) were considered of equal importance by beginner and advanced teams. Finally, DG6 (Active and balanced participation) was considered as a little more important by advanced teams.

According to the feedback obtained from participants, there are several recommendations, such as (i) including a deeper analysis on aspects considered as important by advanced participants, (ii) creating a list of heuristics to evaluate designs, and (iii) generating a shorter version of the guidelines to allow focusing on the most relevant elements of each guideline.

The main contribution of this work is the proposal of a preliminary set of guidelines to facilitate the design of digital games that support elderly-children interaction. According to the validation process, we present necessary evidence about the impact of using the guidelines to design digital games with this aim. In addition, the quantitative results indicate high scores on the perceived usefulness, ease of use and intended use of participant developers towards its use. Directions of future work include: (i) the use of the guidelines to design, develop and evaluate additional digital games in the context of elderly-children interaction; and (ii) the revision and refinement of the guidelines based on the results obtained from these additional evaluation studies.

Acknowledgements. We would like to thank the personnel and students from the ITSON, students from the Maestro Ezequiel A. Chavez Elementary School, the postgraduate students from UABC and CICESE and the elderly people for their participation in the evaluations. Finally, our thanks to CONACYT for the support provided to the first author through scholarship 404782.

References

1. Meza-kubo, V., Morán, A.L.: AbueParty: An Everyday Entertainment System for the Cognitive Wellness of the Worried-well. In: Proceedings of the 5th International Symposium of Ubiquitous Computing and Ambient Intelligence (2011)
2. Romero, N., Sturm, J., Bekker, T., de Valk, L., Kruitwagen, S.: Playful persuapsion to support older adults social and physical activities. Interacting with Computers 22(6), 485–495 (2010)
3. Acosta, C., Davila, M., Iribarren, M.R.: Activities of Daily Living and Successful Aging. In: Human Aging. A Transdisciplinary Vision (in Spanish: Actividades de la Vida Diaria y Envejecimiento Exitoso. En Envejecimiento Humano, Una visión Interdisciplinaria), p. 402 (2010)
4. Baltes, P.B.: On the Incomplete Architecture of Human Ontogeny, Selection Optimization and Compensation as Foundation. American Psychologist 52(4), 366–380 (1997)
5. Dowd, J.J.: Aging as exchange: a preface to theory. Journal of Gerontology 30(5), 584–594 (1975)

6. Bradner, E.: Social Affordances: Understanding Technology Mediated Social Networks at Work. In: Extended Abstracts of the Conference on Human Factors in Computing Systems (2001)
7. Stewart, J., Lizzy, I., All, A., Mariën, I., Schurmans, D., Looy, V., Jacobs, A., Willaert, K., Grove, F.D.: The Potential of Digital Games for Empowerment and Social Inclusion of Groups at Risk of Social and Economic Exclusion: Evidence and Opportunity for Policy. Joint Research Centre, Tech. Rep. (2013)
8. Ijsselsteijn, W., Nap, H.H., de Kort, Y., Poels, K.: Digital game design for elderly users. In: Proceedings of the 2007 Conference on Future Play, Future Play 2007, p. 17 (2007)
9. Ornelas Villagómez, P.: Demographic Aging in Mexico. In: Human Aging. A Transdisciplinary Vision (in Spanish: El Envejecimiento Demográfico en México: Reflexiones en Torno a la Población de Adultos Mayores. En Envejecimiento Humano. Una visión transdiciplinaria), p. 402 (2010)
10. Derboven, J., Van Gils, M., De Grooff, D.: Designing for collaboration: a study in intergenerational social game design. Universal Access in the Information Society 11(1), 57–65 (2011)
11. Grimaldo Martinez, A.I., Morán, A.L., Calvillo Gámez, E.H.: Towards a taxonomy of factors implicated in children-elderly interaction when using entertainment technology. In: Proceedings of the 4th Mexican Conference on Human-Computer Interaction, MexIHC 2012, pp. 51–54 (2012)
12. Alisi, T., Del Bimbo, A., Valli, A.: Natural interfaces to enhance visitor's experiences. IEEE Multimedia 12, 80–85 (2005)
13. Brown, M.: Comfort Zone: Model or metaphor? Australian Journal of Outdoor Education 12(1), 3–12 (2008)
14. Parker, J.R.: Buttons, Simplicity, and Natural Interfaces. In: Proceedings from DiGRA 2005: Changing Views - Worlds in Play (2005)
15. Rogers, Y.: HCI Theory: Classical, Modern, and Contemporary. Synthesis Lectures on Human-Centered Informatics 5(2), 1–129 (2012)
16. Strauss, A., Corbin, J.M.: Basics of Qualitative Research. Sage (1998)
17. Roupa, Z., Nikas, M., Gerasimou, E., Zafeiri, V., Giasyrani, L., Kazitori, E., Sotiropoulou, P.: The use of technology by the elderly. Health Science Journal 4(2), 118–126 (2010)
18. Razzaghi, M., Bayat, A.: Designing for Children. Tech. Rep. (2010), Ryan, R.M., Deci, E.L.: Intrinsic and Extrinsic Motivations: Classic Definitions and New Directions. Contemporary Educational Psychology 25(1), 54–67 (2000)
19. Kozulin, A., Gindis, B., Ageyev, V., Miller, S.: Vygotsky's educational theory and practice in cultural context. Cambridge University Press (2003)
20. Bastien, C., Scapin, D.: Evaluating a user interface with ergonomic criteria, Tech. Rep. (1994)
21. Meza-Kubo, V.: Guidelines for the design of usable cognitive stimulation applications for older adults. Ph.D. dissertation, Universidad Autónoma de Baja California (2013)

Enriching (Learning) Community Platforms with Learning Analytics Components

Tilman Göhnert[1], Sabrina Ziebarth[1], Nils Malzahn[2], and Heinz Ulrich Hoppe[1]

[1] COLLIDE Group, University of Duisburg-Essen, Germany
{goehnert,ziebarth,hoppe}@collide.info
[2] Rhine-Ruhr Institute for Applied System Innovation, Duisburg, Germany
nm@rias-institute.de

Abstract. In this paper we present a generic and extensible analytics workbench and show how it can be integrated with learning environments in order to analyze the learners' activities. As the analytics workbench already supports a wide range of analysis types including network analysis, statistical analysis, and analysis of activity logs, the main effort needed for embedding learning analytics features into a learning platform lies in data exchange for input and output of analysis processes between the learning platform and the workbench. However the analytics workbench is also designed for extensibility so that more specific analysis capabilities can be added to it easily if desired. We report three case studies of such integrations and show the benefits for different target groups.

Keywords: learning analytics, analysis workbench, integration.

1 Introduction

Software tools for supporting learning often do not include support for learning analytics. If there is analytic support (e.g. in larger open source projects like Moodle[1] or commercial solutions), it is mostly integrated into the specific learning environment and cannot be easily reused for other environments. Thus, much effort is invested in implementing even basic analytical features again and again for different learning environments.

There are also first approaches for generic analysis environments like LeMo[2], which provides tools for monitoring of learning processes on arbitrary learning management systems (LMS).

Thus current solutions are either tightly integrated into the learning environment or the analysis results are created and offered through a separate analysis tool. The first approach leads to the development of platform specific analysis modules, the second forces users interested in analysis results to access an additional tool which is not usable directly in the learning environments. The essential challenge is the flexible integration of a generic analysis module with different types of learning environments in such a way that the analysis results are available directly inside of the learning environments.

---

[1] http://moodle.com/

N. Baloian et al. (Eds.): CRIWG 2014, LNCS 8658, pp. 177–184, 2014.

We have developed an analytics workbench[2] [4], which covers a wide spectrum of analyses and can be easily applied to a wide range of data. This includes data from different learning environments, since lots of standard data formats are supported and the internal data representations are simple and generic, thus most data can be mapped. The workbench can be used exploratively by analysts, but it can also perform stored analysis workflows automatically to embed the analysis results directly into the target environment like it is shown in [5]. In this paper, we present the analytics workbench and show how it can be easily integrated with existing learning platforms and used for meaningful analyses regarding different target groups.

2   Background

Learning analytics (LA) covers "the measurement, collection, analysis and reporting of data about learners and their contexts, for purposes of understanding and optimizing learning and the environments in which it occurs" [1]. Relying on quantitative measures, LA augments the understanding of learning processes on the part of different stakeholders, including researchers, teachers, students and educational managers. LA can support and promote evidence-based practices that include assessing student progress [6], identifying successful learning patterns, detecting indicators of failure, or evaluating appropriateness of learning materials. It can also provide material for reflection on teaching and learning practices, and enhance the way teachers communicate the results of their experiences. Using these techniques, educators are nurtured in analytic thinking that helps them to make better-informed decisions and to improve their interventions on students' learning.

3   Analytics Workbench

The analytics workbench was initially created in the SiSOB[3] project as a generic, easily extensible analysis tool that would enable even non-computer experts to access the full analytical power behind the tool and that would also allow reusing and sharing the created analysis workflows.

As depicted in figure 1 the workbench offers a web-based user interface for designing analysis processes. The main area of the user interface is the workspace for building the analysis workflows. It is surrounded by an overview of available analysis modules on the left hand side, an overview of the analysis results on the right hand side, and a menu for storing, loading, and executing analysis processes. The workflows are represented in a visual language based on a pipes-and-filters metaphor, in which modules of the language represent analysis steps and links between these modules describe the data flow.

---

[2] http://workbench.collide.info/

[3] http://sisob.lcc.uma.es/

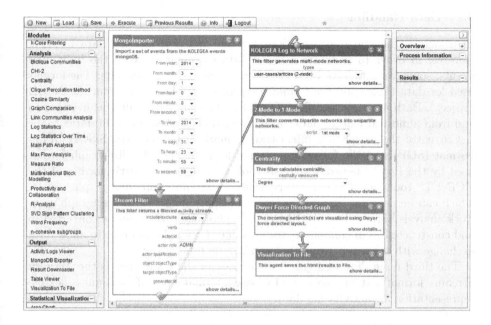

Fig. 1. SiSOB analytics workbench user interface

The workflow presented in figure 1 is used within the KOLEGEA platform (see section 4.2) to get an overview of the community. It generates a user-network based on the interaction with shared artifacts on the platform. The process starts with the *MongoImporter* retrieving all events that occured in March 2014 from the event data base. The following *StreamFilter* is used to exclude the events of users with the role *ADMIN*. Based on the filtered event stream the *KOLEGEA Log to Network* module creates a two-mode network consisting of users and artifacts (cases and articles). This network is folded by the *2-Mode to 1-Mode* module to get a user-user-network. To identify *central* actors, the degree centrality is calculated for each node by the *Centrality* module. To visualize the resulting graph the *Dwyer Force Directed Graph* visualizer is selected. Since the results of this process should not be shared yet, they are saved by the *Visualization To File* module to the local hard drive of the analyst (and not send to the KOLEGA platform).

A wide range of analysis modules are currently available: Apart from quite generic components for putting data into the workbench, retrieving data from the workbench, or duplicating data for splitting a workflow into parallel branches, there are also over ten different modules connected to processing and analyzing graphs (including several modules for community detection and a module offering a wide range of centrality measures), around ten modules connected to processing and analyzing activity logs (including modules for deriving statistical information, creating networks, and doing sequence analysis), and a wide range of different visualizations for both graphs and statistical information.

3.1   Data Handling

In principle, there are no restcitions for the choice of internal data formats. However in order to facilitate an easy data exchange between the individual agents, some formats have been chosen as main exchange formats. The format used for data tables and the format used for graph data stem from the SiSOB project[4]. Both are very flexible JSON based formats which allow converting to and from almost any other data format available for these kinds of data.

Currently the workbench offers transformations between the SiSOB Graph Format (SGF) and the following external formats: the edge list graph format used by the Pajek[5] network analysis tool, the adjacency matrix format of the UCINET[6] tool, and GML[7], a very widespread and expressive graph format, which can be used for example as input and output format of the igraph[8] network analysis library. The SiSOB Datatable Format (SDT) can be converted to and constructed from comma-separated value (CSV) files in order to allow data exchange with almost any tool that can handle tabular data.

The third main data exchange format used in the workbench is the Activity Streams format[9]. It is also a JSON based format and offers a quite flexible representation for activity log data.

3.2   Architecture and Extensibility

The workbench combines a web-based user interface with a multi-agent system in such a way that each of the modules in the visual language corresponds to one agent in the backend. A Node.js[10] based server component acts as access point to the workbench system for external clients (e.g. web browser, web services). It enforces access control so that the analysis system only reacts to authorized requests. The system itself uses an SQLSpaces [7] server as communication and data exchange platform, which is an open source implementation of the "tuple space" [3] concept. The blue parts of figure 2 show an overview of the workbench architecture.

A special internal data repository is the result repository. Most of the output creating agents of the workbench (e.g., visualization agents) are writing their output into this repository. The Node.js based server component accesses this repository for offering analysis results within the web based user interface.

---

[4] see SiSOB deliverable 5.2, http://sisob.lcc.uma.es/repositorio/deliverables/SISOB-D52.pdf

[5] http://pajek.imfm.si/doku.php?id=pajek

[6] https://sites.google.com/site/ucinetsoftware/

[7] http://www.fim.uni-passau.de/fileadmin/files/lehrstuhl/brandenburg/projekte/gml/gml-technical-report.pdf

[8] http://igraph.sourceforge.net/

[9] http://activitystrea.ms/

[10] http://nodejs.org/

3.3 An Architectural Pattern for Embedding the Workbench in Learning Environments

The analytics workbench cannot only be used for exploration by analysts, but it can also be integrated into (learning) platforms to support other target groups like learner, teachers, and decision makers. Advantages of this approach compared to developing platform dependent analysis features from scratch are reduced implementation time, less bugs due to prior testing and usage, and ultimately less costs.

In learning analytics, often three types of users are distinguished: students, teachers and analysis experts (e.g. researchers). While students and teachers will usually just be interested in the results of the analysis process that are fed back into the learning environment, the analysis experts should be able to edit and adapt pre-defined analysis processes and to define new ones.

Usually the resulting software design aims at a system through which the end-users can flexibly select different sets of analysis instruments in a monitoring cockpit (or dashboard) that allows teachers and students to reflect on their learning activity. The requirement to be able to adapt analysis processes implicitly encompasses the option of a growing set of indicators over time, enabling the system to evaluate different sets of analytic measures with teachers and adapting existing ones or creating new ones depending on the evaluation results.

Figure 2 shows how these needs can be mapped to an architecture incorporating the Analytics Workbench. This general architectural approach shows that the only necessary addition to the analysis system is a way of connecting the learning environment and the analysis workbench. There are several possible ways to implement this connection layer (see section 4) but the tasks of this layer are always the same. It manages reading data and writing data from and to the learning environment to feed it into a particular analysis process and write the results back into the learning environment.

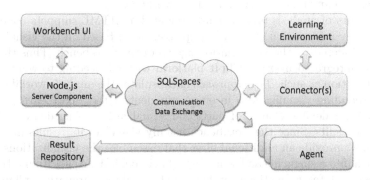

Fig. 2. Generic architecture for coupling the Analytics Workbench (blue) and a learning platform (orange)

In a first step analysis experts design and save the analysis workflows to be applied to the data from the learning environment. In a second step these analysis workflows will be executed independently from the workbench user interface through the connection layer.

To prove the validity and effectiveness of our approach we elaborate on three existing implementations of the architectural approach above.

4   Case Studies

The presented generic architectural approach has already been successfully implemented in the FoodWeb2.0[11] project as reported in ([5]). In this section we present two further case studies of integrating the analytics workbench in ongoing projects.

4.1   JuxtaLearn

The EU project JuxtaLearn[12] aims at fostering learning in different areas of science (or STEM) by stimulating curiosity and supporting learning through performative and productive activities on the part of the learners. Concretely, the students performance is substantiated in the form of creative and participatory video making and video editing in combination with the sharing and commenting of videos in a learning community.

Therefore, a key part of the JuxtaLearn learning pathway is to understand the learning needs of teachers and students through learning analytics.

In the integration of the workbench and ClipIt[13], the JuxtaLearn platform, web services implement the connector layer (cf. figure 2). These web services have been implemented as an extension to the Node.js server component of the basic workbench architecture. The web services allow retrieval of available analysis workflow templates and requesting the analysis of log file data (encoded as activity streams) using a given analysis workflow.

The results of analysis processes are provided as HTML snippets containing the necessary information (e.g. JavaScript code etc.) based on the visualization techniques offered by the visualization agents of the workbench. Thus they can be simply integrated into the ClipIt platform without needing to worry about the presentation as such. This approach encapsulates the analysis workflow with its corresponding visualizations in one component of the JuxtaLearn platform and allows for flexible change of particular visualizations if needed e.g. due to evaluation results. Since the workbench usually visualizes its results in its own interface, we provide an additional filter that assembles the visualizations of the workbench into a single file and sends them back to ClipIt. Apart from retrieving available templates, a functionality that may be used for automatic configuration of ClipIt Dashboard, all calls are asynchronous as the analysis process may

---

[11] www.foodweb20.de

[12] http://juxtalearn.org/

[13] http://juxtalearn.org/joomla/blog-cat/69-clipit-1-0

use some time. To make the connection between the analysis request and the result created by the workbench ClipIt generates a return id as parameter of the request.

4.2 KOLEGEA

The project KOLEGEA[14], which is facilitated by the German Federal Ministry of Education and Research, aims to support doctors specializing in family medicine (general practice (GP)) by providing a platform for collaborative learning in occupational, social communities. It addresses the problem of missing opportunities for networking, which highly restricts occupational knowledge exchange as well as collaborative learning in peer communities among young doctors. Since GP specialty training is based on learning by solving real problems/cases in every day working life, learning is problem-based, self-directed and based on intrinsic motivation. Thus, KOLEGEA's pedagogical approach (see also [8]) is focused on collaboratively working with user-generated cases in the spirit of problem-based learning (PBL). Users can share and discuss cases in self-regulated or mentor-supported small groups and share the results with the community.

A special target group in KOLEGEA are the platform operators, who need input to select a "case of the month" or trainees who have the potential for supporting (or "tutoring") self-regulated groups.

To integrate the analytics workbench with the KOLEGEA system the KOLE-GEA WebServices were extended to provide raw data for the analysis (proprietary log files as well as information on objects and users stored in the data base) as well as to accept and store analysis results. For the workbench agents have been implemented that fetch log data from the KOLEGEA platform, transform it into the Activity Streams format, and store it in a JSON based data base (MongoDB[15]). These agents play the role of the connectors between the workbench and the platform (cf. figure 2).

Besides workflows that collect data for subsequent analysis processes there is a second set of workflows that perform the actual analysis. These workflows may contain additional KOLEGEA specific analysis modules but mainly make use of the analysis features already available through generic workbench modules.

All of these workflows are regularly executed by a so called workflow executor to provide up-to-date analysis results for the users.

4.3 Comparison and Discussion

All three case studies follow the generic architectural approach presented in section 3.3. However there are differences in the concrete implementation especially concerning the trigger for analysis and generation of the visualization of analysis results. In KOLEGEA the execution of analysis processes is periodically triggered from within the workbench whereas in FoodWeb2.0 and JuxtaLearn the

---

[14] http://www.kolegea.de/

[15] https://www.mongodb.org/

analysis is triggered by the learning environment. Regarding analysis results, the KOLEGEA platform and the FoodWeb2.0 platform receive result data (graphs or data tables decorated with analysis results) and the visualization techniques are implemented in the platform. In contrast, the JuxtaLearn platform receives HTML snippets containing the visualization engines as well as the data. Sending only the data means less network traffic and allows the platform designers and possibly the platform users to choose and influence the visualization. Transporting the complete visualization technique together with the data causes more traffic but allows including all visualization techniques keeping together analysis results and its visualization.

References

1. 1st international conference on learning analytics and knowledge, Banff, Alberta, February 27-March 1 (2011), https://tekri.athabascau.ca/analytics/
2. Beuster, L., Elkina, M., Fortenbacher, A., Kappe, L., Merceron, A., Pursian, A., Schwarzrock, S., Wenzlaff, B.: Lemo: An analytics tool for lmss as well as portals. Poster Presentation at the LAK 2013 (2013), http://lemo.htw-berlin.de/public/publication/LAK13_Beitrag.pdf
3. Gelernter, D.: Generative communication in linda. ACM Trans. Program. Lang. Syst. 7(1), 80–112 (1985)
4. Göhnert, T., Harrer, A., Hecking, T., Hoppe, H.U.: A workbench to construct and re-use network analysis workflows – concept, implementation, and example case. In: Proceedings of the 2013 IEEE/ACM International Conference on Advances in Social Networks Analysis and Mining (2013)
5. Malzahn, N., Ganster, T., Sträfling, N., Krämer, N., Hoppe, H.U.: Motivating students or teachers? challenges for a successful implementation of online-learning in industry-related vocational training. In: Hernández-Leo, D., Ley, T., Klamma, R., Harrer, A. (eds.) EC-TEL 2013. LNCS, vol. 8095, pp. 191–204. Springer, Heidelberg (2013)
6. Rodríguez-Triana, M.J., Martínez-Monés, A., Asensio-Pérez, J.I., Dimitriadis, Y.: Script-aware monitoring model: Using teachers' pedagogical intentions to guide learning analytics. In: Towards Theory and Practice of Teaching Analytics (TAPTA) – Workshop in Conjunction with the EC-TEL 2012, Saarbrücken, Germany (2012), http://ceur-ws.org/Vol-894/paper4.pdf
7. Weinbrenner, S.: SQLSpaces – A Platform for Flexible Language-Heterogeneous Multi-Agent Systems. Ph.D. thesis, Universität Duisburg-Essen (2012)
8. Ziebarth, S., Kötteritzsch, A., Hoppe, H.U., Dini, L., Schröder, S., Novak, J.: Design of a collaborative learning platform for medical doctors specializing in family medicine. In: To See the World and a Grain of Sand: Learning Across Levels of Space, Time, and Scale: CSCL 2013 Conference Proceedings, pp. 205–208 (2013)

An Ontology Engineering Approach to Gamify Collaborative Learning Scenarios

Geiser Chalco Challco[1], Dilvan A. Moreira[1], Riichiro Mizoguchi[2], and Seiji Isotani[1]

[1] University of São Paulo, ICMC, São Carlos, SP, Brazil
geiser@usp.br, {dilvan,sisotani}@icmc.usp.br
[2] Japan Institute of Science and Technology, Ishikawa, Japan
mizo@jaist.ac.jp

Abstract. The design of collaborative learning (CL) scenarios that increase both students' learning and motivation is a challenge that the CSCL community has been addressing in the past few years. On one hand, CSCL design (i.e. scripts) has been shown to be effective to support meaningful interactions and better learning. On the other hand, scripted collaboration often does not motivate students to participate in the CL process, which makes more difficult the use of group activities over time. To deal with the problem of motivation, researchers and educators are now looking at gamification techniques to engage students. Gamification is an interesting concept that deals with the introduction and use of game design elements in a proper way to satisfy individual motivational needs. The use of gamification in educational settings is a complex task that requires, from instructional designers, knowledge about game elements (such as leaderboards and point systems), game design (e.g. how to combine game elements) and their impact on motivation and learning. Today, to the best of our knowledge, there are no approaches for the formal systematization of the instructional design knowledge about gamification and its application in CL scenarios. Thus, to address this issue, we have applied ontological engineering techniques to develop an Ontology called OntoGaCLeS. In this paper, we present the main concepts and ontological structure used to represent gamified CL scenarios. In this ontology, we formalize the representation of gamification concepts and explain how they affect motivation in the context of collaborative learning. Particularly, we will focus on the definition of player roles and gameplay strategies. Furthermore, to show the utility of our approach, we illustrate how to use our ontology to define a personalized gamification model that is used to gamify a CL scenario based on motivational needs and individual traits of learners in a group.

Keywords: gamification, ontology, collaborative learning.

1 Introduction

In the field of CSCL (Computer-Supported Collaborative Learning), to create effective collaborative learning (CL) scenarios, researchers and practitioners have used learning/instructional theories and best practices to set up well-thought-out

N. Baloian et al. (Eds.): CRIWG 2014, LNCS 8658, pp. 185–198, 2014.

CSCL scenarios (or CSCL scripts) that increase the occurrence of meaningful interactions [12]. Using well-designed CL scenarios, there is the possibility to increase students' participation and learning during group activities. Despite of these benefits, some researchers have indicated that scripted collaboration may cause, in some situations, demotivation among students, which makes more difficult to use group activities over time [8, 12]. Thus, to support the design of better CL activities, this work intends to combine the design of CL scenarios with a motivational strategy known as *gamification*.

In the last years, many researchers have contributed to the development of the concept of gamification and its application in education [9, 17]. Deterding and colleagues define gamification as "*the use of game design elements in non-game contexts*" [6]. It aims to increase engagement and motivation through the application of game mechanics, such as point system, social connections and so on, in a situation that normally has other purposes than entertainment. The educational benefits that a learner gets through the use of gamification depend strongly on how well game design elements are connected with pedagogical approaches [17]. Thus, in the CL context, we assert that the chances of increasing motivation and educational benefits happen when game design elements and theoretical concepts from CSCL scenarios are correctly linked.

Nevertheless, such a task is not trivial. Some researchers have indicated that many current uses of gamification are incorrect or poorly designed [24]. One of the main reasons for such poor designs is the assumption that all gamified scenarios can share the same game elements (game mechanics, game dynamics and game aesthetics) in different situations. For example, a point system that rewards all learners with the same quantity of points for each lesson does not make the learning more enjoyable. It is most enjoyable for learners, with the psychological need to demonstrate their mastery, to receive more points than other learners.

To deal with this challenge, this paper will describe the development of an ontology that organizes and adequately links knowledge related to CL design and game elements. This ontology is called **OntoGaCLeS** - an *Ontology to Gamify Collaborative Learning Scenarios*. It has been developed using the Hozo Ontology editor [18], and it is available at *http://labcaed.no-ip.info:8003/ontogacles*. Particularly, in this paper, we will focus on describing the representation of a *gamified CL scenario* and its relationship with game player roles and game mechanics.

The next sections are divided as follows: First, we present the related works and an overview of the representation of CL scenarios using ontologies. Next, we define the concepts of a gamified CL scenario. Then, a personalized gamification model is defined to illustrate the utility of our approach. Using this model, we show how to gamify a CL scenario using individual traits and psychological needs of learners. Finally, we present conclusions and future steps in our research.

2   Overview of the Collaborative Learning Ontology

The CL ontology (Collaborative Learning Ontology) [12] has been developed to formally and explicitly describe CL scenarios based on learning theories. It is based on the understanding of the interrelations among different concepts extracted from learning theories, such as interaction patterns, group goal, individual goal, learner's role, and others. Currently, the CL Ontology has been successfully applied to support group formation [15], design of CL activities [12], creation of CL scenarios using the interaction patterns [14], and modeling learner's development [11].

To avoid possible misunderstanding, when using a conventional vocabulary, the CL ontology provides a specific terminology to define a CL scenario. The definition of these terms in the ontology is shown in Figure 1 (which illustrates the basic concepts and their relationships), where:

I-goal is the individual goal that represents what a learner (*I*) is expected to acquire, described as a change of a learner's stage of learning.

I-role is the role played by the person in focus (*I*).

You-role is the role played by the participant (*You*), who is interacting with the person in focus (*I*).

Y<=I-goal is the learning strategy that represents the strategy used by *I* to interact with another learner (*You*) in order to achieve *I-goal*.

W(L)-goal is the common goal for group members (*group goal*).

W(A)-goal is the goal of the rational arrangement of the group's activity used to achieve *W(L)-goal* and *I-goal*.

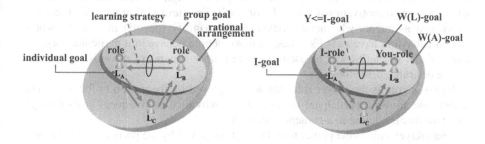

Fig. 1. Concepts and terms defined in the CL Ontology [12]

The learning strategy (*Y<=I-goal*) specifies how a learner (*I-role*) should interact with other members of the group (*You-role*) to achieve an individual goal (*I-goal*). The CL process (*W(A)-goal*) specifies the common goals of process (*W(L)-goal*) and the rational sequence of interactions (interaction pattern) provided by theories. The interaction patterns are represented by necessary and desired interaction activities, among members of a group, as influential Instructional-Learning events [12].

3   Gamifying Collaborative Learning Scenarios

A gamified CL scenario is a CL scenario in which game design elements are applied to make the learning experience more enjoyable and meaningful. In a gamified CL scenario, the learning experience itself intends to be so enjoyable that learners will do the proposed activities, even at great cost, because they are highly motivated, particularly because of the use of different game mechanics (e.g. leaderboards, point system, social connection, etc.). These game mechanics are elements introduced in

CL scenarios that define how these scenarios operate as games. They are the elements that convert specific inputs of a CL scenario into specific outputs (game rewards).

As motivation is the process used to allocate energy and to maximize the satisfaction of needs [21], a circular flow of "needs, behavior and satisfaction" is set in a CL scenario to gamify it, where to fulfill the learner's motivational needs, a learner must be engaged in behaviors that will lead to the satisfaction of those needs using game mechanics. In many cases, the combination of different game mechanics provides the adequate environment to satisfy a person's motivational needs, called human desires by Domínguez et al. [9] and Simões et al. [22]. Thus, to support this fact in CL scenarios, our current formalization of a gamified CL scenario introduces the concepts and terms shown in Figure 2, where:

I-mot goal is the *individual motivation goal* of the person in focus (*I*). Since motivation is circular, at the end of a CL scenario, the needs of a person may change or intensify, and the level of motivation (*motivation stage*) will be increased. Thus, individual motivation goals will be used to represent needs that must be satisfied and motivational stage that will be achieved.

Y<=I-mot goal is the *motivational strategy* that will be used by game mechanics to enhance the learning strategy (*Y<=I-goal*) employed by (*I*). The motivational strategies are guidelines that represent what game design elements are necessary to attain individual motivational goals (*I-mot goal*). Vassileva [23] argues that users can be viewed as agents who act to maximize their utility (payoff) in a world where certain behaviors have payoffs. Thus, to make people behave in a particular way, the motivational strategies will be used to create proper systems of rewards (incentives) for the desirable behaviors.

I-player role is the player role that will be played by the person in focus (*I*). The player role allows a participant to achieve his individual motivational goals through game mechanics define in a gameplay strategy.

You-player role is the player role that will be played by the participant (*You*), who is interacting with the person in focus (*I*).

I-gameplay is the gameplay strategy employed by the person in focus (*I*). The gameplay contains the definition of game mechanics that will be used and the behavior person (*I*) should use when interacting with the run-time using these game mechanics. In this sense, the game mechanics, user rewards and user behaviors are called game dynamics and defined using the guidelines of motivational strategy (*Y<=I-mot goal*). Furthermore, the gameplay (*I-gameplay*) is also used to define the rational arrangement among player roles, motivational strategies and game mechanics.

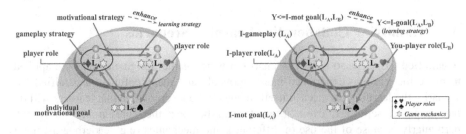

Fig. 2. Concepts and terms defined in a gamified CL Scenario

In the following subsections, we detail the concepts of a gamified CL scenario, showing how they are defined as ontological structures in Hozo. First, the ontological structures for individual motivational goals and player roles are presented because they are used by other structures. Next, the ontological structures for motivational strategy and gameplay strategy are detailed. Finally, we show the ontological structure used to represent a gamified CL scenario.

3.1   Individual Motivational Goals and Player Roles

Our ontological structure of a gamified CL scenario is based on the idea that "to satisfy a set of psychological needs, a learner will perform learning activities if game design elements introduce in them promise the satisfaction of these needs." Furthermore, as a consequence of satisfactory learning results and pleasure experiences, obtained from game design elements, "a learner will increase his liking for the actual learning activities." We call this fact internalized motivation and it consists in the change of the current motivational stage.

In our current version of OntoGaCLeS, employing self-determination theory [5] and Pink Dan Pink's theory [20], we define four psychological needs shown in Figure 3 (a), which are: autonomy, relatedness, purpose and mastery. According to self-determination theory, the motivation stages, shown in Figure 3 (b), include six stages: amotivation, external regulation, introjected regulation, identified regulation, integrated regulation and intrinsic motivation.

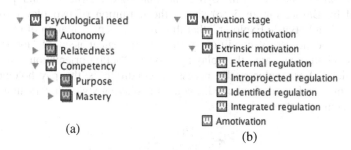

(a)

(b)

Fig. 3. (a) Psychological needs, and (b) motivation stages

Based on these ideas, Figure 4 shows the ontological structures used to represent the concepts of "satisfaction of need" and "internalization of motivation." These two concepts are individual motivational goals (*I-mot goal*) and they have two parts known as the initial stage and goal stage. In the case of "satisfaction of need", the initial stage is a psychological need and the goal stage is "without need," while the initial and goal stages, for the case of "internalization of motivation," are both motivational stages.

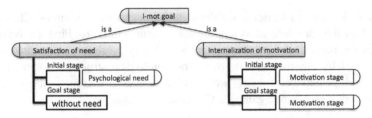

Fig. 4. Ontological structures used to represent individual motivational goals (*I-mot goal*). At the bottom are "satisfaction of need" (left) and "internalization of motivation" (right).

To allow a learner to attain his individual motivational goals, a player role must contains necessary information to define who can play it. This information includes two types of prerequisites defined as necessary and desired conditions. A learner cannot play a player role if he does not fulfill the necessary conditions; and a learner that fulfills the necessary conditions but that does not fulfill the desired conditions, he may play the player role but his individual motivational goals might not be attained.

Figure 5 (a) shows the ontological structure used to represent the concept of player role. In this structure, the current motivational stages and psychological needs are defined as necessary conditions, while the playing styles are defined as the desired conditions. The playing styles represent individual personality traits that define preferences of a learner when he is playing a game. According to Bartle [2], these playing styles represent two preferences: (1) the preference of interacting with other players (*user-orientation*) vs. exploring the game (*system-orientation*); and (2) the preference of unilateral action (*action-orientation*) vs. interaction in the game (*interaction-orientation*). Employing the ontological structure and playing styles, defined by Bartle, Figure 5 (b) shows the definition of two player roles, "Bartle Achiever" and "Bartle Explorer," that illustrate the representation of player roles. For achiever, the necessary condition is the psychological need of "mastery" and the desired conditions are the playing styles "action-orientation" and "system-orientation." For explorer, the necessary condition is the psychological need of "autonomy" and the desired condition are the playing styles "interaction-orientation" and "system-orientation."

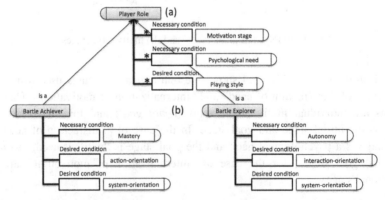

Fig. 5. Ontological structure used to represent player roles (top). At the bottom are the representations for "Bartle Achiever" (left) and "Bartle Explorer" (right).

3.2   Motivational Strategy and Gameplay Strategy

The motivational strategy ($Y<=I\text{-}goal$) is a guideline set that defines how a learner can attain his individual motivational goals (*I-mot goal*). In a gamified CL scenario, a game mechanics uses these guidelines to define the proper way of interacting with learners, and this way depends on each player role. For example, if a learner l_a, with player role achiever, and other learner l_b, with player role explorer, are participants of a CL scenario, the learning experience of l_a will be enhanced using a "point system" (game mechanics) when he obtain more points than l_b. At the end of the CL scenario, the learner l_b also obtains points as rewards, but these do not have the purpose of demonstrating mastery. Their purpose can be to unlock some special gift and/or activity in the system to satisfy learner l_b psychological need for autonomy.

Based on this idea, Figure 6 (a) shows the ontological structure used to represent the concept of motivational strategy ($Y<=I\text{-}mot\ goal$). The person in focus (*I*) is playing the player role *I-player role*. The other learner (You), who is interacting with the learner in focus (*I*), plays a role known as *You-player role*. Finally, the motivational strategy benefits, for learner in focus (*I*), are represented in terms of individual motivational goals (*I-mot goals*).

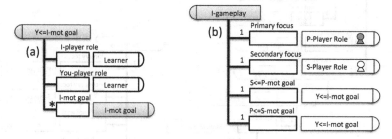

Fig. 6. Ontological structure used to represent motivational strategies (left). Ontological structure used to represent gameplay strategies (right).

As discussed before in the first part of this section, the main goal of a gameplay strategy (*I-gameplay*) is to define the rational arrangement among player roles, motivational strategies and game mechanics. Figure 6 (b) shows the ontological representation used to represent gameplay strategies. The arrangement defined by a gameplay strategy has the purpose of representing how different player roles have the potential to affect each other. Thus, in a particular strategy, the primary focus is a learner (*P*) that plays the primary player role (*P-Player role*), the secondary focus is a learner (*S*) that plays the secondary player role (*S-Player role*), and, for both (*P* and *S*), we define the motivational strategies $S<=P\text{-}mot\ goal$ and $P<=S\text{-}mot\ goal$.

3.3   Gamified CL Scenario

Figure 7 (a) shows the ontological structure developed in this work to represent a gamified CL scenario. It extends [12, 13, 16] and consists in the adequate connections of all the concepts presented in the previous subsections. The concept *Gamified CL*

Scenario adds two parts to the concept *CL scenario*, including the motivation strategy and the gameplay strategy. The motivational strategy (*Y<=I-mot goal*) is related to the learning strategy by the relationship "*enhance*" and each player role (*I-player role* and *You-player role*) defined in this strategy is related with the concept of *Player Role*, Figure 7 (b). The concept of gameplay strategy (*I-gameplay*) showed in Figure 7 (c) is included in the gamified CL scenario to define: the proper game mechanics (*what use*) that can be used by learner (*I*). Each game mechanics includes a set of game dynamics (*gameplay*) in terms of game rewards (*rewards*) that will be used during the scenario execution.

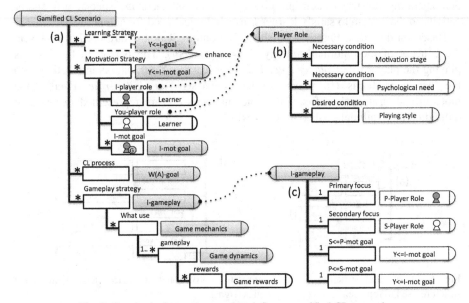

Fig. 7. Ontological structure used to define a gamified CL scenario

4   Illustration: Define a Personalized Gamification Model

To demonstrate the utility of our approach, the ontological structures, shown in previous sections, are used to define a personalized gamification model based on some ideas extracted from Marczewski's[1] blog. He is a leader and expert in gamification.

We begin defining eight player roles, with the information shown in Table 1. These player roles expand Bartle's player roles (socializer, explorer, achiever, and killer) through the addition of intrinsic motivation stages as a necessary condition to play the roles: socializer, free spirit, achiever, and philanthropist.

---

[1] Website: http://marczewski.me.uk

Table 1. Player roles defined in Marczewski's blog

Player role	Necessary and desired condition		
	Psych. need	Motivation stage	Playing style (ind. trait)
networker	relatedness		interacting-orientation, users-orientation
socializer		intrinsic motivate	
exploiter	autonomy		interacting-orientation, system-orientation
free-spirit		intrinsic motivate	
consumer	mastery		acting-orientation, system-orientation
achiever		intrinsic motivate	
self-seeker	purpose		acting-orientation, users-orientation
philanthropist		intrinsic motivate	

In our model, for each player role, a motivation strategy and a gameplay strategy are defined, using the information shown in Table 2. The values *You-player role* and *S-Player* are default values that are defined in the ontological structures for motivational strategy (*Y<=I-mot goal*) and gameplay strategy (*I-gameplay*).

Table 2. Motivation and gameplay strategies for Marczewski's player roles

Motivation strategy		Gameplay strategy	
I-player role / You-player role	Motivational goal (*I-mot goal*)	P-Player Role / S-Player Role	Game mechanics (*what use*)
Networker role / *You-player role*	satisfaction of relatedness, internalize motivation	Networker / *S-Player*	Social Status, Point System, and Badges System
Socializer role / Socializer role	satisfaction of relatedness	Socializer / Socializer	Social Status, and Social Connections
Exploiter role / *You-player role*	satisfaction of autonomy, internalize motivation	Networker / *S-Player*	Point System, Virtual Goods System, and Badges System
Free-spirit / *You-player role*	satisfaction of autonomy,	Networker / *S-Player*	Unlockable System, and Customization Tool
Consumer / *You-player role*	satisfaction of mastery, internalize motivation	Networker / *S-Player*	Virtual Goods System
Achiever / *You-player role*	satisfaction of mastery	Networker / *S-Player*	Quests System, Point System, and Exclusive Reward System
Self-seeker / *You-player role*	satisfaction of purpose, internalize motivation	Networker / *S-Player*	Leaderboard, Badges System, and Exclusive Reward System.
Philanthropist / *You-player role*	satisfaction of purpose	Networker / *S-Player*	Gifting System

Employing the motivation strategies and the gameplay strategies, shown in Table 2, we defined eight ontological structures to represent the gamified CL scenarios (for networker, socializer, exploiter, free-spirit, consumer, achiever, self-seeker and philanthropist), using the structure shown in Figure 7. For example, Figure 8 (a) shows the ontological structure used to represent a gamified CL scenario for a socializer, where the roles of learner on focus (*I*) and of participant (*You*), who is interacting with the learner (*I*), are Socializer roles, Figure 8 (b). Next, the gameplay strategy (*Socializer gameplay*), shown in Figure 8 (c), is used in the gamified CL scenario for a socializer (Figure 8 (a)) to define, as proper game mechanics, the tools: social status and social connections.

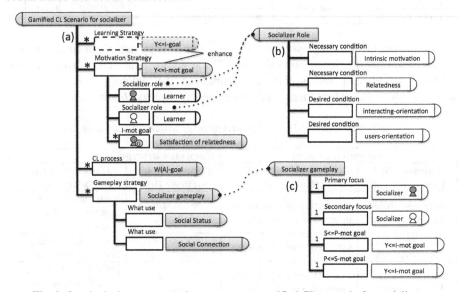

Fig. 8. Ontological structure used to represent a gamified CL scenario for socializers

4.1   Pseudo-algorithm to Define Proper Player Roles and Game Mechanics

Employing a personalized gamified model, the next procedure is defined to identify the proper player roles and game mechanics for all learners in a CL scenario.

1. Match the individual motivational goal, for each learner, by looking at the *I-mot goal* in all gamified CL scenario. The result usually has more than one scenario that can help to internalize motivation and to satisfy basic needs.
2. Check if learners have the necessary conditions to play the game roles for the CL scenarios obtained in step (1).
3. Set the game roles, obtained in the step (2), for each learner according to the priorities calculated using the desired conditions that were satisfied. Learners with all satisfied conditions have high-priority, and learners with only necessary conditions have low-priority.

4. Finally, after the game role definition, set the gameplay for all learners. This task is completed through the selection of proper game mechanics for each learner defined in each gameplay (I-gameplay).

4.2   Gamifying a CL Scenario Using the Personalized Gamification Model

The pseudo-algorithm, defined in the previous subsection, can be used to gamify a CL scenario using the information of individual traits, current motivational stage and motivational psychological needs that is extracted from all learners. Thus, to illustrate how to gamify a CL scenario, the learners' information of a fictional CL scenario, shown in Table 1 is used. In a real scenario, this information can be obtained through a Bartle test [2] and a test of self-determination theory [1].

Table 3. Learners' information in a CL scenario

ID	Gameplay style	Motivation stage	Psychological needs
l1	acting-orientation, system-orientation	Intrinsic	mastery
l2	interacting-orientation, users-orientation	Intrinsic	relatedness
l3	acting-orientation, users-orientation	Amotivation	purpose, relatedness
l4	interacting-orientation, system-orientation	extrinsic (external regulation)	mastery, autonomy
l5	interacting-orientation, system-orientation	extrinsic (identified regulation)	autonomy, purpose
l6	interacting-orientation, users-orientation	extrinsic (external regulation)	relatedness, autonomy

After the execution of the procedural steps, defined in Subsection 4.1, Table 3 is obtained. In step (1), the gamified CL scenarios for socializer and networker can be used by learner l2 to help him in the satisfaction of relatedness needs, while the gamified CL scenarios for self-seeker and networker can be used by learner l3 for satisfaction of purpose/relatedness and internalization of motivation (amotivation) needs. In step (2), the player roles for l2 that satisfy the necessary condition are socializer and networker, and the player roles for l3 that satisfy the necessary conditions are self-seeker and networker. In step (3), the highest role that can be played by l2 is socializer because his playing styles are interacting-orientation and users-orientation, while the highest role that can be played by l3 is acting-orientation and users-orientation. After step (4), the player role for l2 is not socializer, the player role assigned to him is networker. We cannot define a gamified CL scenario for socializer, because does not exist another learner that can play the socializer role. Finally, after the groups are created, the gameplay is defined and game mechanics are defined for each learner, as shown in Table 4.

Table 4. Player roles and game mechanics for learners in a CL scenario

ID	Player role	Game mechanics
l1	achiever	Quests System, Point System, and Exclusive Reward System
l2	networker	Social Status, Point System, and Badges System
l3	networker	Social Status, Point System, and Badges System
l4	exploiter	Point System, Virtual Goods System, and Badges System
l5	Exploiter	Point System, Virtual Goods System, and Badges System
l6	Networker	Social Status, Point System, and Badges System

5   Related Works

In the literature, there are many gamification frameworks [7, 9, 10, 19, 22, 25] that are applied in different contexts, situations and scenarios. In the education field, Zagal [26] have proposed abstracting and cataloguing patterns in order to provide a set of reusable design elements and a language for discussing them. Furthermore, Domínguez et al. [9] and Simões et al. [22] proposed gamification frameworks that help instructional designers select proper game mechanics based in learners' individual traits. These frameworks were developed employing the relationship between game mechanics and human desire, where each game mechanics satisfies a set of human desires.

Our work extends these achievements by proposing concepts in a formal ontology that can be used by humans and computers as patterns and guidelines to define gamified CL scenarios using player roles, game mechanics, psychological needs and individual traits.

Despite the growing number of studies and applications of gamification in the field of education [4], to the best of our knowledge, this is the first ontology that enables humans and computers to find, share, and combine information related to CL scenarios and game design elements.

6   Conclusions and Future Research

In this paper, we presented an ontological structure that enables the representation of gamified CL Scenarios. This structure allows the development of personalized gamified models. The personalization of this models is archived through the rational arrangement between *motivational strategies* and *player roles*. To demonstrate this personalization, in the Section 4, we performed the organization of the knowledge related to eight scenarios. This knowledge allows the selection of proper game mechanics for each learner based in his psychological needs and individual traits.

We believe that the results of this work are the first steps forward for creating new semantic web authoring tools that can provide assistance for the development of more engaging and motivating CL scenarios. With well-grounded instructional designer knowledge about gamification, our ontology will be used to facilitate the inclusion of game mechanics, through the pseudo-algorithm proposed in Subsection 4.2.

In the current version of our ontology, we did not define the game dynamics that personalize the reward systems for each learner. Thus, our next steps will consider how this game element must be formalized according to our ontology. Furthermore, it is also important to identify what is the association between game mechanics and CL interaction patterns defined in [12, 16]. Future research will also consider the inclusion of optimal flow theory [3] and meaningful gamification [19].

Acknowledgements: We thank CNPq and CAPES for supporting this research.

References

1. Araújo Leal, E., Miranda, G.J., Souza Carmo, C.R.: Self-Determination Theory: An Analysis of Student Motivation in an Accounting Degree Program. Revista Contabilidade & Finanças-USP 24(62) (2013)
2. Bartle, R.A.: Designing virtual worlds. New Riders (2004)
3. Csiksczentmihalyi, M., Kolo, C., Baur, T.: Flow: The psychology of optimal experience. Australian Occupational Therapy Journal 51(1), 3–12 (2004)
4. De Sousa Borges, S., Durelli, V.H.S., Macedo Reis, H., Isotani, S.: A Systematic Mapping on Gamification Applied to Education. In: Proceedings of the 29th Annual ACM Symposium on Applied Computing, pp. 216–222. ACM, New York (2014)
5. Deci, E.L., Ryan, R.M.: Self-Determination. Wiley Online Library (2010)
6. Deterding, S., Sicart, M., Nacke, L., O'Hara, K., Dixon, D.: Gamification. Using Game-design Elements in Non-gaming Contexts. In: Extended Abstracts on Human Factors in Computing Systems, CHI 2011, pp. 2425–2428. ACM, New York (2011), doi:10.1145/1979742.1979575
7. Dignan, A.: Game frame: Using games as a strategy for success. Simon and Schuster (2011)
8. Dillenbourg, P.: Over-scripting CSCL: The risks of blending collaborative learning with instructional design. Three Worlds of CSCL. Can we Support CSCL?, 61–91 (2002)
9. Domínguez, A., Saenz-de-Navarrete, J., de-Marcos, L., Fernández-Sanz, L., Pagés, C., Martínez-Herráiz, J.-J.: Gamifying learning experiences: Practical implications and outcomes. Computers & Education 63, 380–392 (2013), doi:10.1016/j.compedu.2012.12.020
10. Duggan, K., Shoup, K.: Business Gamification for Dummies. John Wiley & Sons (2013)
11. Inaba, A., Ikeda, M., Mizoguchi, R.: What learning patterns are effective for a learners growth. In: Proc. of the International Conference on Artificial Intelligence in Education, Sydney, pp. 219–226 (2003)
12. Isotani, S., Inaba, A., Ikeda, M., Mizoguchi, R.: An ontology engineering approach to the realization of theory-driven group formation. International Journal of Computer-Supported Collaborative Learning 4(4), 445–478 (2009)
13. Challco, G.C., Moreira, D., Mizoguchi, R., Isotani, S.: Towards an Ontology for Gamifying Collaborative Learning Scenarios. In: Trausan-Matu, S., Boyer, K.E., Crosby, M., Panourgia, K. (eds.) ITS 2014. LNCS, vol. 8474, pp. 404–409. Springer, Heidelberg (2014)
14. Isotani, S., Mizoguchi, R.: Deployment of ontologies for an effective design of collaborative learning scenarios. In: Haake, J.M., Ochoa, S.F., Cechich, A. (eds.) CRIWG 2007. LNCS, vol. 4715, pp. 223–238. Springer, Heidelberg (2007)

15. Isotani, S., Mizoguchi, R.: Adventures in the Boundary between Domain-Independent Ontologies and Domain Content for CSCL. In: Lovrek, I., Howlett, R.J., Jain, L.C. (eds.) KES 2008, Part III. LNCS (LNAI), vol. 5179, pp. 523–532. Springer, Heidelberg (2008)

16. Isotani, S., Mizoguchi, R., Isotani, S., Capeli, O.M., Isotani, N., de Albuquerque, A.R.P.L., Jaques, P.: A Semantic Web-based authoring tool to facilitate the planning of collaborative learning scenarios compliant with learning theories. Computers & Education 63(0), 267–284 (2013), doi:http://dx.doi.org/10.1016/j.compedu.2012.12.009

17. Kapp, K.M.: The gamification of learning and instruction: game-based methods and strategies for training and education. Pfeiffer, San Francisco (2012)

18. Kozaki, K., Kitamura, Y., Ikeda, M., Mizoguchi, R.: Hozo: an environment for building/using ontologies based on a fundamental consideration of Role and Relationship. In: Gómez-Pérez, A., Benjamins, V.R. (eds.) EKAW 2002. LNCS (LNAI), vol. 2473, pp. 213–218. Springer, Heidelberg (2002)

19. Nicholson, S.: A user-centered theoretical framework for meaningful gamification. Proceedings GLS 8 (2012)

20. Pink, D.H.: Drive: The surprising truth about what motivates us. Penguin (2011)

21. Pritchard, R., Ashwood, E.: Managing motivation: A manager's guide to diagnosing and improving motivation. CRC Press (2008)

22. Simões, J., Redondo, R.D., Vilas, A.F.: A social gamification framework for a K-6 learning platform. Computers in Human Behavior (2012)

23. Vassileva, J.: Motivating participation in social computing applications: a user modeling perspective. User Modeling and User-Adapted Interaction 22(1-2), 177–201 (2012)

24. Webb, E.N.: Gamification: When It Works, When It Doesn't. In: Marcus, A. (ed.) DUXU 2013, Part II. LNCS, vol. 8013, pp. 608–614. Springer, Heidelberg (2013)

25. Werbach, K., Hunter, D.: For the win: How game thinking can revolutionize your business. Wharton Digital Press (2012)

26. Zagal, J.P., Mateas, M., Fernández-Vara, C., Hochhalter, B., Lichti, N.: Towards an ontological language for game analysis. In: Proceedings of the International Digital Games Research Association Conference (2005), http://lmc.gatech.edu/~mateas/publications/OntologyDIGRA2005.pdf

Group Formation Algorithms in Collaborative Learning Contexts: A Systematic Mapping of the Literature

Wilmax Marreiro Cruz and Seiji Isotani

Department of Computer Systems
University of Sao Paulo
Sao Carlos, Brazil
{wilmcruz,sisotani}@icmc.usp.br

Abstract. Group Formation is a complex and important step to design effective collaborative learning activities. Through the adequate selection of individuals to a group, it is possible to create environments that foster the occurrence of meaningful interactions, and thereby, increasing robust learning and intellectual growth. Many researchers indicate that the inadequate formation of groups can demotivate students and hinder the learning process. Thus, in the field of Computer-Supported Collaborative Learning (CSCL), there are several studies focusing on developing and testing group formation in collaborative learning contexts using best practices and other pedagogical approaches. Nevertheless, the CSCL community lacks a comprehensive understanding on which computational techniques (i.e. algorithms) has supported group formation. To the best of our knowledge, there is no study aimed at gathering and analyzing the research findings on this topic using a systematic method. To fill this gap, this research conducted a systematic mapping with the objective of summarizing the studies on algorithms for group formation in CSCL contexts. Initially, by searching on six digital libraries, we collected 256 studies. Then, after a careful analysis of each study, we verified that only 48 were related to group formation applied to collaborative learning contexts. Finally, we categorized the contributions of these studies to present an overview of the findings produced by the community. This overview shows that: (i) there is a gradual increase on research published in this topic; (ii) 41% of the algorithms for group formation area based on probabilistic models; (iii) most studies presented the evaluation of tools that implement these algorithms; but (iv) only 2% of the studies provide their source code; and finally, (v) there is no tool or guideline to compare the benefits, differences and specificities of group formation algorithms available to date. As a result of this work an infographic is also available at: *http://infografico.caed-lab.com/mapping/gf.*

Keywords: Group Formation, Algorithms, CSCL, Systematic Mapping.

1   Introduction

Computer-supported Collaborative Learning (CSCL) is a pedagogical approach in which knowledge construction occurs from social interactions among individuals with

N. Baloian et al. (Eds.): CRIWG 2014, LNCS 8658, pp. 199–214, 2014.
© Springer International Publishing Switzerland 2014

explicit or implicit support from computers and its technologies [3][6][10][12]. Several researchers have highlighted the potential benefits of learning through collaboration [7][11][5]. Nevertheless, learning through interactions does not occur in any situation. According to Barkley and colleagues [2] the design of collaboration is fundamental to achieve desired learning goals.

Among the studies in the design of well-thought-out collaborative learning scenarios, an important task to be conducted is the formation of groups. As indicated by Dillenbourg [3], forming groups without careful considerations (i.e. randomly) often causes problems such as disproportional participation of individuals, demotivation and resistance to group work in futures activities. Isotani et al. [5] also emphasizes that group formation is the first step to design a CSCL scenario where students can learn and participate more effectively. Through the process of selecting individuals to participate in a group, one can analyze and combine characteristics such as cultural background, knowledge, skills, learning styles, roles and so on, to create a positive synergy among participants that will lead to meaningful interactions and better learning situations.

Yet, due to the possibility of using several learner's characteristics and combine them in different ways to form learning groups, this task often requires computational support to be completed successfully. In this context, there are various studies in the literature on computer-supported group formation using different approaches, presenting new algorithms, frameworks, tools, techniques, experiments, and so on. For example, the work of Soh et al. [10], describes the development of an algorithm for group formation using a multi-agent approach and a pedagogical technique known as Jigsaw [1]. In another work, Moreno et al. [7] uses a genetic algorithm approach to consider an arbitrary number of student characteristics to create groups that are more effective. Finally, Ounnas et al. [8] proposed a semantic framework to improve the performance of some existing algorithms reflecting the diversity of approaches, algorithms, inputs and attributes (such as learning style, gender, personality, and so on).

Although, there are several benefits from the use of computing techniques for group formation, to the best of our knowledge, there is not study that summarize and catalogue them in a comprehensive and systematic manner. Because of that, the CSCL community lack a better understanding about how many computational approaches have been developed and used to support group formation, where to find them, how good they are, what the difference between the various existing algorithms, and so on.

To answer some of these questions this work carried out a systematic mapping of the literature to collect the research on group formation in CSCL contexts. We used the method proposed by Petersen [9] to conduct our research as described in Section 2. In summary, first, we defined the research protocol and selected the most important digital libraries in the field of computing and educational technologies. By searching on these digital libraries, we collected 256 studies (i.e. published articles). Then, through a first screening of each study, we verified that 48 of them match our needs (i.e. they are related to group formation in CSCL and met the inclusion and exclusion criteria defined in this work). Finally, we carefully analyzed and categorized these studies as shown in Section 3 to answer the following questions: 1) what are their main research objectives and contributions? 2) What are the algorithms used to support group formation? 3) Is the source code available for analysis and reuse? 4) Is there any study or tool that compare group formation algorithms?

We conclude this paper by discussing the results and practical applications of our findings in Section 4.

2   Method

Systematic mapping is a research method that provides guidelines to conduct literature reviews [9]. It consists of methodical steps to search, interpret, synthetize and analyze the information presented in published papers related to the target domain. The use of this method aims to provide an overview of the field of interest and minimize the chances of errors during the review process. Such a systematic process also gives better control to the review activity and remove possible mistakes that may cause misleading or imprecise conclusions.

In this work, we used the steps proposed by Petersen [9] to carry out the systematic mapping. It consist of five sequential steps as follows: (i) definition of research questions; (ii) conduct search for primary studies; (iii) screening of papers for inclusion and exclusion; (iv) classification scheme; and (v) data extraction and mapping of studies. Each of these steps are presented in the next subsections.

2.1   Definition of Research Questions

The focus of this systematic mapping is to identify and classify computational techniques, particularly algorithms that have been used to assist the formation of groups in collaborative learning environments with computational support. To define our objectives and then search for evidences (i.e. primary studies[1]) on algorithms for group formation, the following research questions were defined:

RQ1: What are the main research objectives and types of contributions from studies on group formation for CSCL?

RQ2: What are the most common computational techniques (i.e. algorithms) used to support group formation in collaborative learning environments?

RQ3: Is the source code of group formation algorithms available for analysis and reuse?

RQ4: Is there any study or tool that compare group formation algorithms in the context of collaborative learning?

---

[1] Individual studies that contribute to provide evidences to answer specific research questions.

2.2   Conduct Search for Primary Studies

The search for primary studies is composed of two steps. In the first step, we define the *search string* considering the most relevant terms related to our topic. In the second step, we select relevant electronic databases to conduct the search.

To create the search string, we used keywords (i) contained in the research questions. (ii) extracted from well-known papers; and (iii) obtained from interview with experts in the field. The result is shown in Table 1. Three main keywords have been defined: "group formation", "collaborative learning" and "algorithms", each keyword form a category that contains their respective synonyms.

Table 1. Categories of keywords and their synonyms

Reference	Category	Synonyms
C1	group formation	group creation
		group design
		group composition
		group organization
		team formation
		team creation
		team design
		team composition
		team organization
C2	collaborative learning	cooperative learning
		cscl
		csgf
		social learning
		group learning
		team learning
C3	Algorithms	Approaches
		methods
		software
		technique

To create the final search string, the categories C1, C2 and C3 were combined by the Boolean operator "AND", and the keywords within each category were combined by the Boolean operator "OR", as shown below:

> *(group formation **OR** group creation **OR** group design **OR** group composition **OR** group organization **OR** team formation **OR** team creation **OR** team design **OR** team composition **OR** team organization) **AND** (collaborative learning **OR** cooperative learning **OR** cscl **OR** csgf **OR** social learning **OR** group learning **OR** team learning) **AND** (algorithms **OR** approaches **OR** methods **OR** software **OR** technique).*

In the second step, to select relevant electronic databases to conduct our search, we started analyzing the results from Dyba et al. [4] that provided a list of important databases in the field of Computer Science and Engineering. Then, to focus our search, we shrink this list to obtain the databases that cover the most important conferences and journals in the field of educational technology. Thus, the following electronic databases were selected:

- ACM Digital Library
- IEEE Xplore
- ScienceDirect – Elsevier
- Scopus
- SpringerLink
- Web of Science

The search engine of each selected database uses different mechanisms and standards. Thus, we adapted the *search string* developed in this work to each database to conduct our search. After that we conduct the search on the titles, abstracts, and keywords of articles to collect the first set of primary studies. The results obtained are shown in Table 2. On the first column, we have the name of the database and, on the second columns, the number of returned papers.

Table 2. Number of primary studies obtained

Database	Quantity
ACM Digital Library	3
IEEE Xplore	23
ScienceDirect – Elsevier	10
Scopus	138
SpringerLink	15
Web of Science	67
TOTAL	**256**

2.3   Screening of Papers for Inclusion and Exclusion

The first screening of the returned papers (Table 2) consist of applying a set of inclusion (I) and exclusion (E) criteria to add or remove papers from our analysis:

I1. If several papers are related to the same study, only the most recent paper is selected;

I2. If the paper describes more than one study, each study is assessed individually;

I3. If there are versions of the same study, a short and a full, the full version must be included.

E1. Papers that do not present studies relating to education;

E2. Papers that do not present studies relating to group formation algorithms;

E3. Papers in languages other than English or Portuguese;

E4. Technical reports and documents that are available in the form of summaries or presentations (gray literature) and secondary studies (i.e., systematic reviews and mapping studies).

After defining the criteria for inclusion and exclusion, we read the titles and abstracts (and sometimes the introduction and conclusion) of each paper in order to identify those considered irrelevant to our work. Then, after the application of these criteria, we carefully read in full the final selection of papers and the data contained in these papers were extracted, analyzed and categorized. The final set of paper and the data analysis is presented in section 3.

2.4   Classification Scheme

To have a better understand of the contributions from each analyzed paper, we used the categories suggested by Wieringa and colleagues [13] to analyze, classify and categorize the types of studies described in the papers. The categories of study types are:

- **Validation Research:** novel techniques that have not yet been implemented in practice. Usually used in experimental settings in laboratory.
- **Evaluation Research:** techniques that are implemented in practice and an evaluation is conducted. This includes analysis of their implementation, benefits and drawbacks.
- **Solution Proposal:** A solution for a problem is proposed, which can be a new solution or an extension of an existing technique. The potential benefits of the solution is presented using case studies (small examples) or other argumentations.
- **Philosophical Papers:** papers that present a new look or direction to the field, often using taxonomies and conceptual frameworks.
- **Opinion Papers:** Studies that express a personal opinion about whether a technique is good or bad and/or how it should be used or implemented.
- **Experience Papers:** Contain the personal experience of the author explaining what and how something has been done in practice.

2.5   Data Extraction and Mapping of Studies

The papers were analyzed and classified according to the steps and categories presented in previous sections. We also created other categories to separate the research contributions of each paper (see Section 3). The data extracted from papers were stored and subjected to qualitative and quantitative analysis. This analysis aimed at finding evidences to answer the questions defined in Section 2.1. To organize the findings, we used a spreadsheet to document the data extraction process, which allowed us to also carry out other statistical analysis such as the number of publications per year, their venue, type, and so on.

3   Results and Analysis

In this Section, we present the results of our systematic mapping. The main purpose is to give an overview on how algorithms for group formation are being developed and applied in collaborative learning environments. This work was conducted over the period of four months between November 2013 to February 2014.

The Figure 1 shows the execution of the steps presented in sections 2.1, 2.2 and 2.3. Initially, by conducting the search in the selected databases 256 papers returned. From these, we identified that 91 of them were stored in multiple databases, thus, we eliminated the duplications leaving only a copy of each paper in our records. Thus, 165 papers remained to be analyzed in the next step. We then applied the inclusion and exclusion criteria in all 165 papers by reading their title, abstract, introduction and conclusion leaving only 65 papers. In the last stage of this process, we carefully read the 65 papers in full, and again applied the criteria for inclusion and exclusion, where 17 papers were eliminated. The final result of this process left 48 papers that were used as evidence to answer our research questions. The list of these papers appears in the end of this document.

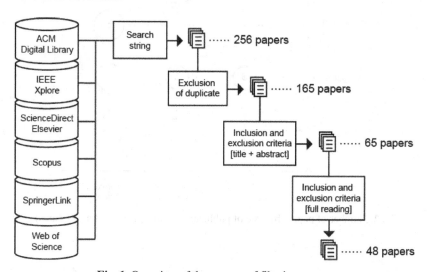

Fig. 1. Overview of the process of filtering papers

Before answering the main research questions of this work, it is also important to give an overview about where and when the 48 papers were published. Figure 2 shows on the *x-axis* the number of papers by publication type (i.e. Journal article, conference proceedings or book chapter) and on the *y-axis* the database where the papers were retrieved. According to Figure 2, the number of papers published in conference proceedings (26 papers) are the most common, followed by Journal articles (19 papers) and book chapters (3 papers). This result highlights the importance of conferences for dissemination of research on the topic of group formation in collaborative learning environments. It is worth to point out that none of the 48 papers is associated with the Scopus database because their primary index came from other databases (e.g. papers published by IEEE are primarily indexed in IEEE Xplorer but also appear in Scopus). Nevertheless, it is also important to emphasize the importance of Scopus database during the validation and calibration of the search string.

Observing the frequency of publications, we found out that 78% of the studies on this topic were published in the last seven years as shown in Figure 3. This trend of gradual increase indicates the growing importance and potential of the area. Regarding the year 2013, in which only two papers were found, it is possible that by the time we run the extraction process some studies had not yet been indexed in the databases.

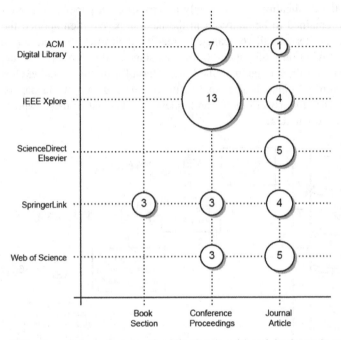

Fig. 2. Number of papers by type of publication (x axis) and databases (y axis)

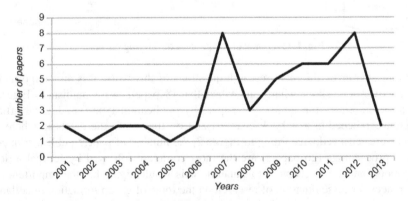

Fig. 3. Number of papers per year

In Section 2.4, we mentioned the use of the categories propose by [13] to cluster papers by type of study. Nevertheless, to answer RQ1 (what are the main research objectives and types of contributions from studies on group formation for CSCL?) we also needed to classify the papers by their research objectives. Thus, after reading all papers we proposed seven categories specific to this work as follows:

- **Tool:** the main objective of the study is to develop or extend a tool that implement a specific algorithm for group formation.
- **Framework:** the objective is to propose a model (technical foundations) that support the creation, or facilitate the use, of group formation algorithms.
- **Investigation:** the purpose of the study is to inspect an existing group formation algorithm.
- **Improvement:** the study focus on propose and implement better solutions for existing group formation algorithms.
- **Methodology:** the study aims to present best practices to use collaborative learning as well as group formation.
- **Script:** the objective of the study is propose guidelines to design CSCL activities, including steps related to group formation.
- **Technical:** the study focus on presenting a computational technique to implement group formation.

We used these categories and the ones present in Section 2.4 to create a bubble chart (Figure 4) to show the distribution of studies in each category. In the *x-axis* we have the study types categories and in the *y-axis* the research objectives. The size and number of each bubble represents the number of studies that fall into a specific x-y situation.

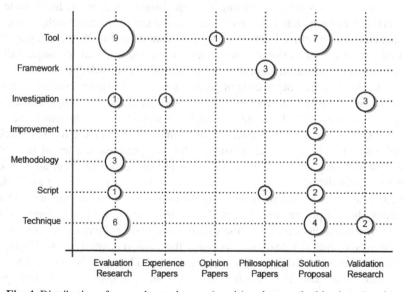

Fig. 4. Distribution of papers by study type (x axis) and research objectives (y axis)

By adding the numbers in the bubbles related to each row of Figure 4, it is possible to verify that the majority of studies, total of 17, fall in the category "Tool" and had as their main research objective to present an application that provides mechanisms to support group formation in CSCL contexts. Furthermore, 7 of these studies evaluated their results through use cases and 9 of them used data from real learning scenarios. The research objective category "*Technique*" is also well explored in the literature with 12 studies. 4 studies in this category aims at presenting a novel algorithm (solution) for group formation; other 2 studies validate these computational techniques in laboratory; and finally 6 studies evaluate these techniques with students in real scenarios. From another viewpoint, by adding the numbers in the bubbles related to each column of Figure 4, it is also possible to verify that most studies concentrate their effort on proposing a solution, total of 17 studies, or evaluating their results in real scenarios, total of 20 studies. This results show the maturity of the field since most studies were conducted with learners and their learning environment.

To answer RQ2 (what are the most common algorithms used to support group formation in collaborative learning environments?), we analyzed each study and found out that 44 from the 48 selected studies proposed or implemented algorithms as a solution to the problem of forming groups in collaborative learning environments. Figure 5 shows the type and number of algorithms utilized by these studies. About 41% of them (18 studies) are based on probabilistic algorithms. 8 are Genetic Algorithms (GA), demonstrating the great interest of researchers in using this technique as a solution to group formation in CSCL due to their applicability to deal with a large number of variables and the possibility to rapidly generate useful (semi-)optimal solutions (i.e. groups). Moreover, 5 studies presented algorithms based on Swarm Intelligence such as PSO (Particle swarm optimization) and ACO (Ant colony optimization) to form groups. In another algorithm category, a data mining approach known as k-means is utilized (4 studies). Among the algorithms listed as "Others" there are many different computational techniques such as the use of semantic web, ontologies, Bayesian Network, machine learning techniques, and so on. Finally the category "Unspecified" include studies that did not specify the computational technique utilized (e.g. ad-hoc group formation algorithms based on authors' knowledge).

To tackle RQ3 (is the source code of group formation algorithms available for analysis and reuse?) we extracted from the studies information about the implementation of the algorithms. We verified that 82% actually developed and implemented the algorithms for group formation in a specific CSCL environment (Figure 6). Nevertheless, although there is a high number of studies that implemented and tested their algorithms, few of them published their source code. In fact, according to our analysis only 2% of the studies did present the source code on the paper or put the code available on the Web. 23% of the studies presented only the pseudo-code and the majority, 75% of the studies, did not present any code of the implemented algorithms (Figure 7). This result is critical and disturbing, because the community do not have access to valuable data that would enable researchers to reuse the achievements (i.e. developed and tested algorithms) obtained in previous studies and build better tools and algorithms on top of that. It is also not possible to check and compare the existing group

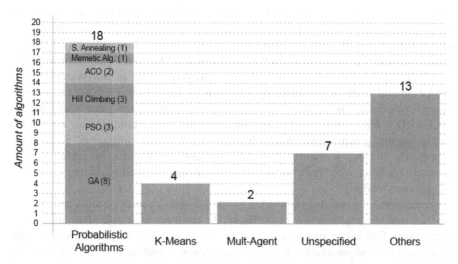

Fig. 5. Amount of algorithms for each type of approach

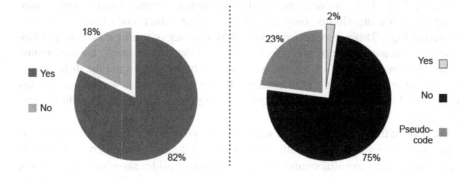

Fig. 6. Percentage of studies developed and implemented

Fig. 7. Percentage of availability of codes used in studies

formation algorithms in several educational settings to acquire more and better knowledge about their benefits for learning; which answers RQ4 (is there any study or tool that compare group formation algorithms in the context of collaborative learning?) – No.

4   Conclusion

Group formation is an important research topic in the field of CSCL due to its potential application to increase learning benefits of individuals when working in groups. Due to the complexity of forming effective groups manually, there are several algorithms to (semi-) automate such a process. In this work, we used a systematic mapping method (Section 2) to collect, analyze and summarize the research achievements on this topic. We initiated analyzing 256 papers and, after a careful inspection, we discarded 208 papers that did not fulfil the defined inclusion and exclusion criteria. The remaining 48 papers were considered as studies that produced evidences to answer the four research questions introduced in section 2.1 - RQ1: What are the main research objectives and types of contributions from studies on group formation for CSCL? RQ2: What are the most common computational techniques (i.e. algorithms) used to support group formation in collaborative learning environments? RQ3: Is the source code of group formation algorithms available for analysis and reuse? RQ4: Is there any study or tool that compare group formation algorithms in the context of collaborative learning?

According to the results presented in Section 3, development of tools and conduction of evaluation research are, respectively, the main research objectives and type of studies identified in the literature (see figure 4) – answering RQ1. Furthermore, the most common computational techniques implemented and used to form groups are probabilistic algorithms (e.g. genetic and swarm intelligence algorithms), followed by data mining techniques (e.g. k-means) and multi-agent approaches (see figure 5) – answering RQ2. Unfortunately, we verified that although many studies implemented group formation algorithms, only 2% provided their source code (see Figure 7) – answering RQ3. This is problematic, since there is a lack of source codes and pseudo-codes to replicate and reuse group formation algorithms. As a result, the community do not have instruments to compare, evaluate and better understand the different approaches to form groups in CSCL contexts. In fact, in our review we did not find any study or tool where the goal is to compare existing group formation algorithms - answering RQ4.

We believe that through the conduction of this systematic mapping, it was possible to provide to the CSCL community an overview of the research on group formation algorithms applied to collaborative learning contexts. Besides showing the increasing number of publications and a variety of computational approaches to deal with the topic, we also identified a critical problem in existing works, namely, the lack of presenting and sharing the source code of developed algorithms. This problem can be exploited to open new and important opportunities for future research.

Finally, to provide a quick visual overview of the findings presented in this paper an infographic was created and it is available at: *http://infografico.caed-lab. com/mapping/gf* . It also contains information that where not fully covered in this paper due to scope and space limitation.

Acknowledgment. We thank FAPESP (Process: 2013/13056-4) and CNPq (Processes: 310204/2011-9; 400481/2013-8 and 470757/2013-2) for providing support for this research.

References

1. Aronson, E., Patnoe, S.: The jigsaw classroom: building cooperation in the classroom, 2nd edn. Addison Wesley Longman, New York (1997)
2. Barkley, E., Cross, K.P., Major, C.H.: Collaborative Learning Techniques: A Practical Guide to Promoting Learning in Groups. Jossey Bass, San Francisco (2005)
3. Dillenbourg, P.: Over-scripting CSCL: The risks of blending collaborative learning with instructional design. In: Three Worlds of CSCL. Can we support CSCL?, pp. 61–91. Open University Nederland, Heerlen (2002)
4. Dyba, T., Dingsoyr, T., Hanssen, G.K.: Applying Systematic Reviews to Diverse Study Types: An Experience Report, 225-234 (2007)
5. Isotani, S., Inaba, A., Ikeda, M., Mizoguchi, R.: An Ontology Engineering Approach to the Realization of Theory-Driven Group Formation. International Journal on Computer-Supported Collaborative Learning 4(4), 445–478 (2009)
6. Isotani, S., Mizoguchi, R., Isotani, S., Capeli, O.M., Isotani, N., de Albuquerque, A.R.P.L., Bittencourt, I.I., Jaques, P.A.: A Semantic Web-based authoring tool to facilitate the planning of collaborative learning scenarios compliant with learning theories. Computers & Education 63, 267–284 (2013)
7. Moreno, J., Ovalle, D.A., Viccari, R.M.: A genetic algorithm approach for group formation in collaborative learning considering multiple student characteristics. Computers and Education 58, 560–569 (2012)
8. Ounnas, A., Davis, H.C., Millard, D.E.: A Framework for Semantic Group Formation in Education. Educational Technology and Society 12(4), 43–55 (2009)
9. Petersen, K., Feldt, R., Shahid, M., Mattsson, M.: Systematic Mapping Studies in Software Engineering. In: Proceedings of the Evaluation and Assessment in Software Engineering, pp. 1–10 (2008)
10. Soh, L.-K., Khandaker, N., Jiang, H.: I-MINDS: A Multiagent System for Intelligent Computer- Supported Collaborative Learning and Classroom Management. International Journal on Artificial Intelligence in Education 18, 119–151 (2008)
11. Strijbos, J., Martens, R.L., Jochems, W.M.G., Broers, N.J.: The effect of functional roles on perceived group efficiency during computer-supported collaborative learning: a matter of triangulation. Computers in Human Behavior 23(1), 353–380 (2007)
12. Wang, D.-Y., Liu, Y.-C., Sun, C.-T.: A grouping system used to form teams full of thinking styles for highly debating, pp. 725–730. World Scientific and Engineering Academy and Society (WSEAS), Stevens Point, Wisconsin, USA (2006)
13. Wieringa, R., Maiden, N.A.M., Mead, N.R., Rolland, C.: Requirements engineering paper classification and evaluation criteria: a proposal and a discussion. Requirements Engineering 11(1), 102–107 (2006)

Further Reading

[RP1] Abnar, S., Orooji, F., Taghiyareh, F.: An evolutionary algorithm for forming mixed groups of learners in web based collaborative learning environments. In: 2012 IEEE International Conference on Technology Enhanced Education (ICTEE), pp. 1–6 (2012)
[RP2] Adán-Coello, J.M., Tobar, C.M., de Faria, E.S.J., de Menezes, W.S., de Freitas, R.L.: Forming Groups for Collaborative Learning of Introductory Computer Programming Based on Students' Programming Skills and Learning Styles. International Journal of Information and Communication Technology Education 7, 34–46 (2011)

[RP3] Ardaiz-Villanueva, O., Nicuesa-Chacon, X., Brene-Artazcoz, O., Sanz de Acedo Lizarraga, M.L., Sanz de Acedo Baquedano, M.T.: Evaluation of Computer Tools for Idea Generation and Team Formation in Project-Based Learning. Computers & Education 56(3), 700–711 (2011)

[RP4] Brauer, S., Schmidt, T.C.: Group formation in elearning-enabled online social networks. In: International Conference on Interactive Collaborative Learning (ICL), pp. 1–8 (2012)

[RP5] Cadavid, J.M., Ovalle, D.A., Vicari, R.M.: A genetic algorithm approach for group formation in collaborative learning considering multiple student characteristics. Computers & Education 58, 560–569 (2012)

[RP6] Cavanaugh, R., Ellis, M., Layton, R., Ardis, M.: Automating the Process of Assigning Students to Cooperative-Learning Teams. In: Proceedings of the American Society for Engineering Education Annual Conference & Exposition (2004),
http://citeseerx.ist.psu.edu/viewdoc/summary?doi=10.1.1.65.682

[RP7] Christodoulopoulos, C.E., Papanikolaou, K.A.: A Group Formation Tool in a E-Learning Context. In: 19th IEEE International Conference on Tools with Artificial Intelligence, ICTAI, pp. 117–123 (2007)

[RP8] Cocea, M., Magoulas, G.D.: User behaviour-driven group formation through case-based reasoning and clustering. Expert Systems with Applications 39, 8756–8768 (2012)

[RP9] Craig, M., Horton, D., Pitt, F.: Forming reasonably optimal groups (FROG). In: Proceedings of the 16th ACM International Conference on Supporting Group Work (GROUP 2010), pp. 141–150 (2010)

[RP10] Daradoumis, T., Xhafa, F., Marques, J.M.: A methodological framework for project-based collaborative learning in a networked environment. International Journal of Continuing Engineering Education and Lifelong Learning 12(5/6), 389–402 (2002)

[RP11] Filho, J.A.B.L., Quarto, C.C., França, R.M.: Clustering Algorithm for the Socio-affective Groups Formation in Aid of Computer Supported Collaborative Learning. In: Collaborative Systems II - Simposio Brasileiro de Sistemas Colaborativos, pp. 24–27 (2010)

[RP12] Fukś, H., Raja Gabaglia Mitchell, L.H., Gerosa, M.A., de Lucena, C.J.P.: Competency Management for Group Formation on the AulaNet Learning Environment. In: Favela, J., Decouchant, D. (eds.) CRIWG 2003. LNCS, vol. 2806, pp. 183–190. Springer, Heidelberg (2003)

[RP13] Gogoulou, A., Gouli, E., Boas, G., Liakou, E., Grigoriadou, M.: Forming Homogeneous, Heterogeneous and Mixed Groups of Learners. In: Proceedings of the Workshop on Personalisation in Learning Environments at Individual and Group Level, pp. 33–40 (2007)

[RP14] Graf, S., Bekele, R.: Forming Heterogeneous Groups for Intelligent Collaborative Learning Systems with Ant Colony Optimization. In: Ikeda, M., Ashley, K.D., Chan, T.-W. (eds.) ITS 2006. LNCS, vol. 4053, pp. 217–226. Springer, Heidelberg (2006)

[RP15] Haake, J.M., Haake, A., Schümmer, T., Bourimi, M., Landgraf, B.: End-User Controlled Group Formation and Access Rights Management in a Shared Workspace System. In: Proceedings of the ACM International Conference on Computer-Supported Collaborative Work (CSCW), pp. 554–563 (2004)

[RP16] Ho, T.-F., Shyu, S.-J., Wang, F.-H., Li, C.T.-J.: Composing High-Heterogeneous and High-Interaction Groups in Collaborative Learning with Particle Swarm Optimization. In: World Congress on Computer Science and Information Engineering (CSIE), pp. 607–611 (2009)

[RP17] Sánchez Hórreo, V., Carro, R.M.: Studying the Impact of Personality and Group Formation on Learner Performance. In: Haake, J.M., Ochoa, S.F., Cechich, A. (eds.)
CRIWG 2007. LNCS, vol. 4715, pp. 287–294. Springer, Heidelberg (2007)

[RP18] Huang, Y.-M., Wu, T.-T.: A Systematic Approach for Learner Group Composition
Utilizing U-Learning Portfolio. Educational Technology & Society 14, 102–117
(2011)

[RP19] Inaba, A., Supnithi, T., Ikeda, M., Mizoguchi, R., Toyoda, J.: Learning goal ontology
for structuring a collaborative learning group supported by learning theories. Electron-
ics Communications in Japan, Part III 86(8), 79–90 (2003)

[RP20] Isotani, S., Mizoguchi, R.: Theory-Driven Group Formation through Ontologies. In:
Woolf, B.P., Aïmeur, E., Nkambou, R., Lajoie, S. (eds.) ITS 2008. LNCS, vol. 5091,
pp. 646–655. Springer, Heidelberg (2008)

[RP21] Isotani, S., Inaba, A., Ikeda, M., Mizoguchi, R.: An Ontology Engineering Approach to
the Realization of Theory-Driven Group Formation. International Journal on Computer-Supported Collaborative Learning 4(4), 445–478 (2009)

[RP22] Khandaker, N., Soh, L.-K.: A Wiki with Multiagent Tracking, Modeling, and Coalition
Formation. In: Proceedings of the 22th Innovative Applications of Artificial Intelligence (IAAI), pp. 1799–1806 (2010)

[RP23] Khandaker, N., Soh, L.-K.: SimCoL: A Simulation Tool for Computer-Supported Collaborative Learning. IEEE Transactions on Systems, Man, and Cybernetics, Part C 41,
533–543 (2011)

[RP24] Kobbe, L., Weinberger, A., Dillenbourg, P., Harrer, A., Hämäläinen, R., Häkkinen, P.,
Fischer, F.: Specifying computer-supported collaboration scripts. International Journal
on Computer-Supported Collaborative Learning 2, 211–224 (2007)

[RP25] Kyprianidou, M., Demetriadis, S., Tsiatsos, T., Pombortsis, A.: Group Formation
Based on Learning Styles: Can It Improve Students' Teamwork? Educational Technology Research and Development 60(1), 83–110 (2012)

[RP26] Layton, R.A., Loughry, M.L., Ohland, M.W., Rico, G.D.: Design and validation of a
web-based system for assigning members to teams using instructor-specified criteria.
Advances in Engineering Education 2(1), 1–28 (2010)

[RP27] Li, Z., Zhao, X.: The Design of Web-Based Personal Collaborative Learning Sys-tem
(WBPCLS) for Computer Science Courses. In: Li, F., Zhao, J., Shih, T.K., Lau, R., Li,
Q., McLeod, D. (eds.) ICWL 2008. LNCS, vol. 5145, pp. 434–445. Springer,
Heidelberg (2008)

[RP28] Lin, Y.-T., Huang, Y.-M., Cheng, S.-C.: An automatic group composition system for
composing collaborative learning groups using enhanced particle swarm optimization.
Computers & Education 55(4), 1483–1493 (2010)

[RP29] Liu, S., Joy, M.S., Griffiths, N.: Incorporating Learning Styles in a Computer-
Supported Collaborative Learning Model. In: International Workshop on Cognitive
Aspects in Intelligent and Adaptive Web-based Education Systems (2008),
http://eprints.dcs.warwick.ac.uk/119/1/liu_joy_
griffiths_ciawes_08.pdf

[RP30] Liu, S., Joy, M.S., Griffiths, N.: iGLS: Intelligent Grouping for Online Collaborative
Learning. In: IEEE International Conference on Advanced Learning Technologies
(ICALT), pp. 364–368 (2009)

[RP31] Liu, S., Joy, M.S., Griffiths, N.: An Exploratory Study on Group Formation Based on
Learning Styles. In: IEEE International Conference on Advanced Learning Technologies (ICALT), pp. 95–99 (2013)

[RP32] Mavrommatis, G., Tsimaras, D.: Forming cliques of collaborating distance learners. In: Proceedings of the 7th Conference on Applied Informatics and Communications, pp. 309–312 (2007)

[RP33] Mehta, D., Kouri, T., Polycarpou, I.: Forming project groups while learning about matching and network flows in algorithms. In: Annual Conference on Innovation and Technology in Computer Science Education (ITiCSE), pp. 40–45 (2012)

[RP34] Messeguer, R., Medina, E., Royo, D., Navarro, L., Juárez, J.P.: Group Prediction in Collaborative Learning. In: International Conference on Intelligent Environments, pp. 350–355 (2010)

[RP35] Muehlenbrock, M.: Formation of Learning Groups by using Learner Profiles and Context Information. In: International Conference on Artificial Intelligence in Education (AIED), pp. 507–514 (2005)

[RP36] Mujkanovic, A., Lowe, D., Willey, K.: Unsupervised Learning Algorithm for Adaptive Group Formation: Collaborative Learning Support in Remotely Accessible Laboratories. In: IEEE International Conference on Information Society, pp. 59–66 (2012)

[RP37] Ounnas, A., Davis, H.C., Millard, D.E.: Towards Semantic Group Formation. In: IEEE International Conference on Advanced Learning Technologies (ICALT), pp. 825–827 (2007)

[RP38] Ounnas, A., Davis, H.C., Millard, D.E.: A Framework for Semantic Group Formation in Education. Educational Technology and Society 12(4), 43–55 (2009)

[RP39] Redmond, M.A.: A computer program to aid assignment of student project groups. In: ACM SIGCSE Conference, pp. 134–138 (2001)

[RP40] Rubens, N., Vilenius, M., Okamoto, T.: Automatic Group Formation for Informal Collaborative Learning. In: Web Intelligence Workshops, pp. 231–234 (2009)

[RP41] Spoelstra, H., van Rosmalen, P., van de Vrie, E., Obreza, M., Sloep, P.B.: A Team Formation and Project-based Learning Support Service for Social Learning Networks. Journal of Universal Computer Science 19, 1474–1495 (2013)

[RP42] Tobar, C.M., de Freitas, R.L.: A support tool for student group definition. IEEE Frontiers in Education Conference (FIE),T3J-7,T3J-8 (2007)

[RP43] Wang, D.-Y., Liu, Y.-C., Sun, C.-T.: A grouping system used to form teams full of thinking styles for highly debating. World Scientific and Engineering Academy and Society (WSEAS), pp. 725–730 (2006)

[RP44] Wang, D.-Y., Lin, S.S.J., Sun, C.-T.: DIANA: A computer-supported heterogeneous grouping system for teachers to conduct successful small learning groups. Computers in Human Behavior 23(4), 1997–2010 (2007)

[RP45] Wessner, M., Pfister, H.-R.: Group formation in computer-supported collaborative learning. In: ACM Conference on Supporting Groupwork, pp. 24–31 (2001)

[RP46] Yannibelli, V.D., Amandi, A.: Forming well-balanced collaborative learning teams according to the roles of their members: An evolutionary approach. In: IEEE 12th International Symposium on Computational Intelligence and Informatics (CINTI), pp. 265–270 (2011)

[RP47] Yannibelli, V., Amandi, A.: A Memetic Algorithm for Collaborative Learning Team Formation in the Context of Software Engineering Courses. In: Cipolla-Ficarra, F., Veltman, K., Verber, D., Cipolla-Ficarra, M., Kammüller, F. (eds.) ADNTIIC 2011. LNCS, vol. 7547, pp. 92–103. Springer, Heidelberg (2012)

[RP48] Yannibelli, V.D., Amandi, A.: A deterministic crowding evolutionary algorithm to form learning teams in a collaborative learning context. Expert Systems with Applications 39, 8584–8592 (2012)

Evaluating Coordination Support Mechanisms in an Industrial Engineering Scenario[*]

Jordan Janeiro[1], Stephan Lukosch[1], Frances M.T. Brazier[1],
Mariano Leva[2], Massimo Mecella[2], and Arne Byström[3]

[1] Delft University of Technology
{j.janeiro,s.g.lukosch,f.m.brazier}@tudelft.nl
[2] Sapienza Università di Roma
{leva,mecella}@diag.uniroma1.it
[3] Bosch Rexroth AB
arne.bystrom@boschrexroth.se

Abstract. Nowadays, industrial engineering collaboration plays a crucial role along product development life cycle, especially for problem-solving and decision-making processes. This paper evaluates the acceptance of two coordination mechanisms for groups when working on a machine diagnosis report collaboratively. The evaluation is organized as a user study and is based on two hypothesis: groups will prefer unstructured over structured coordination, and groups using structured coordination will accomplish their task more efficiently.

Keywords: user evaluation, shared environment, coordination.

1 Introduction

Nowadays, the design and manufacture of innovative and highly specialized machinery goes through a complex life cycle that starts with the understanding of user requirements and the context in which the product is placed in, through design, development, testing, evaluation, maintenance and disposal. Industrial engineering collaboration plays a crucial role along product development life cycle, especially for problem-solving and decision-making processes.

Group participants may choose between different coordination mechanisms, implemented in groupware, to facilitate and support collaboration to achieve goals. Coordination mechanisms vary between providing more or less structured support, such as the concept of scaffolding [8] that provides supporting mechanisms that are gradually removed as users gain practice. The variation of coordination mechanism support for computer-supported cooperative work can be mapped in a spectrum ranging from structured to unstructured extremes [1,9]

---

[*] This work has been partially supported by the EU FP7-257899 project Smart Vortex. The authors would like to thank Serena Carnevale, Emanuela Bauleo and Michele Biancucci for their contributions in the organization and running of the study and the students of Sapienza Università di Roma for their participation.

N. Baloian et al. (Eds.): CRIWG 2014, LNCS 8658, pp. 215–222, 2014.
© Springer International Publishing Switzerland 2014

and containing delimited subspectra that implements a particular coordination mechanism.

Although implementation of coordination mechanisms varies, their major purpose is to support groups achieving a shared goal. For example, in a scenario of remote machine maintenance, both structured and unstructured coordination mechanisms can support a virtual team of engineers to diagnose a machine and avoid its breakdown. The difference relies on the preferences of the team to use one mechanism over the other. One possible factor that influences such preferences is the implementation of a particular coordination mechanism. Some groups may prefer more guidance to achieve their goal by following a process that has been predefined by an expert. Other groups, may prefer more flexibility and use awareness information to coordinate their activities to achieve their common goal.

Although structured as well as unstructured coordination have their advantages, so far there has been no study that compares their effect in a unified environment. This paper thus explores whether groups prefer unstructured coordination over structured coordination. Unstructured coordination mainly assumes that a group can leverage workspace awareness information to coordinate their activities spontaneously. Compared to this, structured coordination prescribes the activities that need to be done to fulfil a shared task. Research has shown that structured coordination can lead to more efficient processes, as overhead for coordination is diminished [3]. Groups, however, may prefer unstructured coordination because this resembles co-located group dynamics, in which most of shared work is unplanned, spontaneous and informal [10]. The paper thus evaluates the following hypotheses:

H1. Groups prefer unstructured over structured coordination.
H2. Groups using structured coordination will accomplish their task more efficiently.

2   Coordination Mechanisms

The major goal of implementing coordination mechanisms in groupware systems is to support users to coordinate tasks to accomplish a common goal. The structured extreme of the spectrum refers to coordination mechanisms that are highly specified and routinised processes, allowing experts in a certain domain to define processes to support and guide group participants.

The unstructured extreme of the spectrum refers to coordination mechanisms that are highly unspecified and dynamic processes, not using structures or processes to guide group participants in their tasks, but conversely relying on awareness mechanisms that allow for an informal understanding of the activities of other participants.

Some groupware systems address more than one part of the coordination spectrum and provide mechanisms from both extremes. In such systems, the decision on a specific coordination mechanism is delegated directly to groups of participants, according to the situation they handle [1,9].

In global software development, Redmiles et al. describe different coordination mechanisms used by teams of developers [7]. They differentiate between two coordination approaches, formal and informal. Formal mechanisms rely on process-based mechanisms that define multiple, independent tasks that are periodically resynchronized. Whereas, informal approaches rely on awareness information that allows for the understanding of activities performed by developers.

For supply chains, Bernstein proposes a system that implements different coordination mechanisms to enable collaboration between users [1]. The system implements four mechanism: guidance through scripts, planning based on constraints, monitoring constraints and context provision. Each mechanism supports users handling unexpected disruption problems in a computer supply chain, such as providing pre-defined supply chain processes or contextual information.

3   Elastic Groupware

Elgar (Elastic groupware) is a groupware that addresses the spectrum of coordination and implements different coordination mechanisms to support groups in collaborative tasks [5]. Elgar was designed and implemented to support the analysis of industrial machines and formulation of diagnosis reports. The current implementation of Elgar offers a unified environment to explore the hypotheses H1 and H2 in an industrial engineering scenario.

Elgar is based on traditional dashboard systems. It was implemented as an extensible web-based groupware that provides a modular infrastructure to integrate software components. In Elgar, such components are called *Elastic Collaboration Components* (ECCs; they are web-based components inspired on portlets. Developers need to implement ECCs according to the architecture described in [5] to use them as part of Elgar. This user study used four ECCs: V2QT, CollPad, Rating and CollPad Categorizer. The Visual Query and Visualization Tool (V2QT) [6] is an ECC that enables users to visually pose continuous queries (CQs) over machine sensor measurements. V2QT handles query requests and responses and displays results in a time-oriented graph. Users gain insights, through the V2QT, over possible anomalous equipment behaviours, e.g., the main hydraulic pressure is above a threshold. All other ECCs were used to support users to share and rate ideas based on lists collaboratively and generate analysis reports.

Table 1. Survey for ad-hoc and prescribed coordination mechanisms

Questions for the prescribed coordination mechanism
qa1: I used workspace awareness information to be aware of if other group members were involved in a group task.
qa2: I used workspace awareness information to join other group member(s) to work on a task.
qa3: I used workspace awareness information to take over on a task in which no other group member was working.
qa4: I used workspace awareness information to divide tasks among other group members.
qa5: I used workspace awareness information to know the progress of the group task.
Questions for the prescribed coordination mechanism
qp1: Elgar coordinated all group tasks for the group members.
qp2: The process helped me to accomplish the group task.
qp3: I had to coordinate the division of tasks with other group members.
qp4: It was important that all group members were working together in one process phase at a time.
qp5: It was important that Elgar prescribed the tools that I needed to use in each process phase.
qp6: I knew the progress of the group task.
qp7: I could identify if other group members were involved in the group task.
qp8: I could follow process phases in Elgar.

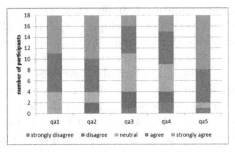

(a) Acceptance of the ad-hoc coordination
mechanism.

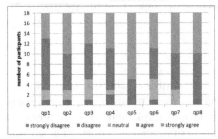

(b) Acceptance of the prescribed coordination mechanism.

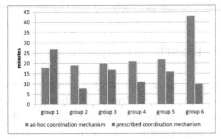

(c) Time spent by groups using Elgar with
ad-hoc and prescribed mechanisms.

Fig. 1. User study results

Currently Elgar supports two coordination mechanisms: ad-hoc and
prescribed. The ad-hoc mechanism enables self-coordination of groups to emerge
naturally and is not guided by prescription. It supports groups that do not need
guidance or prescribed coordination during collaboration. This mechanism relies
on the use of workspace awareness information to divide and synchronize tasks
among group participants. In this mechanism, such information represents all
group participants connected in a collaboration session and the ECCs in which
participants work. Through such information, participants may create an understanding about activities of others, facilitating their coordination to achieve a
common goal, such as in [2,9].

The prescribed coordination mechanism guides groups with processes planned
by professional facilitators. Guidance is described as a collaboration process,
which represents a sequence of collaborative phases, each associated with a set
of ECCs, and explicit instructions for the phase. In the prescribed coordination
mechanism, the collaboration process is described at design-time and provides
guidelines for a group. It is based on process-awareness [4], in which processes are
described to coordinate users in accomplishing established goals. The emphasis
of the prescribed coordination mechanism is to provide guidance to users in
collaborative activities, and not to require a facilitator to lead groups [11].

4   User Study

The goal of this study is to evaluate hypotheses H1 and H2. We used an industrial engineering scenario, in which group participants had to generate a machine diagnosis report for an industrial machine. Group participants had to identify, divide and coordinate tasks that contribute to the accomplishment of the report. Groups had to analyze readings from two sensors (oil pressure in the A-side pump and oil pressure in the B-side pump) of a wood shredder machine and indicate machine problems in the report. For this study, a group of mechanical engineers extracted machine data from a deployed machine, in a remote site in Sweden. They incorporated machine failures to the data to enable a realistic diagnosis process, on purpose. Participants had to identify anomalous values whenever they were above an established threshold at a specific moment. Based on value analysis, they should formulate the problem causing the anomaly describing: $i)$ the name of the sensor, $ii)$ the current sensor reading and $iii)$ the difference between the sensor reading and the established threshold for the sensor. Once group participants describe a list of all possible problems, they should rate the descriptions according to a criticality scale. The higher is the difference between a sensor reading and the value representing the threshold, the higher is the criticality. Rating based on the criticality is important for machine maintenance teams to handle the most critical problems first.

A total of eighteen participants took part in this user study, fourteen males and four females participants. There were three participants working as research fellows and fifteen students enrolled in master's programs. The eighteen participants were divided in six groups of three people indistinctly.

Setup. The user study followed six phases: $i)$ training for diagnosis session using the ad-hoc mechanism, $ii)$ training for diagnosis session using the prescribed mechanism, $iii)$ user preference inquiry, $iv)$ diagnosis session using the ad-hoc mechanism, $v)$ diagnosis session using the prescribed mechanism and $vi)$ fill in the surveys for ad-hoc and prescribed coordination mechanisms. The first two phases were training sessions to familiarize the groups with Elgar. The last three phases refer to the evaluation of these coordination mechanisms.

There were two training sessions per group to instruct participants using the system. In phase 1, group participants used the ad-hoc mechanism and in phase 2, the prescribed mechanism. During the training sessions, users were co-located and had to generate diagnosis reports for two simulated sensors of a wood shredder machine. An instructor was also present and taught the use of Elgar and particulars for each coordination mechanism. In phase 1, the instructor trained users on creating and sharing ECCs dynamically, based on workspace awareness information. In phase 2, the instructor trained the users in following a process that contains pre-defined instructions and shared components. After finishing the training sessions, groups, were asked to report about their preferred coordination mechanism (phase 3). For the user study, each group participant was lead to a separate room, provided with a notebook and connected through a voice channel for communication purposes. First, the instructor requested the group to generate a report for two sensors of a wood shredder machine using Elgar's ad-hoc coordination mechanism (phase 4). Then, the instructor requested the group to generate another report for other readings of machine sensors using Elgar's prescribed support coordination mechanism (phase 5). After the end of the study, the instructor requested groups to answer two surveys that aimed to assess their perception about used coordination mechanisms (phase 6).

Method. The focus of questions listed in Table 1 is to understand the satisfaction of group participants using ad-hoc and prescribed coordination mechanism. The survey assesses whether provided processes were able to coordinate participants' activities to generate a diagnosis report, or workspace awareness information enabled participants to divide tasks among them and collaborate with others.

Possible answers for each survey question ranged between the following values: strongly disagree, disagree, neutral, agree and strongly agree. In addition, each survey had two open-ended questions for group participants to better understand their experience using a particular coordination mechanism, e.g., "*Describe the positive aspects of using the ad-hoc coordination mechanism to accomplish the group task*". The focus of the questions was to understand the positive and negative aspects of each coordination mechanisms, perceived by participants.

5   Results

Figure 1a summarizes the answers of group participants to questions related to ad-hoc coordination mechanism. As illustrated by the bars qa1, qa2 and qa5, the majority of group participants agree that workspace awareness information was important for task coordination to write a diagnosis report. However, as illustrated by bars qa3 and qa4, current features implemented in this mechanism were not sufficient to divide and assign tasks among participants, or to enable participants to take over tasks. Instead, they had to coordinate the division of tasks spontaneously, through the voice communication channel.

Figure 1b summarizes the answers of group participants to questions related to the prescribed coordination mechanism; it shows that in general the majority of group participants accept the coordination imposed by a predefined process. It is important to remark that all users agreed (or strongly agreed) with the question qp5, reinforcing the preference for the prescription of ECCs to accomplish a group task. However, the majority of participants (13 participants) agree that they had to coordinate the division of their tasks, in addition to the ones prescribed in the process. One possible reason for this is that the used process coordinated the group for major abstract activities, e.g., analyse sensor data, formulate problems and generate report. It did not coordinate groups for specific and less abstract activities, e.g. user 1 shall analyse sensor pump B-side and user 2 should analyse sensor pump A-side.

6   Discussion

There are two measures collected during this user study: completion time and coordination mechanism preference. The first one measures the time used by a group to write a diagnosis report. Figure 1c summarizes the time spent by all groups using a coordination mechanism. The majority of the groups (5 groups) were quicker to perform the group task using the prescribed mechanism confirming hypothesis H2, in which groups will be more efficient using the prescribed coordination mechanism. A main reason for this result is the prescription of all tasks and components. In the prescribed mechanism, Elgar coordinates group participants guiding them through tasks and sharing ECCs for them automatically. Therefore, participants did not need to spend time discussing their strategy to handle a problem or sharing ECCs manually. The second measure, coordination mechanism preference, reports on the number of groups that preferred one coordination mechanism over the other. Group participants answered to this enquiry after the training sessions with Elgar. The analysis of this enquiry shows that five groups preferred to use of the prescribed coordination mechanism, whereas one group preferred to use the ad-hoc mechanism. Such analysis contradicts the first aforementioned hypothesis in this paper (H1), in which groups will prefer the use of unstructured coordination mechanisms over structured.

We also briefly discuss the answers of group participants to open-ended questions. According to the analysis, process descriptions facilitated more the accomplishment of the group task by guiding participants through phases and sharing ECCs automatically. In response to the question *"Describe the positive aspects of using the prescribed coordination mechanism"*, a group participant answered *"Since it is more driven than the ad-hoc type, it is quite difficult to do mistakes"*, referring to the guiding aspects of a process. Another user answered *"Having the process already defined helps very much. The best thing is that all the ECCs are already shared"*, referring to the prescription of components. With regard to the ad-hoc mechanism, group participants mentioned that ad-hoc mechanism is more suitable for experienced users that do not need guidance from the system. In response to the question *"Describe the positive aspects of using the ad-hoc coordination mechanism."*, a group participant answered *"It's best suited for expert users"*. Another participant answered *"It is customizable and good for a pro user"*. A possible reason for such answers is that experienced participants would develop their own work practices, creating continuously new processes and not using existing ones.

7   Conclusions and Future Work

This paper evaluated the acceptance of two different coordination mechanisms to generate a machine diagnosis report collaboratively. The evaluation was organized as a user study and focused on two hypothesis: H1) groups prefer unstructured over structured coordination, and H2) groups using structured coordination will accomplish their task more efficiently.

The study shows that hypothesis H1 was disproved whereas H2 was confirmed. The hypothesis H2 fulfils our expectations. Indeed, the large majority of group participants used less time to accomplish group task using a prescribed coordination mechanism. One major reason is here that the prescription reduced the time for coordination. However, the study disproves hypothesis H1. We assumed that users, after training sessions to become familiar with Elgar, would prefer ad-hoc mechanisms because of its support to informal and spontaneous group interactions. However, in this study group participants did not have enough expertise on the subject to coordinate their task spontaneously.

The use of engineers and students from different domains represents a preliminary study to assess the overall acceptance of the system. However, it is necessary to run the same study with experts in industrial machine diagnosis. The goal is to understand whether their preferences on coordination mechanisms remain the same or change. Future work also includes an extension in Elgar to enable the continuous change of coordination mechanisms during a collaboration session. Through this extension, group participants can explore coordination mechanisms dynamically and identify the most suitable for them.

References

1. Bernstein, A.: How can cooperative work tools support dynamic group process? bridging the specificity frontier. In: Proc. ACM CSCW 2000 (2000)
2. Biehl, J.T., Czerwinski, M., Smith, G., Robertson, G.G.: Fastdash: a visual dashboard for fostering awareness in software teams. In: Proc. CHI 2007 (2007)
3. Briggs, R.O., Kolfschoten, G.L., de Vreede, G.-J., Albrecht, C.C., Lukosch, S.G.: Facilitator in a box: Computer assisted collaboration engineering and process support systems for rapid development of collaborative applications for high-value tasks. In: Proc. HICSS 2010 (2010)
4. Dumas, M., Van Der Aalst, W., Ter Hofstede, A.: Process-Aware Information Systems. Wiley (2005)
5. Janeiro, J., Lukosch, S., Radomski, S., Johanson, M., Mecella, M., Larsson, J.: Supporting elastic collaboration: integration of collaboration components in dynamic contexts. In: Proc. EICS 2013 (2013)
6. Malagoli, A., Leva, M., Kimani, S., Russo, A., Mecella, M., Bergamaschi, S., Catarci, T.: Visual query specification and interaction with industrial engineering data. In: Bebis, G., et al. (eds.) ISVC 2013, Part II. LNCS, vol. 8034, pp. 58–67. Springer, Heidelberg (2013)
7. Redmiles, D., Van Der Hoek, A., Al-Ani, B., Hildenbrand, T., Quirk, S., Sarma, A., Filho, R., de Souza, C., Trainer, E.: Continuous coordination-a new paradigm to support globally distributed software development projects. Wirtschafts Informatik 49(1), 28 (2007)
8. Rogoff, B.: Apprenticeship in thinking: Cognitive development in social context. Oxford University Press (1990)
9. Sarma, A., Redmiles, D., van der Hoek, A.: Categorizing the spectrum of coordination technology. Computer 43(6) (2010)
10. Whittaker, S., Frohlich, D., Daly-Jones, O.: Informal workplace communication: What is it like and how might we support it? In: Proc. CHI 1994 (1994)
11. Knoll, S.W., Horning, M., Horton, G.: Applying a thinkLet-and thinXel-based group process modeling language: A prototype of a universal group support system. In: Proc. HICSS 2009 (2009)

Virtual Operating Room for Collaborative Training of Surgical Nurses

Nils Fredrik Kleven[1], Ekaterina Prasolova-Førland[2], Mikhail Fominykh[2],
Arne Hansen[3], Guri Rasmussen[3], Lisa Millgård Sagberg[4], and Frank Lindseth[5]

[1] Department of Computer and Information Science,
Norwegian University of Science and Technology, Norway
[2] Program for Learning with ICT, Norwegian University of Science and Technology, Norway
[3] Faculty of Nursing at the Sør-Trøndelag University College, Norway
[4] Department of Neurosurgery, St. Olav's University Hospital, Norway
[5] Department of Medical Technology, SINTEF, Norway
nilsfrk@stud.ntnu.no,
{ekaterip,mikhail.fominykh,lisa.millgard.sagberg}@ntnu.no,
{arne.hansen,guri.rasmussen}@hist.no,
frank.lindseth@sintef.no

Abstract. In this paper, we present the first results of a study exploring how to support collaborative learning of surgical nursing students in a 3D virtual world. A Virtual Operating room, resembling the one at St. Olav's University Hospital in Trondheim, Norway was created in Second Life to accommodate an educational role-play. In this role-play, the operating nursing students could practice communication with patients and cooperation in the team while preparing patients for surgery. At the first stage of the evaluation, the virtual simulation has been tested among nine postgraduate nursing students. The participants gave their evaluation and opinions in the form of questionnaires and discussion after the role-plays. Following the analysis of the data, we present a summary of the most important results in this paper. This study provides a number of suggestions for improving the learning process when role-playing in a virtual environment. We demonstrate that an educational simulation can be implemented with limited resources, and yet be practically useful in education of health personnel. Further research with medical and nursing students is highly applicable and feasible, and should include a larger group of participants. In the next stage of our work, the evaluation of the Virtual Operating room has been conducted with nurses, who are on an earlier stage of their study program, as well as anesthesia nurses and non-medics.

Keywords: 3D collaborative virtual environments, medical training, collaborative learning, virtual operating room, educational role-play.

1 Introduction

At the core of educational activities of health professionals at all levels is the patient. The ultimate goal of basic research is the treatment offered to patients. Traditionally,

N. Baloian et al. (Eds.): CRIWG 2014, LNCS 8658, pp. 223–238, 2014.

the bulk of the student contact with the patient has been through placement mainly in hospitals. However, there is a major challenge to this – the availability of time for contacting patients. One aspect is the increase in number of students. This is in part ameliorated by increasing the number of patients. The other aspect is not so easily remedied, the fact that patients spend less and less time in hospitals. As the hospitals improve their effectiveness, e.g. by increasing the number of day patients, there is dramatically less time for the contact between students and patients. Consequently, students get less time on the task. Thus, there is a need for solutions that give the students more time on tasks or make the time with the patients more effective.

In addition, in today's hospitals, a patient is treated not by a single practitioner, but by a team of specialists, with complex collaborative procedures and practices within the team. That means that a student needs to practice not only on patient interaction, but also on complex interactions within a team of professionals, such as when preparing a patient for surgery.

One has to look for alternatives to facilitate practice and explorative learning experiences to meet these challenges. Addressing the knowledge needs requires that students are provided with flexible online educational solutions which must be embedded in a holistic system. Therefore, the idea of an online virtual university hospital has emerged, to be a venue for learning, research, and development. The idea is to make a virtual mirror of St. Olav's University Hospital (St. Olav) as it is one of the most modern university hospitals in the world. It has a state of the art technological platform and modern clinical buildings with a unique feature: the Faculty of Medicine at the Norwegian University of Science and Technology (NTNU) has integrated its teaching and research facilities within the hospital. In addition, NTNU and St. Olav have developed 'Kunnskapsportalen', a portal for distributing knowledge and information to patients and the general public, as well as to students, staff, and researchers.

In this paper, we present a pilot study within the Virtual Hospital project. The goal of this study is exploring educational role-playing in a 3D virtual environment as a method for training communication, cooperation within a team, and other practical skills of surgical nurses. The study was conducted at the NTNU, involving several other organizations. The major learning objective of the training simulation developed is to obtain in-depth knowledge of communication and interaction with patients and their families. In particular, we focused on the procedures of receiving and delivering patients on their way to complicated operations, something that requires coordination between ward and surgical nurses. The teachers from the Faculty of Nursing at the Sør-Trøndelag University College (HiST) have contributed to providing learning goals and designing scenarios for the study. Another group of subject experts from the Department of Neurosurgery at St. Olav was involved to provide requirements and feedback for the design of the virtual environment.

2   Background

As identified in recent relevant studies in the field, "as available teaching time in anatomy and surgery are expected to continue to decline, the adoption of unique instructional methods such as virtual learning may serve not only to attract more tech-nologically inclined candidates but also improve the efficacy of the relatively fewer opportunities that will remain" [1]. This motivates the exploration of different modes for virtual learning, i.e. flexible low-cost 3D virtual simulations, 3D virtual environ-ments, and associated infrastructure accessible over the Internet. Recent studies in the field indicate that this technology can "[...] effectively replicate clinical PBL scena-rios", with the potential "to considerably augment, if not eventually, revolutionize medical education" [1].

Many studies report the potential of three-dimensional virtual worlds (3D VWs) for educational activities [2]. This technology can benefit educational process due to low cost and high safety, three-dimensional representation of learners and objects, and interaction in simulated contexts with a sense of presence [3,4]. Possibilities for synchronous communication and interaction allow using 3D VWs by various colla-borative learning approaches [5], as well we facilitate situated learning [6] and project-based learning [7] approaches. Nowadays, 3D VWs can be used in combina-tion with other VR technologies, such as motion tracking and head-mounted displays, to increase the sense of immersion and, therefore, improve the experience, making it more believable and transferable to the real life.

There have been several cases where 3D VWs have been used in the health care domain, including both desktop-based VWs and other VR applications. Examples include training facilities for nurses [8,9] and doctors (e.g., in palliative care units [10]), health information centers, and 3D visualizations of internal organs. Such train-ing is, on several occasions, reported to provide a cost-efficient and user-friendly alternative to real-life role plays and training programs [10]. As demonstrated in sev-eral studies, "virtual worlds offer the potential of a new medical education pedagogy to enhance learning outcomes beyond that provided by more traditional online or face-to-face postgraduate professional development activities" [11].

Desktop-based environments have been augmented with VR elements for treat-ment of various neurological and psychiatric disorders such as autism, phobias, and post-traumatic stress syndrome, the latter especially in military settings. For example, Virtual Afghanistan/Iraq system has undergone successful clinical trials in using ex-posure therapy for treatment of combat-related post-traumatic stress syndrome among veterans [12]. VR is being increasingly used for developing educational medical visu-alizations, for example, to be used in anatomy classes [13].

Several leading world universities and hospitals, especially in the US, UK, Austral-ia, and New Zealand have adopted 3D virtual simulation as a part of their educational programs. Examples include virtual hospitals/medical faculties at University of South Florida, Imperial College of London, and Auckland University Hospital. Such envi-ronments typically include an array of different facilities, such as emergency room, intensive care unit, nursing simulation, and general information for the public. Other examples include Maternity Ward at Nottingham University and Emergency Prepa-redness Training at University of Illinois in Second Life (SL). Our own Virtual Hos-pital initiative is inspired by these projects but seeks to achieve a more holistic and coherent approach to the development of a virtual hospital.

3   Learning Objectives and Collaborative Scenarios

In this project, we focused on communication with patients. Surgical nurses often have to inform, prepare, and even calm both the patient and their relatives before a complicated operation. These elements can be practiced through role-playing different scenarios in the virtual environment in SL.

The learning objectives below are based on the teaching plan for postgraduate surgical nurses at the Faculty of Nursing at HiST and have been used to develop the scenarios for the simulation. The major learning objective is to obtain in-depth knowledge of how to communicate and interact with patients, and their families, in particular when receiving or delivering patients who are on their way to a complicated operation. The major learning goal is split into several sub-goals:

- Learning objective L1: Reassuring a patient in advance of an important and complex operation
- Learning objective L2: Dealing with relatives
- Learning objective L3: Communicating / dealing with patients with immigrant background, especially women
- Learning objective L4: Reassuring / dealing with children in advance of an operation
- Learning objective L5: Dealing with a seriously ill and potentially dying patient
- Learning objective L6: Performing basic medical tasks prior to the operation (e.g., moving the operating table and disinfection)

In order to address the learning objectives given above, four typical scenarios have been developed, including a variety of medical cases and a variety of patient and relatives groups, as well as both males and females, children and patients with immigrant background:

- Learning scenario 1
  - Actors: A woman (patient), surgical nurse, and ward nurse
  - Plot: A 35-year-old woman is admitted to the gynecological department. She is on her way to a surgery for an abscess that will be operated in spinal anesthesia. She lies on a patient bed transported by a nurse from the ward who delivers her to the surgical nurse.
- Learning scenario 2
  - Actors: An immigrant woman (patient), husband (relative), surgical nurse, and ward nurse
  - Plot: An immigrant woman is going to a scheduled hip operation. She will have general anesthesia during the procedure. Her husband comes with her along with the nurse from the ward. She speaks poor Norwegian, and her husband must therefore be there to translate. She is concerned with keeping her hijab on and wishes to be treated by female personnel only.
- Learning scenario 3
  - Actors: A young boy (patient), his mother (relative), surgical nurse, and ward nurse

- Plot: A five-year-old boy arrives to the sluice with his mother and nurse from the ward. He is going to recto- and gastroscopy.
- Learning scenario 4
  - Actors: A man (patient), surgical nurse and ward nurse
  - Plot: A man in the age of 40 is going to surgery due to a malignant brain tumor. He has two teenage children that he alone is responsible for back home.

4   Requirements, Design, and Implementation

4.1   Requirements for the Virtual Operating Room and Avatars

The requirements for the virtual operation room were acquired from the Department of Neurosurgery at St. Olav, including photographs and textual description. However, no formal set of requirements was made, therefore, we employed Scrum methodology for the development. We divided the requirements into two parts: the virtual environment and the avatars.

The requirements to the virtual environment describe three rooms, their size, structure and artifacts to fill them (such as equipment and furniture). According to the final requirements, it consists of a hallway leading to three rooms, described below:

- *Waiting room* is an ordinary waiting room that you find within all ordinary hospital clinics, consisting of a reception desk, sitting chairs for patients and relatives, and a table with magazines and papers.
- *Sluice* in this context is a room that health personnel use for the delivery of patients on their way to an operation.
- *Operating room* is a place where the surgeries are conducted. The room is usually equipped with operation lamps, different medical equipment, and an operating table for the patient.

The avatars are divided into three types: patients, relatives, and nurses. The avatars for the relatives had to match the description in the scenarios, e.g. a 'mother' or a 'person with immigrant background'. The patient and the nurse avatars had more details, and they should have been dressed in accordance with the standards adopted at the Norwegian hospitals.

4.2   Design and Implementation

The process of designing and implementing the virtual environment for conducting role-plays was performed in two iterations with feedback sessions in between. The first one was based on the initial requirements received from St. Olav's hospital. The second iteration was based on the feedback from the Faculty of Nursing teachers which generated new requirements and suggestions for improvements to be implemented. We were, however, not able to implement all the requirements (e.g., certain artifacts/equipment) from the initial set, and there were refinements that could have

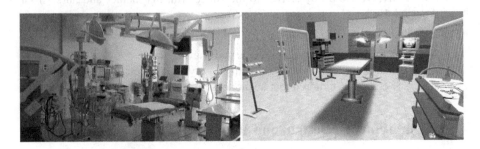

Fig. 1. Real life and Second Life operating room

Fig. 2. Nursing avatars

been done in more details (e.g., images and instruments). The goal of this process was to create a virtual environment (Fig. 1, right) that is realistic enough to give a feeling of being in a real operating room at the Department of Neurosurgery at St. Olav's hospital (Fig. 1, left). Since the environment was meant for conducting role-play with the focus on communication (not on interaction with virtual artifacts), the operating room was not evaluated (e.g., considering the functionality of the medical equipment).

We designed and implemented the building, its rooms, and some details from scratch, but most of the more complex artifacts, such as the operating table, anesthesia machine, and other similar equipment were purchased at the SL marketplace. The negative side of purchasing items created by others is that some minor conflicting details may follow. For instance, the patient bed on wheel was supposed to be used to animate the transportation of patients from the ward or sluice to the operating room. We purchased the only bed with that function available at that time on the SL market-place. However, it was an 'emergency bed', with an integrated animation of blood infusion into the patient. This last detail would not be correct in the situations given in the scenarios, and the students were instructed to ignore such details.

Nine avatars were required according to scenarios. Customization of the avatar appearance to fit some of the more detailed role descriptions was a time-consuming task. We created 11 avatars, including two spare ones. The resultant nursing avatars had pale green or white clothing where the former is used for the surgical nurses and the latter for the ward nurses (Fig. 2). SL does not have default avatars matching the description of some of the patients and relatives, such as a young boy and a Muslim person. Therefore, we purchased 'skins' and related accessories (e.g., a hijab and jewelry) at the SL marketplace. All patient avatars were required to wear a plain patient gown with a front opening. We purchased a plain white open shirt for the adult patients and a gown with a back opening for the child.

5   Study Settings and Results

5.1   Role-Playing Settings

After completing the virtual environment and a set of avatars, we recruited nine post-graduate nursing students from the Faculty of Nursing at HiST, who were on their last year to become surgical nurses. These students already had a bachelor degree in nursing and at least two years of professional experience in addition. In other words, they were well experienced and should already have acquired skills given by the learning objectives (section 3). Therefore, as we could not expect them to learn much from the role-playing, we used their knowledge and experience to evaluate the teaching method and the environment developed. Instead of assessing the learning improvement, we rather asked them to evaluate the simulation and provide feedback on how well it might be suited in the nursing education at an earlier stage.

Low computer competence and little experience with 3D VWs were expected for such a group. In order to address this, a tutorial on the gameplay, such as camera controls and avatar navigation, and a one-hour practice session were conducted before the role-playing session. In order to accommodate role-playing with four players, four fully equipped computers were set up in two different offices at our university (Fig. 3). The third location with a computer, a large screen, and speakers was prepared for the teachers and those students not playing to observe the role-play.

Fig. 3. Surgical nursing student participates in the role-play (photo by Anne Midling)

In the beginning of the role-playing session, the students chose one of the roles from each scenario and received "role-play cards" describing their characters. The cards contained information about name, role description, and a description of the situation. In all scenarios, the students were told to improvise their role as best as they could using their knowledge and earlier experience. The teachers and other students were observing the role-play and discussing questions that have been emerging during the play. After completing a scenario, its players joined this group to have a discussion before going to the next scenario.

5.2   Data Collection

The data in this study was collected from several sources. The role-play in SL was recorded as a screen capture (with sound), while the subsequent discussions were recorded with written notes and sound capture. In addition, a questionnaire consisting of 28 questions was given the students after the role-playing session. It included multiple-choice questions using a five-level Likert scale, 'check-box' questions allowing to select multiple answers from a list of options and open questions.

The questionnaire was divided into four main topics. The first one covered the competences of the participants in use of computers and their previous experience with SL or similar VWs. The second topic included questions about the use of SL during the role-play, including the process of getting used to the navigation, realism of the environment, and game experience. The third topic contained more subject-specific questions on believability of the simulation and its suitability for providing knowledge and skills described in the learning objectives. The fourth topic contained open questions where the students were asked to input proposals for changes and improvements, describe what learning outcomes a student may gain from the simulation, and suggest other areas within medicine they thought could benefit from it.

The group discussion was conducted to supplement the questionnaire and let the participants express ideas immediately after the role-playing. We used five questions to engage the students in a discussion. However, several other questions appeared too. The major topics were the general impression, possible educational value, level of engagement, alternative solutions, and application domains within medicine.

5.3   General Issues

The teachers reported after the role-playing session that not all the students were prepared for the settings. We observed some hesitations every time we asked who wants to go next in playing a scenario. Some of the nurses said that they did not feel that comfortable role-playing in the VW when they knew the other students and the teachers watched the play on a large screen. "One should like to role play, or else the role will become limited", a student mentioned in the questionnaire. One student asked if it was possible to identify the person behind an avatar. Even though role playing while being remotely observed by others made some of them uncomfortable, they all tried to play at least once and they came through all the scenarios. It was also observed that the role-playing became smoother for each new scenario as they started to get hold of the navigation in SL and more immersed into acting. This trend was also reflected in the open questions of the questionnaire were one student answered: "The role plays got better eventually. Got more comfortable after some practice".

5.4   Previous Experience and Technical Issues

The questionnaire reveals that none of the students had any previous experience with SL or other VWs and/or games. Only three out of nine also describe themselves as having good or better computer competence. This is likely to be the reason to why some of the participants were not so steady when moving around and interacting with objects in the virtual world. Six of them answered "neutral" on the question regarding the difficulty in moving the avatar to different places. This may be interpreted that it has been easy to move the avatar as we did not observe any issues there, but challenges came when they were to move the avatar while interacting with objects, for instance, operating the rolling patient bed.

Half of the participants agreed that it went quickly to learn the interface of SL, while the other half answered neutral to this question. We observed that fewer questions were asked as we proceeded, and the participants even started to inform and teach one another of how to manage different SL controls. The students experienced some echo inside the game while playing. This was most likely due to having two computers with microphones located in the same room with a distance not greater than three meters from each other.

Even though the students got one hour of training before conducting the role play, the avatar and camera movement inside the game were still challenging. The majority of the students expressed (both during the role play and during the discussion afterwards) that it was difficult to focus both on role-playing and movement simultaneously. This was also observed, for instance, when one of them lost control over the patient bed, which turned from side to side when trying to reach the operating room. The student tried to get control of the bed in the game while at the same time trying to role-play. It often resulted in laughter from the audience, which appeared interrupting.

5.5   Collaboration

We observed that when the immersion and acting part became better, the collaboration between the participants in SL improved as well. The reason was probably that while the students became more confident in playing 'difficult' patients or relatives, the students playing the nurses had to work harder to get things in order. For example in Scenario 2, the Muslim woman refused to take off her hijab, and the two nurses had to talk their way around to get the patient and her husband understand the rules. Another example is from Scenario 3, where the 'boy' resisted a while before positioning his avatar on the bed. His mother, the ward, and the surgical nurse together tried to convince the anxious boy to cooperate and to calm him down.

At some point in every scenario, the role playing started to halt because of a missing part. It was explained that surgical nurses collaborate most often with anesthesia nurses at the operating room. We observed during the role play that anesthesia nurses/doctors were frequently mentioned in the dialogue between the surgical nurses and the patients. The students and teachers mentioned during the discussion that it would have been natural for an anesthesia nurse to take over some of the interaction, as the surgical nurse could not proceed with their tasks without them doing their job. Therefore, it was concluded that anesthesia nurses should be included in future role-plays. This is implemented at the subsequent evaluation as mentioned in the Future work.

5.6   Evaluation of the Environment and the Play

Even though role-playing is not currently used as a tool in their study program, all students except one reported that they felt engaged in the virtual role-plays and the same people found the experience fun and motivating.

The majority of the students were neutral to the question about how realistic or representative the virtual environment in SL was in comparison to St. Olav. Seven students answered that it was difficult to read and interpret information from the body language and facial expressions of the avatars. The ability to read nonverbal cues or signs such as these is an important aspect of being a nurse [14]. The avatar appearance was easier to interpret, as four of them agreed.

We asked the students about what elements they thought gave the most and least information in the simulation. The most informative elements included room design/environment, avatar clothing and their positioning, while the least informative elements were sound and body language. Six of the students agreed that the simulation in general gave them enough information so that they were able to understand the patients, relatives, and their situation. The rest were neutral on this question.

5.7   Value of Using Virtual Simulation

Evaluating the educational value of the simulation, the participants used what they learnt at the college as well as their professional experience. We gave them nine questions asking how well they thought the simulation would contribute to enhancing skills in collaboration and communication with relatives, other health personnel, and different types of patients.

The positive choices on the Likert scale were more frequent. Therefore, we assume that using a simulated environment may be a positive supplement to the education of practical nursing skills. Eight out of nine students answered positively to the question if the use of role-play in a VW can be a supplement to help the surgical nurse students in communicating with patients and relatives.

We listed six other medical areas that might be suitable for practicing inside a VW and asked the students to choose the areas they believed to be most relevant. Four most popular areas selected by seven or more students were collaboration in teams in an operation room or emergency department, procedural training, anatomical visualization, and diagnostic training. Other popular areas included education of patients and relatives to improve understanding of the treatment procedures and disseminating health-related information to the public in general.

6   Discussion

6.1   Environment Design

The fact that only three of nine students agreed that the virtual environment was realistic and representative of an operating room in real life could be partly attributed to the fact that not all of them have been to the real Neurosurgery operating room that was replicated in this study. From the questions regarding "potential for improvements", some of them suggested that we should have "more equipment to interact with". One student mentioned during the discussion that "more items to interact with needs to be included if surgical nurses are to treat and communicate with a patient". This was explained by that the surgical nurses often communicate their actions to their patients in order to inform the patient and to keep the patient's thoughts occupied. The lack of interactive items may be an explanation to why most of the students did not find the virtual environment realistic. Such an item could for instance be a blanket to put over the patient, or other items such as suited clothes for the relatives to put on if they were to follow the patient and the nurse to the operating room. However, tasks such as changing avatar clothing during the role-play would require additional training, and therefore, these details were skipped.

The replies to the question on what aspects gave most information show, however, that six of the students did use information given from the "surroundings" which is also reflected in the recordings of the role plays. For example, when one of the surgical nurses tells the patient's husband to step away from the instrument table when he gets too close, as these instruments are sterile and prepared for surgery (Fig. 4). Another example is the nurse asking the same person not to be in the way, but stand beside the "machine with the screen" located at a distance from the operating table.

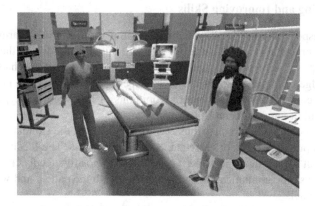

Fig. 4. Role-playing, Scenario 2

Another factor contributing to inhibiting the feeling of realism amongst the students might be the lack of experience working with VWs. All nine students answered in the survey that they had no earlier experience with either SL or other VWs. This lack of experience might have made it difficult to feel immersed inside a VW.

6.2   Using the Technology

There are at least three aspects to look at when considering simulation in a VW as a suitable tool in the education of nurses. The first being how much time and effort it takes for both the students and the teachers to gain enough technical knowledge before it becomes useful to practice inside a VW. Our students had no experience with SL or VWs in general, and they got only one hour of training before they were set to role-play the scenarios. The evaluation reveals that is was challenging for the students to focus on their actual task when they simultaneously had to interact with a few set of objects (such as rolling the patient bed and using the operating table). Observing the role play and the recordings afterwards also reveals several disruptions caused by object handling in game. For instance, some students did lose control over the rolling bed, or those playing patients often misclicked and sat down at other objects rather than lying down at the operating table as supposed to. This often led to some laughter amongst the students, and the need for some help before continuing role-playing.

Even though the study was exploring communication with patients, the role play became to a certain extent challenging considering the number of objects the participants had to handle. One hour with tutorial would not have been enough if the students were to interact more and still conduct role-playing fluently. The students from our discussion remarked themselves that 'it does take a while to learn', however, another uttered that 'a day more with role-playing, and they would be skilled'. A third one said that 'it was difficult to role-play when you had to concentrate on the technical part such as navigation, interaction and camera movement'. They all agreed on that a larger quantity of training would have been necessary in order to forget the technical aspect completely and focus on their actual task during the role-play.

6.3   Learning and Improving Skills

The second aspect to consider is whether the simulation supports learning and improvement of practical skills. We did not measure this directly during the evaluation, as our participants were postgraduate students on their last year and had both sufficient knowledge of the procedures and work experience. However, we engaged them as evaluators contributing their subjective opinions of how well this type of simulation would benefit nurse students at an earlier stage.

The survey contained nine questions regarding how well the simulation would improve various practical skills of early-stage nursing students. The feedback was positive overall, but we divided it to three types. The most positive feedback was received in two questions on communication and interaction with the patient and with the other medical personnel.

In five other questions, the feedback was also positive, but with one or two neutral answers. Such moderately positive feedback was given to the educational value of the simulation in calming down the patient, communicating and interacting with the patient from a different culture, the child patient, the relatives of the patient, and relatives of the child patient.

Two answers to the remaining two questions were also positive, but had more neutral and some negative answers. These questions were evaluating the usefulness of the simulation for training communication and interaction with the patient who needs an interpreter and the patient who is visibly under stress.

During the discussion, several aspects were appreciated and many suggestions were made by the participants. The students reported that it would be much more appreciated if they could get some feedback during the role-plays from either a teacher or a more experienced nurse. This could for instance be through an avatar standing in the corner and contributing with inputs, functioning as an expert guide. Another suggestion made was to let teachers play the patients, as they often have more experience and they would be more suitable to play them, as it was done e.g., in [11]. We also observed that it was easier for participants to put themselves into the role of a young boy than of the Muslim woman and her husband. However, the results from the two questions regarding their opinion of how well such simulation would improve the skills in communication and interaction were the same for the two patient types.

The students gave several suggestions for improving the scenarios. They agreed that more strict guidelines for the role-plays would be helpful, as the scenarios were quite open only providing the students with some background information before they were asked to improvise. One of the student mentioned "the scenarios would have been too difficult if you do not have any real experience, one would not know how to proceed". This is important feedback when it comes to using the simulation with students at earlier stages in their studies later.

The last suggestion was to integrate different scenarios in a single story with game elements. In such a game, completing one scenario would allow a team of participants to advance to the next level, i.e. to the next scenario.

6.4   Motivation, Engagement, and Fun

The third aspect concerns whether the role-playing activity was motivating, engaging, and fun. This is important if simulation is to be accepted by the students as a part of their educational program. As mentioned earlier, laughter was frequently recorded during role-play, especially when the participants made small mistakes, such as crashing the rolling bed with the patient or making their avatar sit at odd places. This was indeed a new experience to them, but we observed that many were eager to immerse themselves in the role-play when they started to be more comfortable with it. Eight out of nine answered either "agree" or "strongly agree" to the two questions "if the game experience was fun/motivating" and "did you feel immersed in the role-play?". The fun factor is, however, not enough to make the students want to use the simulation in their study program, as they need to have a stronger reason and motivation.

An important point was raised when discussing changes needed in order to increase the educational value for the surgical nurse students. Surgical nursing is a practical profession and the procedure needed to prepare a patient for surgery is relatively complex. Many of the students stated that it would be natural to do more practical things in the virtual operating room, such as washing hands, covering patients with blankets, position the operating table, disinfect, prepare instruments and similar tasks and routines, which is an important part of their job when dealing with a patient. As one of the students puts it, "use our hands more". Most of our discussion centered on how we can improve the virtual operating room to make empathy and interaction to feel more like in real life. Therefore, it appears that some of the motivation amongst the students also lies in the possibility to practice on medical procedures as a surgical nurse and working on associated tasks in the VW to gain early experience.

7   Conclusion and Future Work

This study gave valuable information that can be used as guidelines for further development of virtual operating facilities and, on the longer term, of a Virtual Hospital. There is also a need for further exploration of the different ways to support collaborative team training and practical medical training with innovative technologies. The role-playing facilities in a virtual operating room have the potentials to provide the students with a safe, realistic, and accessible environment for practicing their nursing tasks. In order to achieve this goal, the virtual operating room should be equipped with interactive objects related to the tasks and routines of a surgical nurse. This could be simple tasks, such as putting a blanket over the patient or preparing instruments before surgery. Surgical nurses use their hands a lot, and the students should be able to do this within the VW in order to immerse themselves properly into the role. It also follows that interactivity in general, such as navigation, object manipulation and interaction with co-players should be improved and made more intuitive.

Another implication of the study is the need for guidelines and methods for scenario development, including definition of roles. For example, a student might not have experience with certain types of patients and certain medical conditions. Therefore, he/she would not be able to give the correct feedback to co-players. This would require formalized guidelines for how to play this type of patient, and probably some more information and preparation for the student in advance of the role play. A related aspect is interaction between different actors in the nursing team. An efficient learning experience requires receiving cues and other sorts of feedback frequently during the role play. Our participating students got these cues mostly from the dialogue with each other as nurses and patients, though they participated in a reflection round afterwards. The students mostly agreed that this should be improved for the role play to have an educational value. Types of feedback suggested were an avatar guide/game leader played by an experienced person who can join them and give inputs along the way. Another suggested solution was to let a more experienced nurse or teacher play the patient, as well as to implement tasks to be done where you either fail and stay put or complete to advance to the next level with a new set of tasks.

The major limitation of the study is that for practical reasons we were able to recruit only nine postgraduate students at their last year of studies. The small number of participants does not give statistically significant data, but, at the same time, they provide some useful indications for further development of the system. For example, all but one of the participants have been positive towards the use of role-play in a VW as a supplement to surgical nurse training in communicating with patients and relatives. Therefore, despite the limitations of this study, the results provide a motivation for further development of the Virtual University Hospital a basis for further elaboration of design principles for a collaborative virtual training environment.

The ongoing second stage of evaluation is based on our experience and feedback from this study. Both anesthesia and surgical nursing students at their first year were recruited to role play together in the Virtual Operating room at this stage. We attempted to make the gaming experience more intuitive and engaging and enhance the feeling of immersion by using head mounted displays (Oculus Rift). The teachers acted as patients, providing correct feedback and giving the students the possibility to focus on their roles as anesthesia or surgical nurses, teamwork and interaction with the patients. We have also conducted an evaluation among non-medics, as the operating room could potentially be used for informing general public and preparing patients for a surgery. Apart from evaluating their experience in the operating room, both user groups, medics and non-medics, have been asked to suggest services and features to be included in the Virtual University Hospital.

We are currently working on analyzing the motivation, degree of immersion, and learning experience amongst the participants at the second evaluation stage, comparing the findings with the results from the study presented in this paper and outlining requirements for the Virtual University hospital as an arena for health education.

Future evaluations will include a larger group of students, but also different scenarios and groups of patients and medical professionals. One example is language and cultural awareness training for nurses from other countries (e.g., Philippines and Eastern and Southern Europe) recruited to work in Norway. We will continue developing the Virtual University Hospital, both conceptually and technologically. In the longer run, such an environment will be enhanced with other features and facilities for collaborative work and learning, such as library of medical resources, anatomical visualizations, meeting facilities, patient information facilities and so on, normally present in a real university hospital.

Acknowledgements. We would like to thank the students from the Faculty of Nursing at HiST who participated in the evaluation.

References

1. Spooner, N.A., Cregan, P.C., Khadra, M.: Second Life for Medical Education. eLearn Magazine. ACM, New York (2011)
2. de Freitas, S., Rebolledo-Mendez, G., Liarokapis, F., Magoulas, G., Poulovassilis, A.: Developing an Evaluation Methodology for Immersive Learning Experiences in a Virtual World. In: 1st International Conference in Games and Virtual Worlds for Serious Applications (VS-GAMES), Coventry, UK, March 23-24, pp. 43–50. IEEE, New York (2009)
3. Warburton, S.: Second Life in higher education: Assessing the potential for and the barriers to deploying virtual worlds in learning and teaching. British Journal of Educational Technology 40(3), 414–426 (2009)

4. Mckerlich, R., Riis, M., Anderson, T., Eastman, B.: Student Perceptions of Teaching Presence, Social Presence, and Cognitive Presence in a Virtual World. Journal of Online Learning and Teaching 7(3), 324–336 (2011)
5. Lee, M.J.W.: How Can 3d Virtual Worlds Be Used To Support Collaborative Learning? An Analysis Of Cases From The Literature. Society 5(1), 149–158 (2009)
6. Hayes, E.R.: Situated Learning in Virtual Worlds: The Learning Ecology of Second Life. In: American Educational Research Association Conference, pp. 154–159. AERA (2006)
7. Jarmon, L., Traphagan, T., Mayrath, M.: Understanding project-based learning in Second Life with a pedagogy, training, and assessment trio. Educational Media International 45(3), 157–176 (2008)
8. Johnson, C.M., Vorderstrasse, A.A., Shaw, R.: Virtual Worlds in Health Care Higher Education. Journal of Virtual Worlds Research 2(2), 3–12 (2009)
9. Rogers, L.: Simulating clinical experience: Exploring Second Life as a learning tool for nurse education. In: Atkinson, R.J., McBeath, C. (eds.) 26th Annual Ascilite International Conference Same Places, Different Spaces, Auckland, New Zealand, December 6-9, pp. 883–887 (2009)
10. Lowes, S., Hamilton, G., Hochstetler, V., Paek, S.: Teaching Communication Skills to Medical Students in a Virtual World. Journal of Interactive Technology and Pedagogy (3) (2013)
11. Wiecha, J., Heyden, R., Sternthal, E., Merialdi, M.: Learning in a Virtual World: Experience With Using Second Life for Medical Education. Journal of Medical Internet Research 12(1), e1 (2010)
12. Rizzo, A.S., Difede, J., Rothbaum, B.O., Reger, G., Spitalnick, J., Cukor, J., McLay, R.: Development and early evaluation of the Virtual Iraq/Afghanistan exposure therapy system for combat-related PTSD. Annals of the New York Academy of Sciences 1208(1), 114–125 (2010)
13. Jang, S., Black, J.B., Jyung, R.W.: Embodied Cognition and Virtual Reality in Learning to Visualize Anatomy. In: Ohlsson, S., Catrambone, R. (eds.) 32nd Annual Conference of the Cognitive Science Society, Portland, OR, August 12-14, pp. 2326–2331. Cognitive Science Society (2010)
14. Wright, R.: Effective Communication Skills for the 'Caring' Nurse, Tertiary Place, pp. 1–3. Pearson Education, Upper Saddle River (2012),
    http://www.pearsonlongman.com/tertiaryplace/pdf/
    ros_wright_effective_comm_skills_for_the_caring_
    nurse_aug2012.pdf

JEMF: A Framework for the Development of Mobile Systems for Emergency Management

Marcus F.T. Machado, Bruno S. Nascimento,
Adriana S. Vivacqua, and Marcos R.S. Borges

Graduate Program on Informatics (PPGI), Institute of Mathematics (IM),
Federal University of Rio de Janeiro (UFRJ),
Rio de Janeiro – RJ – Brazil
{marcus.machado,bruno.nascimento)@ppgi.ufrj.br,
{avivacqua,mborges}@dcc.ufrj.br

Abstract. In recent years, Emergency Management has become the target of multiple research efforts. This domain is characterized by collaborative aspects and several researchers discuss the use of mobile applications to complement traditional forms of information sharing. However, there is little software reuse in proposed solutions, as very few projects have focused on developing reusable components. Reusing components is desirable for it harnesses the strengths of existing systems, speeds up development and increases system reliability. In this paper, we present a framework to support software reuse in mobile systems development, based on specifications from the emergency domain. It provides guidance for the construction of new systems and simplifies their development, as required by the developers.

Keywords: Emergency Management, Software Reuse, Mobile Devices.

1 Introduction

Emergencies happen with a certain frequency and often result in material and human losses. In 2011, 332 natural disasters killed 30,773 people and caused the largest economic damage registered to date: around U$ 366,100 billion [18]. In view of these occurrences, many countries strive to put national Emergency Management (EM) plans in place, to address flaws in existing response strategies. The conduction of studies and development of new technologies are an essential part of these efforts.

EM is characterized by its collaborative aspect: it involves the work of different people and organizations involved in responding to crisis situations. During disaster response, there should be constant information exchange between the command team, responsible for planning and managing response tasks, and the operations team, responsible for the execution of said tasks, such as victim rescue and firefighting. Generally, the device most widely used for this purpose is UHF radio. However, radio channels typically are insufficient (2 channels) to support the response processes and can get congested. This limitation becomes evident when the command team has to coordinate simultaneous emergencies. Usually, the main channel must be shared

N. Baloian et al. (Eds.): CRIWG 2014, LNCS 8658, pp. 239–254, 2014.
© Springer International Publishing Switzerland 2014

among all participants and this may delay information sending/receiving. When channels are busy, the information sources may get mixed up, confusing messages. The sheer volume of information presents a challenge to individuals handling it. These delays sometimes force teams to improvise their response actions, because they are not able to wait for further information. In addition, text messages, maps, photos and videos are relevant information that helps with the understanding, organization and coordination of the emergency response process. However, there is no possibility of their delivery through radio systems [8].

These characteristics highlight the complexity of communication, collaboration, information management and decision making in critical situations. EM information systems have been designed to help these teams through the integration of modern technology such as mobile and wearable devices. Several recent studies discuss mobile software solutions and experiments, in light of their resources and capabilities. As these software solutions become more complex, greater importance is placed on software development techniques to facilitate their production. Software reuse is one of the main goals pursued by developers and it is essential to simplify software development in the EM domain. The problem this paper seeks to address is the need for better support for new systems development, and to better take advantage of advanced resources offered by modern devices. We present the Java Emergency Management Framework (JEMF) to contribute with the construction of mobile systems through software reuse to support emergency management.

This paper is structured as follows. Section 2 presents a literature review with a set of EM information systems developed few years ago and classifies these EM information systems based on functionalities, technological and collaboration aspects. Section 3 presents the main specifications of the framework. Section 4 discusses related work and concludes the paper.

2   Literature Review

2.1   Emergency Management Information Systems

In the past decade, the necessity of providing technological devices for emergency teams to improve their access to information has become evident. Among other things, the use of technology helps with the development of a common understanding, among the people involved, of the tasks being undertaken, which makes emergency response more efficient. Several EM projects emerged in order to support collaboration, decision making and increase situational awareness of response teams. Among the projects discussed in the literature, some have adopted the use of mobile computing. We discuss some of these mobile projects below.

NOAH [20] is a system that supports physicians in the initial patient data documentation at emergency site, such as vital signs and injuries. The system connects the ambulance and physicians with the command office and hospital. In this way, an electronic record of the patient is processed and transmitted from emergency site to hospital, improving patient prognosis. This causes a reduction in radio communication and in errors during victim transfer. Furthermore, the hospital may be selected based on their availability of beds and equipment.

DUMBONET [11] combines satellite network, wireless ad hoc network and conventional internet to provide its users coordination, communication and information management tools where no network infrastructure is available. The system offers a real-time multimedia information exchange in search and rescue operations on areas affected by disasters. It enables connecting many affected areas and the command office via satellite links. Each of these areas has its own mobile communication network between firefighters and each mobile unit can be a laptop or a Personal Digital Assistant (PDA).

DMT [1] is a system to assist reconnaissance missions during the emergency response phase that offers a set of features for task monitoring in subsequent phases as well, such as the coordination of multiple rescue teams at the disaster site. DMT is basically composed of (i) a simulation module, establishing functions for possible damage and casualty estimation; (ii) an analysis module that involves the damage, tasks and resources allocation obtained from the disaster site; (iii) a communication module for the integration of firefighters in managing information between them.

CCCMS [12] helps coordinate response in emergency situations after a car accident. Through the system, it is possible to dispatch the human resources to the crash site, and reassess the response strategy when new information about the incident is captured. Therefore, CCCMS supports the response process by orchestrating the communication and coordination efficiently between all parties: it centralizes all information related to the accident. This increases the flow of shared information, facilitates task management and allows a faster transportation of victims to hospitals.

MIKoBOS [14] is an emergency response system to extend the communication of the fire department for the mobile domain. It allows field personnel to take orders from commanders and access information that would otherwise only be available to them through the exchange of voice messages via radio transmissions. These agents are allocated to response tasks and can interact electronically with colleagues, superiors or experts to share sensor data, status reports or equipment available. Additionally, victim's information may be transferred to hospitals.

The emergency management mobile system GeoBIPS [13] allows firefighters collect dynamic data from the incident scene and combine it with static data stored on local or remote servers, providing quick and easy access to relevant information. The system is used by the operations team to share data, and the incident commander is responsible for coordinating response tasks. Moreover, this system enables the transmission of updated information to the crisis center for decision-making and receives other useful information, such as operational plans and instructions, from this center.

EMERGENCY project [10] aims to explore the allocation of resources through a map-based interface to support the decisions of the Police incident commander. Tasks, policemen and their context, in other words, the human resource management during emergencies can be interpreted more intuitively. This system allows the commander to maintain an overview of the crisis response process by presenting map-based data, containing symbols and metaphors instead of just text. Thus, the manipulation of resources at the scene becomes easier and the information is understood quickly, requiring less concentration.

SisC2Celular [16] is an information system designed to manage the process of emergency response based on command and control systems. The mobile phone is used as a platform to capture, transmit and receive information between the command and operation teams, to achieve common understanding and better situational awareness faster. Decisions made by the command officer are sent to the field response teams through operational tasks. Following these decisions, these firefighters forward requests and reports to the command center, simplifying information sharing.

The WORKPAD project [6] provides a communication infrastructure to support the firefighters in emergency situations. This system plays a key role in work team's objectives achievement, with a focus on task management, support decision making and planning, and an easy user interface for mobile devices. Furthermore, WORKPAD allows users to have an overview of the affected area and viewing points of interest displayed on a map, facilitating the execution of planned tasks and efficient monitoring by the crisis managers.

ProRad [21] is a radiological responder support system that uses a simplified interface and specific data sources in the mobile device, providing information, calculating and classifying risk categories. Additionally, it enables the recording of important information and provides overviews of radiological emergency situation in short amounts of time. ProRad supports radiological responder's tasks by facilitating the investigation of the disaster current situation, for example, measure an isolation area tailored to the disaster severity. Consequently, reduces agent exposure to radiation.

MobileMap [8] is a collaborative mobile system that uses GPS data to locate the emergency site, and the human and material resources available. This system provides interaction between the firefighters and command center, in order to retrieve and report information about an incident. It provides communication and data sharing about the threats and affected area, the firefighting coordination and the collection and recording of field information. MobileMap improves firefighters' learning process with the results obtained in previous events, so they'll make better decisions and feel more confident.

Araujo developed a system for managing information crisis that supports collaboration and decision making between the operations team and incident commander, named Sis.Emergência [2]. This system aims to record the main information related to the firefighters involved in the response process, as rescuers relevant data, their specialty and tasks, and the victims data found at the crisis scene. Also, it is possible to exchange messages between the firefighters, take photos and view main information collected in digital maps, conducting more efficiently the emergency response and enabling a more detailed assessment by incident managers.

2.2 Systems Overview

Some of these approaches have been created to reduce information sharing problems found in radio communications at the command post office, the operation team or between these two groups. Nilsson et. al. [15] identified common needs among the actors involved in midrange emergency response operations, characterizing them into

11 types of information systems functionalities that support different actor's tasks, such as firefighters, police officers and physicians. We performed an analysis of the functionalities provided by each of the projects described in the previous section, classifying them into the following categories. A summary can be seen in Table 1.

1. Operational picture: maintains information about the emergency and visualization of response situation.
2. Resource management: supports the management of personnel and equipment available.
3. Actions & plans: supports the planning and execution of tasks.
4. Monitoring: keeps health status record of responders or victims.
5. Communication: supports responders in the use of information sharing mechanisms.
6. Information services: supports access to useful information supply services, such as reports and documents.
7. Transmission: supports image transfer among responders.
8. Special interaction: supports with different interactions through unusual interfaces, such as augmented reality.
9. Incident details: maintains information about affected organizations and people.
10. Automatic reasoning: supports the automatic analysis and information retrieval of the tasks coverage area.
11. Logging: keeps track of events or actions performed by responders.

We can also note that the adopted technology evolved over time (Table 1). Nowadays, mobile computing offers a number of features such as microphone, multitouch screen, camera, GPS, which can facilitate the capture and exchange of information between people. Some of the aforementioned projects adopt these resources.

In addition, we analyzed these systems using the based on the 3C Collaboration Model [4]. This model references the collaboration within systems. It divides collaboration into 3 aspects: communication, coordination and cooperation, as seen in Fig. 1 (a). When working in groups, individuals communicate through messages to negotiate and make decisions. During coordination, group members deal with conflicts and organize tasks and resources, trying to avoid wasting communication and cooperation efforts. While they work together to achieve a common goal, renegotiation is needed to handle unforeseen situations, and demand communication and coordination to rearrange tasks [17]. The EM domain is characterized by collaborative work and every aforementioned project emphasizes a combination of the 3C aspects. We present a synthesized classification of these systems in relation to the 3C Collaboration Model aspects in Fig. 1 (b).

The analysis of these projects provides a solid basis for the identification of requirements for systems to support crisis management. These solutions define a set of specifications and present the system description through use case and class diagrams and requirements lists.

Table 1. EM projects classification according to functionality system categories [15] and characteristics

Functionality	Project											
	NOAH	MIKo-BOS	Geo-BIPS	DUM-BONET	DMT	WORK-PAD	SisC2-Celular	CCCMS	Mobile-Map	EMER-GENCY	Sis.Emergência	ProRad
Operational picture	Yes	Yes	Yes		Yes		Yes		Yes	Yes	Yes	Yes
Resource management			Yes			Yes	Yes	Yes	Yes	Yes	Yes	
Actions & plans			Yes			Yes	Yes	Yes				Yes
Monitoring			Yes						Yes			Yes
Communication	Yes	Yes	Yes	Yes		Yes	Yes		Yes		Yes	
Information services	Yes	Yes			Yes	Yes						Yes
Transmission	Yes	Yes	Yes	Yes					Yes		Yes	
Special interaction								Yes				
Incident details	Yes	Yes	Yes	Yes	Yes	Yes	Yes		Yes		Yes	Yes
Automatic reasoning			Yes	Yes	Yes	Yes	Yes		Yes	Yes	Yes	Yes
Logging									Yes			
Characteristic												
Year	2000	2006	2006	2007	2009	2009	2010	2010	2011	2011	2012	2012
Country	Germany	Germany	Belgium	Thailand	Germany	Italy	Brazil	Canada	Chile	Norway	Brazil	Brazil
Device	Portable Computer, PDA	PDA	Portable Computer, PDA	Portable Computer, PDA	Portable Computer	PDA	Cell Phone	PDA	PDA	Smart-Phone, Tablet	Smart-Phone, Tablet	Smart-Phone, Tablet
Platform	Windows	Windows	Windows	Windows	Windows	Windows	Windows Java ME	Windows	Windows	Android	Android	Android

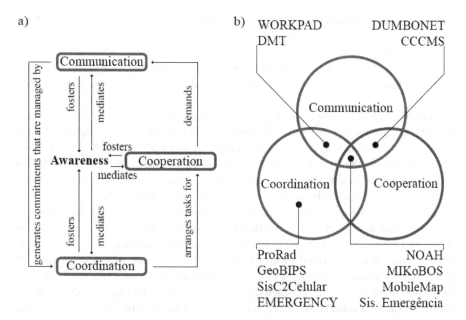

Fig. 1. 3C Collaboration Model instantiated for group work (a) [adapted from 17] and EM projects classification according to 3C Collaboration Model (b)

However the projects details, the development process is usually not fully described or even mentioned, making them hard to replicate. There is little on how solutions should be implemented. Moreover, these approaches usually do not consider the reuse of common specifications or other prototypes from the same domain. Some projects discuss how to overcome systems issues, such as heterogeneous network problems, but do not discuss specifications and software design similarities with other projects [10]. According to [2], one of the main challenges in software development is to define requirements and functionalities that are useful and easy to handle during the use of new devices in the EM domain. Project design and software development to meet all specifications are a challenge that needs to be met.

3   JEMF: A Software Framework for Emergency Management

3.1   Software Reuse

A framework is usually composed of a set of classes that implements an abstract design that addresses a family of related problems [9]. It allows the reuse of both code and design and contains a set of cooperating classes that represent objects of a particular domain. The design of a system is usually described in terms of components, how they interact and collaborate with each other, the responsibilities each one has, and their control logic and information flow. A framework defines these elements so as to make it easy to use individual components, following the full structure of relationships, patterns and design decisions identified in an application domain. When a framework is developed, the generic aspects of the domain are captured and different applications can be developed to make use of those concepts present in the framework, thus saving time and effort. Only application specific aspects need to be modeled, since the generic requirements have already been modeled and implemented.

According to [5], benefits of using a framework include modularity, reusability and extensibility. Modularity happens through encapsulation of volatile implementations behind stable interfaces. It improves the quality of software due to the localized impact of changes in the design or implementation. This reduces the effort required for understanding and maintaining existing software. These stable interfaces increase reusability through the definition of generic components that can be reused to create new applications. This takes advantage of the domain knowledge and effort employed by experienced developers when developing the framework, and avoids the need to rebuild and revalidate common applications requirements and recurring design solutions. Extensibility is provided by an adaptive architecture that allows applications to specialize their classes and aggregate new features and services. Over time, software reuse leads to an increase in productivity and in product reliability, and contribute to cost reduction. To enable reuse, a previous step is needed, where the common features in a target domain are identified, analyzed, designed and implemented.

3.2   Framework Sub-phases

The methodology for the development of JEMF is based primarily on four sub-phases, following the literature [23]:

1. Analysis: conducting a domain requirements survey and modeling use cases.
2. Framework design: developing class and sequence diagrams.
3. Implementation and instantiation of a test application.
4. Unit and integration tests.

The analysis sub-phase comprises (i) specifications discussed in previous projects in the EM domain and (ii) specifications discussed in the literature for related areas (e.g., medical domain). Requirements were defined through an analysis of the basic aspects of development in mobile devices, as seen on next subsection. In design sub-phase, Unified Modeling Language (UML) and Object-oriented programming (OOP) are used for modeling and implementation. A framework typically uses several design patterns in its construction, which benefits future expansions. The invariant elements of a domain are implemented in the framework and reused in instantiations. The framework also reduces technical complexities, since a partially implemented architecture has already been defined and implementation details encapsulated [17]. JEMF has been developed for Android, as it is one of the most widespread mobile operating systems today. The test sub-phase is important to validate the software architecture implemented.

A framework supporting the development of mobile applications with generic specifications enables software reuse for the construction of efficient and maintainable solutions, streamlining the development process. The systematic use of frameworks for modeling, specification and implementation of systems aims to improve the software development process, ensuring less effort and higher quality of the final product. The framework should allow the artifacts of a software creation process (requirements, logical structure, code, etc.) to be reused for the development of new mobile systems. Additionally, it should enable the use of a variety of resources available on mobile devices.

Analysis and Design Sub-phases. In this paper we present a set of JEMF artifacts generated during analysis and design sub-phases. JEMF design requirements were obtained primarily from the initiatives discussed in the second section. Through analysis of previous computational solutions, we can extract a compilation of the common requirements for this domain. Specific requirements definition is left to each new project, such as usability requirements, so they will most likely vary according to different characteristics and objectives of the organizations within this domain and preferably determined closely by stakeholders.

- Communication: framework should support information sharing between the command and operation teams. Data access and collection about disasters helps organize response operations, increases collaboration and improves personnel safety.
- Persistence: framework should support for storing, updating, and accessing all data on both resolved and on-going disasters.
- Integrity: decision making depends on data integrity, which involves source reliability, credibility of strategic information and data transfer.
- Interoperability: data must be open for sharing among different devices or systems, using a common standard for different platforms.
- Availability: data should be available to everyone, everywhere and anytime, allowing operational teams to work in synchronous or asynchronous mode.

An emergency may require response actors with varying levels of authority, functions and specialties. EM information systems should include features for each profile and enable them to execute their responsibilities with the support of these systems. Emergency victims and witnesses are also present in this context, but unlike the responders, they usually may be part of the rescue task scope and they are not systems users, Fig. 2 (a). Below, we describe the characteristics of the main actors defined in JEMF.

- Commander: Actor responsible for managing the emergency response. This actor defines and assigns tasks to other users, concentrates most of the relevant information during response and performs the decision making with an overview of the current situation. Examples: incident commander, police officer.
- Responder: Actor responsible for executing response tasks. The responder receives tasks assigned to him/her by the commander and executes them. This actor also captures contextual information on site and shares it with other actors. Their main objective is to stabilize the incident scene and rescue victims. Examples: firefighter, policeman, vehicle driver, helicopter pilot, etc.
- Medical: Actor responsible for executing health care tasks to victimized people. This actor has the ability to assess and stabilize the health of people who were

injured in the incident. Examples: physician, paramedic, first aid specialist, nurse and other professionals related to health care.

- Volunteer: Actor responsible for executing response tasks with lower priority or risk. This actor supports the responder and commander in tasks that require greater amount of rescuers or may assist with specialized knowledge. Examples: humanitarian staff, analysts, technicians, light, water and telecommunications specialists.
- Victim: Represents the entity that was affected by an incident, comprising a single person or a community. Victims are affected by responder, physician or volunteer actions. Example: citizen.
- Witness: Represents an actor who saw the incident. They can provide relevant information about events that occurred. Examples: citizen or observer.

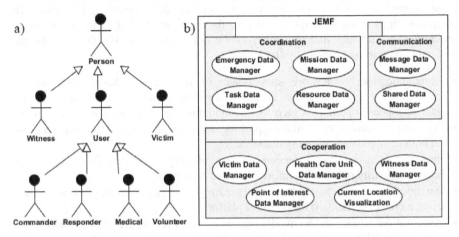

Fig. 2. JEMF main actors (a) and features (b)

Relevant features provided by existing projects were incorporated into JEMF and can be seen in Fig. 2 (b). They represent a set of fundamental functions that EM information systems should support for proper decision making by actors in disasters response. Below, we detail these features that allow data management (Create, Read, Update and Delete operations), which encompass each actor's tasks:

- Emergency Data Manager: manages the main event data of a disaster. It registers the type, intensity, location and date of an emergency. For example, a wildfire in Canada or landslide in Rio.
- Mission Data Manager: manages the macro tasks data of a responder group that can be registered by crisis managers. For example, firefighting in region A, search and rescue of victims in region B, citizen removal in region C, etc.
- Task Data Manager: manages the individual task data of each responder allocated in a mission. For example, agent 1 responsible for the aftermath task, agent 2 responsible for the stabilizing victims task, agent 3 responsible for the equipment or rescue vehicles operation, etc.

- Health Care Unit Data Manager: manages data about health care entities that are part of response effort. For example, ambulances, care tents and hospitals that can receive disaster victims.
- Victim Data Manager: manages the data relating to each victim rescued in the response process.
- Witness Data Manager: manages witness data about the event, acquiring information that may be important to responders.
- Resource Data Manager: manages data about the tools, equipment and supplies available to support response efforts.
- Point of Interest Data Manager: manages data about strategic locations, such as infrastructure damaged by an event.
- Message Data Manager: manages the data exchange between each responder, including text messaging, image or audio.
- Shared Data Manager: manages the common shared data considered relevant to all responders during crisis response, such as action plans, documents, etc.
- Current Location Visualization: allows the responder to view your current location on a map.

These JEMF features are organized into general and specific functions. The general functions, such as Task, Resource, Point of Interest, Message and Shared Data Management, are meant for all those involved in the response process and these are key functions within this domain. The specific functions are meant for each responder according to their responsibility level within the response team. The Commander uses Emergency, Mission and Health Care Unit Data Management. The Responder uses Mission, Victim and Witness Data Management. The Medical uses Mission, Victim and Health Care Unit Data Management. The Volunteer uses Witness Data Management.

After defining these requirements, we developed class diagrams to build the supporting framework. Some studies indicate that a framework should be evolved starting from general classes, and use inheritance in the individual applications [9, 19]. Inheritance allows developers to construct new classes and change code in an object-oriented environment, simply inheriting the desired behavior from an existing class and overriding only the methods that is different in the subclass. It may require many new subclasses for specifying business unit differences, but key functional areas may be abstracted out and reused in an expedient way. Design patterns like Template Method and Factory Method were applied to increase the amount of reusable code in the super classes.

JEMF is divided into two packages: Core and Mobile. The Core package is where we find the fundamental classes and represent entities that involve an emergency response (Fig. 3). These classes should not go through many changes, since they have a higher level of abstraction, and new concrete classes should be extensions that form the set of features previously described.

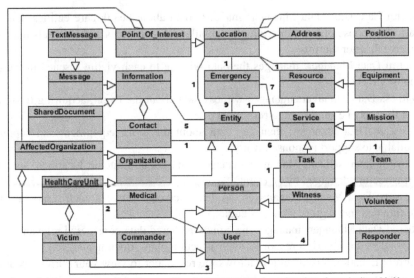

Legend: 1. has; 2. managed by; 3. rescued by; 4. interviewed by; 5. send/receive; 6. required/provided by
7. response to; 8. used by; 9. affected by.

Fig. 3. JEMF Core package

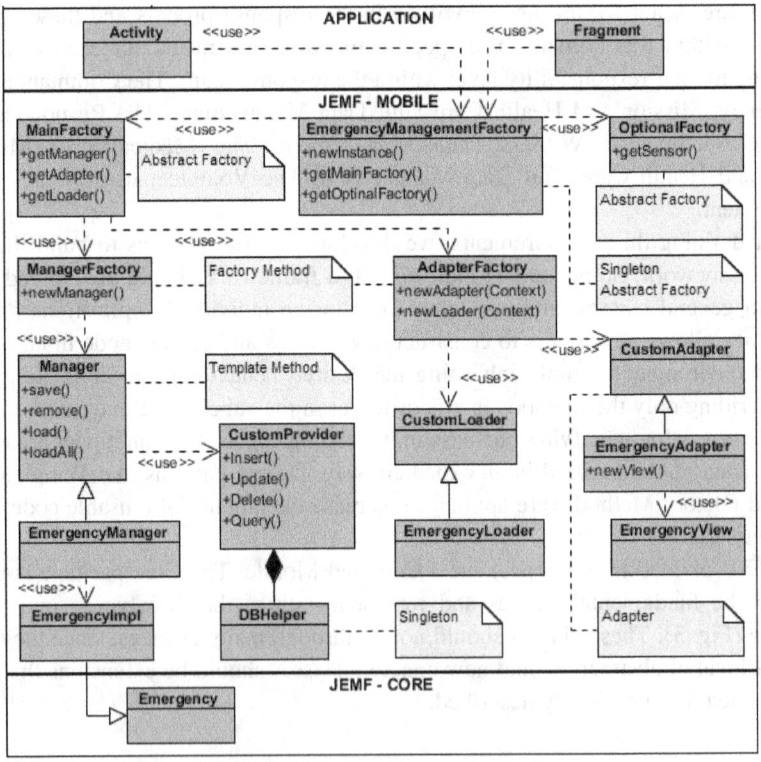

Fig. 4. JEMF Mobile package illustrating Emergency Data Manager feature

The Mobile package is aimed for concrete classes that are linked with the mobile operating system, such as patterns, components and resources, and will be reused to integrate the software architecture of a new mobile system (Fig. 4). This package contains the basic data management operations and allows developers modify and increment methods offered by these concrete classes or aggregate with other Android classes.

Fig. 3 shows that an *Emergency* needs response *Services* (Missions or Tasks). These *Services* may be provided by response *Entities* (Organizations or Personnel) or may be required by affected *Entities*. Response *Services* and *Entities* may use available *Resources* (Equipments). Response *Entities* may also exchange information (Messages or Shared Documents) between them or manage others *Entities* (Health Care Units). An *Emergency*, a *Service*, a *Resource* or an *Entity* has a *Location*. Fig. 4 is an example for Emergency Data Manager feature and shows how an application can access main methods to insert, update, delete *Emergency* data through *Emergency-Manager* class and retrieve stored data through *EmergencyLoader* class.

Implementation and Test Sub-phases. Programming-by-difference [9] is how we implemented JEMF classes and other developers may use it with JEMF code in the same way. When a new class is needed and it is similar to a JEMF developed class, they should create a subclass and override the methods that are different. So, the main abstract classes contain the common code portions and original methods can be defined once, enabling all subclasses override it to write specific portion. For example, based on Fig. 4 if a new *MissionManager* class is needed then just inherit from Manager and write the methods save, remove and load for this new class. Both Core and Mobile packages can be customized to support new applications because all classes and code are available for developers, once this framework is defined as open source.

Below we present some code snippets demonstrating how to instantiate JEMF and call Emergency Data Manager basic operations by an Android application project. We started on the second line of the example with the instantiation of an object *EmergencyImpl*. This object represents an *Emergency* and we can assign values to its attributes through getter and setter methods. Then, we instantiate the framework to perform save, remove and load operations (line 3). On lines 4, 5 and 6, we present each of these operations to the emergency object, respectively. In these operations it is necessary to inform the framework which feature we are working through a parameter. When it is necessary to perform these operations for other Data Manager feature, for example Mission Data Manager, we should instantiate a *MissionImpl* object and change the parameter in the methods presented, detailed on lines 7 and 8. The framework code is not limited to these methods, but it offers a larger features and functions coverage to work and integrate with the application in a proper way.

During the test sub-phase, we developed a set of unit test cases to catch and decrease problems and errors. We also developed examples to test the framework in real implementation scenery. These examples are distributed with JEMF code to help developers jump start development. Both the framework and its examples are fully documented so they understand JEMF properties easily.

Example of Emergency and Mission Data Manager feature within the *onClick* event of a button.

```
public void onClick() {
  EmergencyImpl emergency = new EmergencyImpl();
  EmergencyManagementFactory emf = EmergencyManagamentFac
tory.newInstance();
  emf.getMainFactory(EMERGENCY).getManager().save(emergen
cy);
  emf.getMainFactory(EMERGENCY).getManager().remove(emerg
ency);
  emf.getMainFactory(EMERGENCY).getLoader();
  MissionImpl mission = new MissionImpl();
  emf.getMainFactory(MISSION).getManager().save(mission);
}
```

We are investigating JEMF's usefulness for EM system development, using sample problems and replicating existing systems. The goal of this initial testing phase is to stabilize the framework so more in-depth testing can be conducted. An experiment is being prepared to obtain a better analysis of frameworks' effectiveness, previous programming experience required and others evaluation questions.

4   Related Work and Conclusion

Along with the vast expansion in the usage of information systems in EM domain, the community already has worked towards addressing software reuse challenges with toolkits and frameworks. For this reason, we briefly describe the main previous works related to the theme.

ESCAPE is a context information management framework for emergency situations [22]. This framework supports multiple levels of context information, such as individual, team, situation and response. Aggregation of data from different sensors and clients is a limitation because the information may not conform to the model proposed. An improvement to handle the data generated by the different mobile devices in this context is needed. Furthermore, ESCAPE is a framework designed for Service-Oriented Architecture (SOA) and involves the web services implementation for the WORKPAD project [6].

Crisis Management Information Systems must deal with information sharing between government entities and the full range of emergency-related organizations. The research presented by [7] constitutes a framework that covers the high-level features and services provided by such a system. Moreover, the main challenges and requirements are addressed and how technologies should meet them. However, their focus is based on SOA and PDA devices. The specifics of the resources provided by the mobile computing are not detailed. Collaboration and decision making aspects during an emergency response are not fully depicted, which we seek to do with JEMF.

Sahana is a free and open source EM web system to hold actions during and after a disaster [3]. It covers common coordination problems experienced during a disaster such as: helping find missing persons, volunteer management, response management, monitoring actions among government groups, civil society and the victims themselves. Sahana is not a framework or middleware, but its main feature is to be an open source system. This allows other nations to reuse and adapt this system to meet the specifics of their locality and support rescue organizations that do not have sufficient investments offered by their government. It is freely shared in order to create opportunities for others to build and maintain response systems with low cost, thus saving lives in countries that have lack the expertise to develop collaborative and decision making systems.

JEMF differs from these other frameworks due to the chosen scope for the development of software architecture, Web technology. This project seeks to contribute with the development of a software architecture for mobile technology. JEMF will act in systems building for current mobile devices such as smartphones and tablets. Through reviews and specifications presented in this paper, we contribute to guide new projects that support the response processes. We defined some essential principles for developing the emergency response mobile systems and introduced some appropriate technologies to remove complexity of programming them. JEMF aims to improve productivity of mobile EM projects, bringing the benefits of reusability, maintainability and provide detailed modeling and implementation features. Another difference from the papers already cited is that future projects may help to expand this software framework because new generic specifications may share goals and characteristics within a common research community. Therefore, new evaluation analysis may be performed to obtain the corrections, perspectives of use and contributions to the systems development. Future collaborations in the development and consolidation of this framework by other developers can be made or it could attend to similar areas. Additionally, this project scope would be expanded to server-side features to offer proper back-end support. We expect to contribute with this software framework to new mobile EM system construction by software reuse.

References

1. Angermann, M., Khider, M., Frassl, M., Lichtenstern, M.: DMT - An integrated disaster management tool. In: 1st International Conference on Disaster Management and Human Health Risk, United Kingdom (2009)
2. Araujo, F.C.S., Borges, M.R.S.: Support for systems development in mobile devices used in Emergency Management. In: IEEE 16th International Conference Computer Supported Cooperative Work in Design, pp. 200–206 (2012)
3. Currion, P., Silva, C., Van de Walle, B.: Open source software for disaster management. Communications of the ACM 50(3), 61–65 (2007)
4. Ellis, C.A., Gibbs, S.J., Rein, G.L.: Groupware - Some Issues and Experiences. Communications of the ACM 34(1), 38–58 (1991)
5. Fayad, M., Schmidt, D.C.: Object-oriented Application Frameworks. Communications of the ACM 40(10), 32–38 (1997)

6. Humayoun, S., Catarci, T., de Leoni, M., Marrella, A., Mecella, M., Bortenschlager, M., Steinmann, R.: Designing Mobile Systems in Highly Dynamic Scenarios: The WORKPAD Methodology. Knowledge, Technology & Policy 22(1), 25–43 (2009)
7. Iannella, R., Robinson, K., Rinta-Koski, O.: Towards a framework for crisis information management systems (CIMS). In: 14th Annual TIEMS Conference (2007)
8. Ibarra, M., Monares, A., Ochoa, S.F., Pino, J.A., Suarez, D.: A mobile collaborative application to reduce the radio traffic in urban emergencies. In: IEEE 16th International Conference Computer Supported Cooperative Work in Design, pp. 358–365 (2012)
9. Johnson, R., Foote, B.: Designing Reusable classes. Journal of Object-Oriented Programming (JOOP) 1(2), 22–35 (1988)
10. Joshi, S.G.: Exploring map-based interfaces for mobile solutions in emergency work. Master's Thesis, University of Oslo, Norway (2011)
11. Kanchanasut, K., Tunpan, A., Awal, M.A., Das, D.K., Wongsaardsakul, T., Tsuchimoto, Y.: DUMBONET: A multimedia communication system for collaborative emergency response operations in disaster-affected areas. International Journal of Emergency Management 4(4), 670–681 (2007)
12. Kienzle, J., Guelfi, N., Mustafiz, S.: Crisis Management Systems: A Case Study for Aspect-Oriented Modeling. In: Katz, S., Mezini, M., Kienzle, J. (eds.) Transactions on Aspect-Oriented Software Development VII. LNCS, vol. 6210, pp. 1–22. Springer, Heidelberg (2010)
13. Luyten, K., Winters, F., Coninx, K., Naudts, D., Moerman, I.: A situation-aware mobile system to support fire brigades in emergency situations. In: Meersman, R., Tari, Z., Herrero, P. (eds.) OTM 2006 Workshops. LNCS, vol. 4278, pp. 1966–1975. Springer, Heidelberg (2006)
14. Meissner, A., Wang, Z., Putz, W., Grimmer, J.: MIKoBOS - A Mobile Information and Communication System for Emergency Response. In: 3rd International Conference on Information Systems for Crisis Response and Management, pp. 978–990 (2006)
15. Nilsson, E.G., Stølen, K.: Generic functionality in user interfaces for emergency response. In: 23rd Australian Computer-Human Interaction Conference, pp. 233–242. ACM, New York (2011)
16. Padilha, R.P., Borges, M.R.S., Gomes, J.O., Canós, J.H.: The Design of collaboration support between command and operation teams during emergency response. In: IEEE 14th International Conference on Computer Supported Cooperative Work in Design, pp. 759–763 (2010)
17. Pimentel, M., Fuks, H.: Collaborative Systems. Elsevier, Rio de Janeiro (2011)
18. Ponserre, S., Guha-Sapir, D., Vos, F., Below, R.: Annual disaster statistical review 2011: the numbers and trends. Université Catholique de Louvain, Brussels, Belgic (2012)
19. Roberts, D., Johnson, R.: Evolving Frameworks: A Pattern Language for Developing Object-Oriented Frameworks. In: 3rd Conference on Pattern Languages and Programming (1996)
20. Schächinger, U., Kretschmer, R., Röckelein, W., Neumann, C., Maghsudi, M., Nerlich, M.: NOAH – A Mobile Emergency Care System. European Journal of Medical Research 2000(5), 13–18 (2000)
21. Silva, A.J.D.: A mobile computing system to support first responder in radiological emergency. Master's Thesis, Federal University of Rio de Janeiro, Brazil (2012)
22. Truong, H.-L., Juszczyk, L., Manzoor, A., Dustdar, S.: ESCAPE – An Adaptive Framework for Managing and Providing Context Information in Emergency Situations. In: Kortuem, G., Finney, J., Lea, R., Sundramoorthy, V. (eds.) EuroSSC 2007. LNCS, vol. 4793, pp. 207–222. Springer, Heidelberg (2007)
23. Yang, Y.J., Kim, S., Choi, G.J., Cho, E.S., Kim, C.J., Kim, S.D.: A UML-based object-Oriented Framework Development Methodology. In: Asia Pacific Software Engineering Conference, Taiwan (1998)

The Semantic Web as a Platform
for Collective Intelligence

Leandro Mendoza[1], Guido Zuccarelli[1],
Alicia Díaz[1], and Alejandro Fernández[1,2]

[1] LIFIA, Facultad de Informática, Universidad Nacional de La Plata, Argentina
[2] CIC, Comisión de Investigaciones Científicas, Argentina

Abstract. The *Semantic Web* constitutes a promising platform for the development of computer support for cooperative work. However, the maturity of the related technologies and available datasets poses new challenges. Knowing what these challenges are, and assessing their impact in advance can save effort and reduce the chance of failure. In this article we discuss the specific challenges in the development of an application that integrates collaborative product reviews available in the *Semantic Web*. The challenges we identify, if not tackled, translate to an additional effort in the integration process, the need to discard available data, and potential inconsistencies and lack of data-quality in the final product.

1 Introduction

The World Wide Web is currently an ecosystem where users contribute and consume content. Part of this content serves as input for collaborative decision making. Such is the case of collaborative reviewing sites for movies, books, and other products, where users share and discuss their opinions. Our ability to build systems that empower users' to exploit this socially created content is limited by our capacity to find and interpret the users' opinions. If users provide their opinions in natural language (i.e., plain English) our systems need to apply NPL techniques. If users publish their opinions in different web-sites our systems must retrieve, interpret and integrate these opinions. The *Semantic Web* [13] proposes methods and technologies to transform the current web in a *Web of (Linked) Data* that programs can more easily interpret and act upon.

Berners-Lee et al. [1] defined the *Semantic Web* as an extension of the current Web in which information is given well-defined meaning, better enabling computers and people to work in cooperation. A more programming-oriented definition given by the author Yu in [13] conceives the *Semantic Web* as a collection of technologies and standards that allows machines to understand the meaning of information on the Web. Two closely related concepts to the *Semantic Web* vision are *Linked Data* and *Web of Data*: while *Linked Data* refers to a set of best practices for publishing and connecting structured data on the Web [2], the term *Web of Data* can be viewed as the result of applying *Semantic Web* technologies

N. Baloian et al. (Eds.): CRIWG 2014, LNCS 8658, pp. 255–262, 2014.

to make *Linked Data* possible. The foundation of the *Semantic Web* is RDF[1], a set of standards published by the W3C that define a data model consisting of resources, properties and statements (triples that connect resources through properties), and the means to publish and access them.

We believe the *Semantic Web* constitutes a promising platform for the development of computer support for cooperative work. In this article we report on the challenges that the *Web of Data* poses on the development of CSCW applications that retrieve, integrate and interpret users contributions available on-line. To illustrate the discussion we introduce *"Collective opinions"*, an application that aggregates *Reviews* and *Ratings* on the *Web of Data*. It is based on the architecture proposed by the LDIF project [12]. LDIF is the "Linked Data Integration Framework", a mature initiative in the *Semantic Web* community.

Following, we position our work in the context of existing research at the intersection of *Semantic Web* and CSCW. Then, we present the requirements for the *"Collective opinions"* application and the principles in its design. The combination of requirements, design approach, and nature of available data on the *Semantic Web* result in a set of challenges that we discuss in section 4. Finally, we summarise our findings, and provide an outlook.

2   Related Work

Within the context of CSCW, the most commonly explored contribution of the *Semantic Web* focuses on its power to model and store knowledge through the use of ontologies. Santos et. al. [11] show how *Semantic Web* technologies add quality to crowdsourced, geo-spacial annotations on maps. Ontologies and ontology modelling languages such as OWL[2] provide the formal semantics that allow for the verification of data consistency, increase interoperability, and enhanced information retrieval. Most importantly, ontologies represent a common, well defined language for users to contribute annotations.

In [7], Tom Gruber argues that the current web (the web 2.0, the *Social Web*) provides "collected intelligence" instead of "collective intelligence". That is, the value of the current web is that it collects the contributions of users and aggregates them into community- or domain- specific sites such as Flicker or Youtube. However, to attain real collective intelligence new levels of understanding on this content should emerge. He presents *RealTravel.com*, an example of a collective knowledge system for the domain of travel, based on the *Semantic Web* principles.

Di Noia and Mirizzi [4] argue that, although the web of data provides tons of data, only few applications exploit this potential. They implemented a content based, movie recommender system that leverages the data available in the *Semantic Web*. They focus on three popular dataset; DBPEDIA [10]; Freebase [3], and LinkedMDB [8]. They construct a content based recommender algorithm

---

[1] http://www.w3.org/standards/techs/rdf#w3c_all - Last accessed on May 1st, 2014.

[2] http://www.w3.org/TR/owl-features/ - Last accessed on May 1st, 2014.

that performs well in terms of precision and recall on the dataset of the Lenskit project [5]. Their work experimentally shows that these three datasets are mature enough and rich in high quality data to serve as the main data source for the proposed recommender system. Moreover they conclude that combining information from various datasets improves recommendations and does not add noise.

Summarising, the work of Santos and colleagues shows how *Semantic Web* technologies support the collaborative construction of data models within the boundaries of a single system. Gruber illustrates how the *Semantic Web* supports the emergence of new knowledge from the contributions of users in a collaborative system. Dinoia and Mirizzi, demonstrate that well curated semantic datasets can be combined to build effective recommender systems to support decision making. Our goal is to understand what is involved in implementing collective intelligence systems that can cope with the open, distributed, variable in quality, large scale nature of the Web of Data.

3   Collective Opinions - The Case Study

Our goal is to illustrate, with a concrete example, the challenges that currently face those that attempt to build groupware applications that exploit the potential of the *Semantic Web* as a repository of "collectively" constructed knowledge.

"Collective opinion" is a system that crawls the (semantic) Web to harvest what users say about books, movies and products to construct a "collective opinion". It processes textual reviews and numeric ratings, focusing on the requirements that pose the most interesting and diverse problems.

- **R1: Trending opinions** - rank the items that users are talking about the most these days. Identify reviews that were published recently, and aggregate them by the item they refer to. That a review was discovered recently is not enough to infer that it is a recent review. It might be the case of a dataset that was recently published to the *Semantic Web* with reviews from the past year.
- **R2: Ranking surprises** - list the items whose aggregated rating plummeted or skyrocketed in the last days. Correctly calculate the rating taking into account that: a) reviews could come in different scales, b) there are individual reviews and aggregated reviews.
- **R3: Associations** - provide associations in the form "users who liked this, also liked ...". A user might express an opinion on several sites; match reviews and ratings to users, and identifying the same user on various sites.

We adopt the architecture proposed by the LDIF project [12]. It models a *Semantic Web* application as consisting of a sequence of phases:

- **Access and retrieve data:** Linked data is published in various forms. RDF documents (using, for example, RDF/XML serialisation[3]) publish a

---

[3] http://www.w3.org/TR/REC-rdf-syntax/ - Last accessed on May 1st, 2014.

collection of RDF triples that systems retrieve through http requests – complex data models are split into various RDF documents. Sparql end-points provide SQL-like query access to (normally large) RDF datasets. It is also possible to embed RDF statements within HTML documents using RDFa[4], Microformats[5] and Microdata[6].

- **Translate to a common vocabulary:** Various models, called schemas or vocabularies, can represent the same data in terms of resources and properties (much like various entity-relation models can represent the same data in the relational database world). They differ, for example, in the level of detail they provide. Vocabularies emerge and evolve independently which means that, at a given movement, several vocabularies for the same domain might coexist. To exploit this data we first need to translate it to one common vocabulary.
- **Resolve identities:** In the *Semantic Web anyone can say anything about anything.* Statements about a resource can be distributed in multiple datasets, in multiple locations. Moreover, there is no central authority to ensure the existence of a unique identifier for each resource. Applications must realise when two statements refer to the same resource, and act accordingly.
- **Fuse data and assure quality:** Once data is represented in a common vocabulary, and identities are resolved, only statements that comply with predefined quality criteria get fused into an integrated dataset.
- **Exploit data:** The final application (in our case, "Collective opinions") works on the resulting dataset to exploit the available, integrated, curated data.

4   Collective Opinion - The Challenges

Building *"Collective opinions"* confronted us with challenges inherent to the nature and maturity (or lack thereof) of the *Semantic Web* and the LDIF architecture, and challenges specific to the datasets available in the domain of our case study. Next, we report on those we faced when selecting the input and common vocabularies. Then we discuss the challenges in finding useful data. Finally, we discuss the challenges for data retrieval and fusion.

4.1   Challenges for the Selection of Vocabularies

A prerequisite to build an application that uses the LDIF framework is to select the vocabularies accepted during retrieval (input vocabularies), and the common vocabulary to use for the fused dataset. We evaluated each vocabulary's popularity in existing datasets as well as the nature of its supporting community of users, to select only those that added more value. To decide on the common

---

[4] http://www.w3.org/TR/xhtml-rdfa-primer/ - Last accessed on May 1st, 2014.

[5] http://microformats.org - Last accessed on May 1st, 2014.

[6] http://www.w3.org/TR/microdata/ - Last accessed on May 1st, 2014.

vocabulary we analysed how well it modelled our domain (*coverage* [14]), and how good it could map data published in the remaining vocabularies (*mappability* [14]). This challenging selection process involved extensive review of scientific publications, technology web-sites, and technical, specialised discussion forums.

The are three main alternatives to publish data about reviews. The **Review Vocabulary**[7] (also known as Review Ontology) was one of the earliest vocabularies to publish reviews and ratings using RDF. It can be traced back to the work of Heath and Motta [9] in the Revyu.com system for collaborative rating an reviewing. The "Microformats community" puts forward its own vocabulary for marking up reviews. **hReview**[8], is a simple, open format, suitable for embedding reviews (of products, services, businesses, events, etc.) in HTML, XHTML, Atom, RSS, and arbitrary XML. **Schema.org**[9] is a *Semantic Web* initiative led by Google, Bing, Yahoo and Yandex to help authors embed semantics into HTML pages. It concentrates on simplicity and on a well understood set of abstractions (including Reviews) that these big search companies think can have special treatment in their search engines, for example showing rich snippets. Microdata is the recommended mechanism so publish Schema.org data within HMTL pages, although RDFa and Microformats are also applicable.

Through API queries to two widely used semantic search engines, *LOD Cloud cache (LODC)*[10] and *Sindice (SIND)*[11], we observed that, in their search indexes, the three vocabularies appeared frequently enough to justify including them as input. There were also traces of a predecessor of Schema.org called **datavocabulary** that we decided to ignore as most sites should eventually upgrade.

We compared the three vocabularies to assess coverage of the data needed to implements the application requirements, and to establish alignments or mappings [6] between equivalent concepts (with similar meaning). The three vocabularies describe the person who creates a review (requirement R3), the date of creation (R1 and R2), a personal opinion in form of text, and a rating that corresponds to a numeric value within a given range (R1, R2, and R3). All of them also foresee a mechanism to associate a review with the resource being reviewed (R1, R2, and R3), the difference being that for the Review Vocabulary, this is achieve through a relationship from the resource to the review (i.e., backwards).

We conclude that Review Vocabulary, hReview, Schema.org are mappable to one-another. Moreover, Review Vocabulary is formally defined in RDFS[12]. It can be used to publish not only within HTML (with RDFa) but also in RDF documents, and in Sparql endpoints. There are already tools that map hReview to Review Vocabulary. Based on this observations, we choose **Review Vocabulary** as the base vocabulary to represent our integrated data.

---

[7] http://purl.org/stuff/rev# - Last accessed on May 1st, 2014.

[8] hReview http://microformats.org/wiki/hreview - Last accessed on May 1st, 2014.

[9] http://schema.org/ - Last accessed on May 1st, 2014.

[10] http://lod.openlinksw.com - Last accessed on May 1st, 2014.

[11] http://sindice.com/ - Last accessed on May 1st, 2014.

[12] http://www.w3.org/TR/rdf-schema/ - Last accessed on May 1st, 2014.

4.2   Challenges for the Selection of Data Sources

Most semantic information about reviews and ratings is currently embedded in HTML documents. Semantic search engines are the standard mechanism to find these documents. Using semantic search engines is challenging in terms of *availability* [14] of these services and the *timeliness* [14] of their responses. The alternative (impracticable for most scenarios) is to implement an ad-hoc web crawler. Both, SIND and LODC, provide a SPARQL endpoint to query its RDF datasets. SIND provides a search API that we can use by calling it programmatically. During our experiments, SIND suffered frequent shutdowns which spanned weeks. LODC remained accessible for the whole duration of our study (two months). In order to assess the amount of available data, we performed a query to retrieve all those documents that contains data about Reviews, using the *Review Vocabulary* as the baseline. The results showed that LODC reports 5,014,468 documents using the *Review Vocabulary*, whereas SIND reports 10,216,632 documents. It is important to note that each document could contain information about more than one Review. To assess data *timeliness*, we searched for any document containing information about Reviews (using any of our input vocabularies). We took a random sample of size 1000 from the results obtained in each engine. We immediately downloaded those documents and inspect them. The percentage of documents from the result set that was still available on-line was 69% for SIND and 54% for LODC. Moreover, 72% of the documents in SIND's result set that were on-line, still had relevant semantic content; in comparison only 40% of those in LODC's result set did, which indicates that search engine's indexes are largely outdated.

4.3   Challenges for Data Retrieval and Fusion

Using URIs to identify resources is a key principle of the *Semantic Web*. Our input vocabularies foresee that the subject of the review and its author are resources. Our application depends on this principle to uniquely identify items and persons for all three requirements. However, we found multiple cases where a string (the name of the person) is used to specify `author` when the expected value for this property is a resource (i.e., a URI) typed as `Person` or `Organisation`. Our approach in these cases was to discard the data. The same problem was present when the item of the review was a string (e.g., the title of the movie) instead of a URI. If we knew the domain was restricted to books or movies, we could guess the identity of the item via comparison to labels of known books or movies in curated datasets such as DBPEDIA. However, this approach would require additional effort and is error prone and of limited applicability.

Our input vocabularies define that the rating value should be numerical and must be in the range defined by the *min* and *max* values. The use of non-numerical values for rating (or rating range) is a recurring problem in the available data. For example, a rating value described using a string such as "rating 1 of 5" instead of a numerical value and a valid range (rating value: 1, Min rating value:1, Max rating value: 5). Reviews that presented this problem were discarded as they are of no use to implement our requirements. For other scenarios they might still be valuable.

In RDF, a properties can take a literal value (instead of a URI that identifies a resource). The name of a person and the numeric value of a rating are typed literals. Typed literal values consist of a string (the lexical form of the literal) and a datatype (identified by a URI). Knowing how dates are represented is critical for our application (specially R1 and R2). It is common practice to use XML schema data types; and the convention[13] to represent dates and times is to use the ISO 8601 Date and Time Formats. When the date was not available or did not follow the conventions, we considered for the total ratings (thus R3) but not for trends and surprises (R1, and R2).

Our input vocabularies have a mechanism to indicate the type of resource being reviewed (i.e., a movie, a book, a restaurant). In Schema.org and hReview the type of resource is a property of the review itself. In the Review Vocabulary, the type is a property of the resource that identifies the item. Being able to tell the type of the reviewed item lets us implement R3, suggesting only resources of the same type (i.e., users who liked this *book*, also liked these *books*). The data we obtained varied widely regarding this aspect therefore we had to resort a more open version of R3.

Search engines indicate that web sites might be more prominently displayed in search results if they provided semantic markup for their content. This situation motivated web-sites creators to indiscriminately copy and republish content from others sites (particularly movie reviews). There are currently no consistent mechanisms to identify and discard exact content replicas. If the date and time of the review, and its author are not available we cannot tell if two reviews are the same or not.

5   Conclusions

The *Semantic Web* can foster the creation of CSCW applications that exploit users' generated content. Existing work shows that, in controlled scenarios, these technologies support the emergence of new knowledge from the contributions of users in a collaborative system. In this work, we discuss some of the challenges of implementing collective intelligence systems that can cope with the open, distributed, variable in quality, large scale nature of the Web of Data. Focusing on the development of an application that integrates product reviews we learnt that: a) the distributed and collaborative nature of the *Semantic Web* originated a variety of alternative vocabularies (and supporting communities) to model the same domain, that developers must find, evaluate, select and combine, which demands considerable effort; b) semantic search engines, a common mechanism to identify data sources, lack stability and timeliness – therefore, alternative mechanism are called for; and c) many existing datasets lack quality and cannot

---

[13] XML Schema: `http://www.w3.org/TR/xmlschema-2/` - Last accessed on May 1st, 2014.

be effectively used in applications that aim at doing rich integration – strategies for the early identification of quality problems and for data curation are needed.

We continue studying the domain of user generated semantic content and the implications of using it for collective intelligence. Next steps are the compilation of a formal model for quality in *Linked Data* that can serve as the basis for automated evaluation of datases, and potentially, automated curation.

In this work we focused on CSCW applications that take the *Semantic Web* as a source of data. In an additional line of research, we explore the potential of RDF as a flexible modelling framework to enable Group Decision Support Systems.

References

1. Berners-Lee, T., Hendler, J., Lassila, O., et al.: The semantic web. Scientific American 284(5), 28–37 (2001)
2. Bizer, C., Heath, T., Berners-Lee, T.: Linked data-the story so far. International Journal on Semantic Web and Information Systems 5(3), 1–22 (2009)
3. Bollacker, K., Evans, C., Paritosh, P., Sturge, T., Taylor, J.: Freebase: a collaboratively created graph database for structuring human knowledge. In: Proceedings of the 2008 ACM SIGMOD International Conference on Management of Data, pp. 1247–1250. ACM (2008)
4. Di Noia, T., Mirizzi, R., Ostuni, V.C., Romito, D., Zanker, M.: Linked open data to support content-based recommender systems. In: Proceedings of the 8th International Conference on Semantic Systems, I-SEMANTICS 2012, p. 1 (2012)
5. Ekstrand, M.D., Ludwig, M., Kolb, J., Riedl, J.T.: Lenskit: a modular recommender framework. In: Proceedings of the Fifth ACM Conference on Recommender Systems, pp. 349–350. ACM (2011)
6. Euzenat, J., Shvaiko, P.: Ontology matching, 2nd edn. Springer, Heidelberg (2013)
7. Gruber, T.: Collective knowledge systems: Where the social web meets the semantic web. Web Semantics: Science, Services and Agents on the World Wide Web 6(1), 4–13 (2008)
8. Hassanzadeh, O., Consens, M.P.: Linked movie data base. In: LDOW (2009)
9. Heath, T., Motta, E.: Revyu.com: a reviewing and rating site for the Web of Data. In: Aberer, K., et al. (eds.) ASWC 2007 and ISWC 2007. LNCS, vol. 4825, pp. 895–902. Springer, Heidelberg (2007)
10. Lehmann, J., Isele, R., Jakob, M., Jentzsch, A., Kontokostas, D., Mendes, P.N., Hellmann, S., Morsey, M., van Kleef, P., Auer, S., et al.: Dbpedia-a large-scale, multilingual knowledge base extracted from wikipedia. Semantic Web Journal (2013)
11. Gonzalez, A.L., Izidoro, D., Willrich, R., Santos, C.A.S.: OurMap: Representing Crowdsourced Annotations on Geospatial Coordinates as Linked Open Data. In: Antunes, P., Gerosa, M.A., Sylvester, A., Vassileva, J., de Vreede, G.-J. (eds.) CRIWG 2013. LNCS, vol. 8224, pp. 77–93. Springer, Heidelberg (2013)
12. Schultz, A., Matteini, A., Isele, R., Mendes, P.N., Bizer, C., Becker, C.: Ldif-a framework for large-scale linked data integration. In: 21st International World Wide Web Conference (WWW 2012), Developers Track, Lyon (2012)
13. Yu, L.: A developer's guide to the semantic Web. Springer (2011)
14. Zaveri, A., Rula, A., Maurino, A., Pietrobon, R., Lehmann, J., Auer, S.: Quality assessment methodologies for linked open data. Submitted to SWJ (2012)

Engineering Peer-to-Peer Learning Processes
for Generating High Quality Learning Materials

Sarah Oeste[1,*], Matthias Söllner[1], and Jan Marco Leimeister[1,2]

[1] Universität Kassel, Chair of Information Systems, Kassel, Germany
{sarah.oeste,soellner}@uni-kassel.de
[2] University of St. Gallen, Institute of Information Management, St. Gallen, Switzerland
janmarco.leimeister@unisg.ch

Abstract. Organizations are facing the challenge of transferring knowledge from experienced to novice employees and are seeking for solutions that avoid the loss of knowledge with retiring experts. A possible way for overcoming this challenge is having employees develop learning materials for their novice colleagues. Based on insights from both, education and collaboration research, designing structured collaborative peer-creation-processes seems a promising approach due to several reasons. Within a peer-creation-process participants are guided to knowledge acquisition, transfer as well as documentation for others. By developing learning materials through collaboration with people at different level of knowledge, e.g., the tacit knowledge of the expert gets codified and is ready for being used by novices. Furthermore, the collaborative creation will create learning effects even among participants and should further increase their knowledge, and the quality of the learning materials. Unfortunately, little research has addressed reusable didactically driven processes of systematically documenting knowledge that can be used by others as learning material. In order to bridge this gap we identify requirements from educational and collaboration literature and conceptualize educationally driven changes in the layer model of collaboration, e.g., to consider learning objectives in the goals layer or to integrate peer review as mechanisms for quality control in the procedures layer. This paper opens up a promising field for collaboration research and provides future research directions for reusable structured peer-creation-processes with focus on learning. This research-in-progress paper closes with a conceptual framework with requirements of a collaborative peer creation process.

Keywords: Layer Model, Collaboration Engineering, Peer Creation, Peer Learning, Knowledge.

1 Introduction

People with different level of knowledge are working together in organizations. Besides this, knowledge leaves organizations as a consequence of demographic change. So mechanisms for a cooperative knowledge generation become important [1].

---

* Corresponding author.

N. Baloian et al. (Eds.): CRIWG 2014, LNCS 8658, pp. 263–270, 2014.

In order to perpetuate competitive capacity, it is necessary for organizations to save and to document explicit and tacit knowledge. Also research shows that collaboration of people with heterogeneous knowledge can lead to a gain in productivity [2, 3]. Therefore focusing on educational research is important. In this context peer learning (PL) and peer creation (PC) comprise the integration of learners in creative PL activities or paradigm changes from a learner as consumer to a producer of learning content [4]. The challenge is to codify and document explicit and tacit knowledge in a way that it is accessible for others as the basis for knowledge acquisition. This constitutes a complex and recurring task of knowledge documentation. So far, little to no research addresses structured reusable processes for such a purpose. Focusing on structured and reusable group processes, research in collaboration engineering (CE) provides useful mechanisms. However didactical elements are necessary, but not anchored in mechanisms of CE. Thus, the integration of people with heterogeneous knowledge in a collaborative process for documentation can be a purchase. We assume that combining mechanisms of educational and CE research is an appropriate approach. Consequently, the creation of a structured group collaboration process for reusable documentation of knowledge in form of high quality learning material LM needs to be addressed. We therefore propose the following research question: What requirements need to be considered in a conceptual framework for designing a peer-creation process (PCP) for developing high quality learning material (LM)? Thereby theoretical contribution of PCP research constitutes an improvement with a level 2 contribution type in form of a first framework for designing a PCP according to a design science point of view [5]. We therefore provide basics for research in PL for developing LM and basics in CE for designing collaborative group processes in section 2. In section 3, we conceptualize a framework with requirements for designing a PCP. Beginning with the discussion of the guiding idea of PCP, we provide a description of requirements and conclude with the presentation of a framework. In section 4, we describe next steps of research and end with the expected contribution.

2   Related Work

2.1   Educational Basics in Peer Learning

Following the assumption that knowledge documentation in form of LM provokes a learning process, didactical basics are necessary. A characteristic of learning is a change in behavior based on experiences [6] like conversations and discussions [7]. Thus, someone is learning on basis of its own experiences and connects this with previous knowledge. In this context PL provides a suitable approach. A group of people learn or attempt to learn something together through social interactions [8]. These interactions, like discussions with peers, foster reflection and cognitive processes [9]. That results in positive effects for the learner, called peer: e.g. knowledge gain which leads to learning success or improvement of communication skills and the peer learns to become responsible for his activities [10],[11],[7],[12]. In addition a peer learning process focuses on the learner and permits interactions between learners on the same level of knowledge [12, 13]. In most cases a lecturer prepares basic conditions and assists the process [14]. Table 1 shows the concept of PL and different kinds. PL, peer tutoring and cooperative learning provide insights on how learning processes are structured and conducted, but do not focus on LM development. For documenting explicit and tacit knowledge through interactions between people, PC provides mechanisms for LM development. This output can be used by an extended group of people. PC comprises mechanisms of co-creation [7] which indicate first insights on how people create artifacts together with the help of collaborative technologies. The peers add value to the LM by yielding their own knowledge in form of learning content [7]. Until now, structure and learning objectives are open or predetermined by the lecturer [15], only sequences of a learning process are addressed, e.g. generating a short multiple-choice-task, and LM development is not reusable [7]. For developing LM [7] identified key principles as shown in table 1, which lead to first insights on how to design processes for documenting knowledge in a standardized and productive way. Nevertheless development of such learning processes often has a strong reference on specific content and context. So, reproducibility and assignability are difficult [16].

Table 1. Different kinds of peer learning [7], [11], [17]

	Goal	Audience	Setting	Principle / Mechanism	Type of task
Peer Learning	Learning from and with others.	Learner (peer) Instructor	Groups of 2-5 people up to >100 people	Social interaction (e.g. discussion) provoke reflexion of knowledge and cognitive processes.	
Peer Tutoring	Practising learning content and basic skills.	Tutor Tutee Instructor	Groups divided in dyads	Assignment of tasks is not completely new. Participants with high-capacity assist other participants. Learner change the role of tutor and tutee after a defined time. Solving of substasks takes place together.	Strong structure.
Cooperative Learning	Common development for a solution of a problem (task) with maximum learning success.	Learner (peer) Instructor	Small groups with 3-6 people	(1) Positive interdependence. (2) Face-to-Face communication and reciprocal support.(3) Individual accountability for actions. (4) Interpersonal skills. (5) Reflexive group processes.	Substasks build up on each other.
Peer Creation	Development of learning material from peer for other peers.	Learner (peer) Instructor	Peer develops material alone or with another peer.	(1) Clear assignment of tasks. (2) Peers are accountable for developed learning material. (3) Ensure expert knowledge of participants. (4) Train participants in didactics and communication. (5) Formal assignments by instructor. (6) Ensure peer interaction	Lightweight learning material development (e.g. multiple-choice tasks).

2.2 Basics in Collaboration Engineering

Knowledge documentation as a recurring task requires a structured and reusable collaborative process. Hence, CE research provides an approach for designing and conducting such processes to solve complex and recurring tasks. The more a task occurs, the more efficient it is to design it as a collaborative process and solve it with same process flow. Thereby a group of people works together towards a common goal while group activities are characterized by communication, cooperation and coordination [18, 19]. These structured activities lead towards an additional benefit which cannot be attained by individual endeavor [3]. CE differentiates between three roles. A collaboration engineer designs and documents a collaborative process. A facilitator is able to design a non-recurring collaborative process. He disposes expert knowledge and moderation skills, so that he is able to conduct a process. A practitioner can act as facilitator or as participant of a collaborative process and is an expert on task and owns expert knowledge. The layer model of collaboration provides a framework for designing collaborative processes [20]. These layers are hierarchical and depend on each other: Goals as the first layer focus on defining a desired state or outcome as a group goal. The product layer addresses tangible or intangible artifacts as the outcome produced by a group. Defining and acquiring sub products in a collaborative process lead to one common product. Activities as the next layer describe particular subtasks a group must do to achieve defined products to fulfill the common goal. The subsequent layer addresses procedures. These are methods, strategies and tactics a group uses to execute work. So called patterns of collaboration - generate, reduce, clarify, organize, evaluate and build consensus - characterize how activities become structured and are observable regularities for the defined activities. The next layer contains tools and describes several technologies to support the execution of the collaborative process. Scripts as the last layer address documentation of behavior people say and do as they collaborate [20, 21]. Thus, in context of knowledge documentation CE provides guidelines for designing a collaborative process with LM as collaborative product. Nevertheless educational claims are not anchored in CE so far.

3 Framework for Developing a Peer-Creation-Process

3.1 Guiding Idea for a Peer-Creation-Process

Based on the benefits of educational and CE we convey the idea of a PCP for documenting knowledge in form of LM. Thereto we deviate following assumptions starting with collaboration, the work of two or more people on common material, which is characterized by coordination, communication and cooperation [19]: (1) Collaboration enables exchange of heterogeneous knowledge; (2) Such exchange of heterogeneous knowledge provokes recapitulation of knowledge; (3) This in turn provokes a learning process for all involved people; (4) A learning process is the basis for correct documentation of knowledge from people at different levels of knowledge. (5) Documentation of knowledge comes to an end when quality of developed LM is high enough. Elsewise recapitulation, learning process and documentation are passed through again. Figure 1 depicts these assumptions.

Fig. 1. Research assumptions

Subsequently, collaboration is the basis for structuring a PCP. Thus, solutions become important which enable a collaborative exchange of knowledge while improving learning success of participants and documenting their expert knowledge and knowhow by generating LM. Hence, we use mechanisms from CE research enriched with educational research for designing a PCP for developing LM. Figure 2 demonstrates the intention of a PCP and points out the demand for two perspectives of analyses. (1) PCP for developing LM and (2) the distribution of developed LM used by an extended group of people. Focusing on the PCP, input is given by participants in form of peers at a different level of knowledge, so-called practitioners. A further input comes from a facilitator with moderation skills. The throughput addresses structured collaboration between practitioners guided by a facilitator. Though reciprocal interactions between practitioners take place which cause individual acquisition as well as transfer of knowledge and documentation of knowledge in a didactical reasonably way. This has two positive effects for the output of the PCP. Social interaction and the assignment of task in structuring knowledge in a didactical reasonable way provoke individual reflection of knowledge. Hence, high quality LM arises as main output whereas individual learning success by practitioners arises as a side effect. They are learning from and with each other during their collaboration. Thus, from an educational point of view, cognitive process dimensions like applying, analyzing, evaluating and creating knowledge [22] will be addressed. Focusing on the distribution of LM, an extended group of people can use it for improving individual learning success. It is expected that this kind of learning will address lower cognitive process dimensions like remembering or understanding [22].

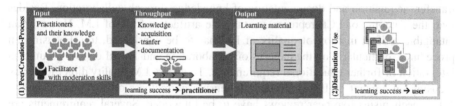

Fig. 2. Guiding idea for a peer-creation-process

3.2   Requirements from Educational Research

In order to design a PCP for developing LM some formal educational requirements need to be considered addressing PL and PC to mediate individual learning success and the development of LM. For that purpose we differentiate between the process and the output in form of LM. By focusing on the way the process has to be conducted, participants of the process should learn something. Through social interactions and collaboration with others they improve their own knowledge and document exchanged knowledge. Thus, knowledge creation, acquisition and transfer take place. This should be anchored in a didactical reasonable way, e.g. with useful learning objectives [22]. Besides that reciprocity in social interaction has to be provided between participants and direct feedback has to be conveyed to participants [14]. Further a lecturer should prepare basic conditions for peer learning activities, assist participants and communicate explicit expectations [14]. Continuing requirements arise from content of knowledge. Hence, the need for different process design depends on knowledge complexity. Therefore educational requirements influence the design of the PCP. By focusing on the output of PCP in form of LM, quality is necessary. This includes the following requirements. Conducted by the intention that an extended group of people can learn with this material didactical requirements like learning objectives [22] and structural requirements like coherent and logical content presentation should also be respected. This refers to the length and design. Further indicators of quality arise from the type of LM. So this could be a textual explanation, a learning exercise or an explanation video. In addition correctness of content should be ensured as a further indicator for LM quality [23]. Therefore controlling mechanisms like peer reviews can be a solution, which should be integrated into the design of PCP.

3.3   Requirements from Collaboration Engineering

A central requirement from CE focuses on the success of a collaborative process with predictable and repeatable results. According to the layer model of collaboration [21] all layers will set requirements to the PCP, but several layers will become important from an educational point of view for expanding CE research. A common collaboration goal, which is congruent to individual goals of practitioners, should be anchored with clear learning objectives. So, a clear description of cognitive process dimension must be considered in process design [22]. In the context of product layer, tangible and intangible artifacts are existent like enhancing individual learning success by transfer, acquisition and documentation of knowledge through collaboration. Besides that, the outcome of PCP represents a tangible artifact in form of LM. However an intangible artifact in form of quality of LM has to be considered, too. Focusing on procedures, particularly on the patterns of collaboration and ThinkLets, we assume an integration of further educational requirements. To ensure LM quality direct feedback focusing on content of LM is needed. To ensure correctness of documented knowledge mechanisms like peer reviews have to be integrated. Several requirements are expected from collaboration scripts. All necessary expertise like moderation know-how of a facilitator or pedagogical skills in form of hints for correct behavior in teaching situations, e.g. how to communicate feedback towards participants, should be contained in process design.

3.4   Conceptual Framework

In order to design a PCP for reusable development of LM we developed a framework in dependence on the established layer model of collaboration. Hence, figure 3 picks up the guiding idea of PCP, consolidates educational requirements for process design and highlights expected educational expansions in the layer model of collaboration.

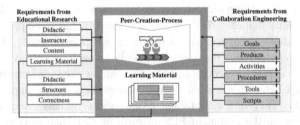

Fig. 3. Conceptual framework of integrating learning into group collaboration processes

The centre of figure 3 indicates the purpose of research with PCP and its output in form of LM. On the left, figure 3 visualizes educational requirements like didactic, instructor and content of knowledge influencing design of PCP with focus on how practitioners have to collaborate to achieve an individual learning success and work towards a common goal in form of LM. Besides that the figure visualizes the influence of LM type. This leads to deduction of quality requirements addressing didactic aspects, structure and correctness of LM in the design of PCP. These educational facts provide important inferences on the assignment of tasks respectively the way the collaborative goal and product should be achieved. On the right requirements from CE are depicted and important layers with an expected educational influence are highlighted. In a structured process design requirements from educational research should be combined with CE especially the layers addressing goals, products, procedures and scripts. These are central elements for anchoring educational requirements, like peer reviews in the procedures or hints for pedagogical skills in the scripts.

4   Next Steps and Expected Contribution

As our next step, we currently plan a study addressing process design, learning success of participants and quality of LM in a large scale university lecture to evaluate our general idea. For that purpose we design and evaluate a PCP where participants develop a storyboard for an explanation video. Thus for our process design, we consider individual learning objectives in the goals layer. We consider intangible learning success as well as tangible LM in the product layer and ensure correctness of LM by integrating lightweight peer-reviews for several subtasks in procedures layer. In the scripts we integrate pedagogical techniques and advices. In the experimental setting we plan with about 300 participants which will randomly be assigned to the treatment (using the PCP) or control group. Within each group, we will randomly build another 6 groups so that roughly 25 students collaborate in each group. A quantitative inquiry of the participants will be conducted for gathering insights on process design and individual learning success. During a focus group qualitative interviews with experts in CE and education should display insights on the process design and the quality of developed LM. After a first successful evaluation of our approach with students, the PCP will be used by at least of our partners from industry, to develop LM supporting the knowledge transfer in the partner organization. In this paper we developed the vision of reusable PCP with the help of CE routines enriched with educational requirements. Therefore, we showed the concept of a PCP, and its potential for challenges in knowledge transfer and documentation. We identified several challenges for designing a PCP that require extension in the layer model of collaboration and conveyed this in a conceptual framework. Such processes have the potential to make organizations independent from educators and standardize inexplicit pedagogical methods and routines. Thus, it allows reusability and the execution by facilitators and practitioners with moderation skills. The output of the PCP can be used as knowledge base for acquisition of factual knowledge and thus, bring about a second opportunity for knowledge transfer towards an extended group of people. With the described research we expect extensions for the body of knowledge of collaboration. Further, the results provide first insights of a structured and reusable way for overcoming challenges in knowledge transfer and documentation.

References

1. Fuchs-Kittowski, F.: Interaktionsorientiertes Wissensmanagement. In: Krcmar, H. (ed.) Interaktionsorientiertes Wissensmanagement. Peter Lang GmbH, Frankfurt am Main (2013)
2. Ries, B.C., Diestel, S., Shemla, M., Christina, L.S., Jungmann, F., Wegge, J., Schmidt, K.-H.: Age Diversity and Team Effectiveness. In: Schlick, C.M., Frieling, E., Wegge, J. (eds.) Age-Differentiated Work Systems, Springer, Heidelberg (2013)
3. Bittner, E.A.C., Leimeister, J.M.: Creating Shared Understanding in Heterogeneous Work Groups: Why It Matters and How to Achieve It. JMIS 31, 111–144 (2014)
4. Johnson, L., Adams Becker, S., Estrada, V., Freeman, A.: NMC Horizon Report: 2014 Higher Education Edition. The New York Media Consortium, Austin (2014)
5. Gregor, S., Hevner, A.R.: Positioning and Presenting Design Science Research for Maximum Impact. MIS Quarterly 37, 337–355 (2013)
6. Gagne, R.M.: Learning Outcomes and Their Effects. American Psychologist 39, 377–385 (1984)
7. Wegener, R., Leimeister, J.M.: Peer Creation of E-Learning Materials to EnhanceLearning Success and Satisfaction in an Information Systems Course. In: 20th ECIS, Barcelona (2012)
8. Dillenbourg, P.: What do you mean by collaborative learning. In: Dillenbourg, P. (ed.) Collaborative Learning, pp. 1–19. Elsevier, Oxford (1999)
9. Arbaugh, J.B.: Online and Blended Business Education for the 21st Century: Current research and future directions. Woodhead Publishing (2010)
10. Damon, W.: Peer education. Jn. of Applied Developmental Psychology, 331–343 (1984)
11. Topping, K.J.: Trends in Peer Learning. Educational Psychology 25, 631–645 (2005)
12. Geer, J., McCalla, G., Collins, J., Kumar, V., Meagher, P., Vassileva, J.: Supporting Peer Help and Collaboration in Distributed Workplace Environments. International Journal of Artificial Intelligence in Education 9, 159–177 (1998)
13. Hua Liu, C., Matthews, R.: Vygotsky's philosophy: Constructivism and its critisms examined. International Education Journal 6, 386–399 (2005)
14. Harris, A.: Effective Teaching. School Leadership & Management 18, 169–183 (1998)
15. Auvinen, A.-M.: The challenge of quality in peer-produced eLearning content. eLearning Papers 17, 1–11 (2009)
16. Kollar, I., Fischer, F., Hesse, F.W.: Collaboration Scripts - A Conceptual Analysis. Educational Psychology Review 18, 159–185 (2006)
17. Büttner, G., Warwas, J., Adl-Amini, K.: Kooperatives Lernen und Peer Tutoring im inklusiven Unterricht. Zeitschrift für Inklusion 1-2 (2012)
18. Kolfschoten, G.L., Vreede, G.-J.: A Design Approach for Collaboration Processes: A Multimethod Design Science Study in Collaboration Engineering. JMIS 26, 225–256 (2009)
19. Leimeister, J.M. (ed.): Collaboration Engineering - IT-gestützte Zusammenarbeitsprozesse systematisch entwickeln und durchführen. Springer Gabler, Heidelberg (2014)
20. Briggs, R.O., Kolfschoten, G.L., Vreede, G.-J., Albrecht, C., Dean, D.R., Lukosch, S.: A Seven-Layer Model of Collaboration. In: 30th ICIS. Phoenix, Arizona (2009)
21. Briggs, R.O., Kolfschoten, G.L., de Vreede, G.-J., Albrecht, C., Lukosch, S., Dean, D.L.: A Six-Layer Model of Collaboration. In: Nunamaker Jr., J.F., Romano Jr., N.C., Briggs, R.O. (eds.) Collaboration Systems, pp. 221–228. Advances in Management Information Systems, New York (2014)
22. Krathwohl, D.R.: A Revision of Bloom's Taxonomy: An Ovierview. Theory Into Practice 41, 212–218 (2002)
23. Leacock, T.L., Nesbit, J.C.: A Framework for Evaluating the Quality of Multimedia Learning Resources. Educational Technology & Society 10, 44–59 (2007)

Start-Smart as a Support for Starting Interaction in Distributed Software Development

Ramón R. Palacio[1], José Ramón Martínez[1], Joaquín Cortez[2],
Luis Adrián Castro[2], and Alberto L. Morán[3]

[1] Unidad Navojoa, Instituto Tecnológico de Sonora, Navojoa, México
ramon.palacio@itson.edu.mx, joseramonmg26@gmail.com
[2] Unidad Nainari, Instituto Tecnológico de Sonora, Cd. Obregón, México
joaquin.cortez@itson.edu.mx, luis.castro@acm.org
[3] Facultad de Ciencias, Universidad Autónoma de Baja California, Ensenada, México
alberto.moran@uabc.edu.mx

Abstract. Distributed software development is a collaborative activity characterized by frequent interactions among the members of the workgroup. However, interruptions that arise when interacting can adversely affect the work of a developer. This work aims at designing a tool to provide support while initiating interactions to indicate the extent to which a colleague can be interrupted, based on a software industry work context. For this, based on the literature of software engineering and the concept of Collaborative Working Spheres, in this study we defined the information elements that must be provided. As a result, an application was derived providing support for initiating interactions in software development working groups, which will be evaluated in different software factories.

Keywords: Interruptions, Distributed Software Development, Collaborative Systems.

1 Introduction

An interruption is an event that requires a user to redirect her attention to another task, suspending her primary or current activity. Often, interruptions can adversely affect users' work. Therefore, it is important to identify the reason for interrupting (i.e., information in an interruption) so as to justify the temporarily stop of the user's current task and attend the requirement right away [1,2]. To improve user experience against disruptions, research groups have made considerable efforts to obtain an efficient way to measure costs derived from interruptions [3-5]. These efforts usually impact ideas or design guidelines that software development teams can consider to diminish the impact of an interruption at work. The aim of these techniques is avoiding disruptions that negatively affect the completion of tasks [3,4], in the error rate or impact [2,6], decision making [1,7], and affective state of people [6].

One of the main problems caused by interruptions is recalling what needs to be carried out at a particular task (prospective memories failure), which is exacerbated

N. Baloian et al. (Eds.): CRIWG 2014, LNCS 8658, pp. 271–278, 2014.

when both the number of interruptions and pending tasks increase [4,8]. With revolutionary advances in mobile computing, for instance, smartphones have now multiple features that can be a source of user interruptions at work such as SMS, telephone calls, and all kinds of social networks, which help individuals stay in touch but also increasingly interrupting their everyday activities. One of the reasons for this is that conventional mobile phone contact manager apps are very limited. Also, Distributed Software Development (DSD) is a work style where interruptions are commonplace and needed. This type of work is carried out in geographically separated locations in a coordinated way, with real-time and asynchronous participation [9]. Therefore, the main challenge that must be faced in DSD is communication. Given geographical and time zone differences, telephone calls, instant messages, and emails are frequently used, which result in a work environment highly pervaded by interruptions.

In this paper, we propose the creation of a mobile device mechanism that supports initiating an interaction in an informed fashion, as proposed in [10], through the concept of Collaborative Working Spheres (CWS). A CWS is a "collection of units of work, resources and people sharing a common goal or objective in a certain period of time, providing knowledge of the activity of the members of the working group to assist initiating an informed interaction".

The CWS concept promotes the distribution of information elements that pertain to the activity carried out by members of a workgroup (e.g., activity schedule, role, project activities). In this way, if the time is appropriate to initiate an interaction, contacting a colleague can be facilitated. Based on this, it would be interesting to determine the information elements enabling the development of a mobile phone app that supports initiating an informed interaction.

2   Using Collaborative Working Spheres for Initiating Interactions

The CWSs concept [10] is fundamental for the Selective Availability (SA) criterion. In this respect, we propose the introduction of the SA criterion, which considers whether a user is selectively available to colleagues whose activity is related to the work unit she is currently dealing with, and may be less (or not) available to other colleagues. Therefore, we propose a mobile phone application (smartphone) providing an augmented contact list (i.e., informed vs. conventional) with an availability indicator. This indicator must be meaningful for the user so that it can be used to determine at a glance the degree of interest in an interaction that a potential contact can have.

In order to determine the appropriate time to initiate an interaction, data or information elements that are found in the work context developer are required. The elements described in Table 1 were the basis for developing the proposed application. These elements were a basis for determining whether a colleague is available. To further illustrate, in Figure 1, we exemplify the context of a DSD worker, where Developer 1 (D1) requires initiating an interaction with Developer 2 (D2). For this, D1 must have a list of colleagues that shows their statuses (*presence*), based on the different elements in the context of individual work (*identity*), such as events, actions, roles and resources, in this case D2. In addition, it is possible to determine the geographic location (*location*) where the workgroup members are located, which is important given the potential time zone differences among the developers.

Table 1. Information elements for the Start-Smart

Element	Description
Presence	A list of potential contacts showing their degree of availability is needed, similar to that of [11,12].
Identity	Identifying the role of a contact and her name can be a rapid way of knowing who is the most appropriate person to contact [5,12]
Location	Knowing the current location of the person to be contacted is needed. This influences decision making when trying to interact, if there is a major time zone difference [9].
Operation	Knowing the current activity in which the workgroup members are working on is one of the most relevant aspects for determining the appropriate time for an interaction [13].
Expectative	Knowing today's activities of the person to be contacted. The caller may detect some spare time or when the colleague being contacted is not under a heavy workload, as similarly reported in [13,8].

The elements of the individual working sphere (IWS) provide information about the activity that must be carried out by developers in certain time periods, as defined by the project manager (*operation*). In the same way, in the calendar there is a description of the activities that developers must perform. This provides a different perspective on what a developer will be performing at a certain time (*expectative*). Based on the CWS elements (Figure 1), it is possible to determine the different situations of the workgroup by reading the following user status: *available, reachable, but busy and busy*.

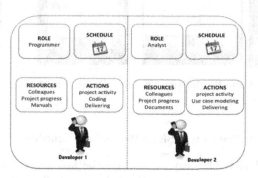

Fig. 1. CWS Elements

3   The Start-Smart Tool

Start-Smart is a tool designed to support initiating interactions in DSD. Considering that CWSs have been reported to be a mechanism to determine "selective availability" [10], we designed the Start-Smart tool to mainly have the following features: i) It provides *useful information about colleagues*, focusing on facilitating personal data (e.g. current project, current activity, role). Besides, it provides mechanisms for finding the *time zone and city* where the colleague is located, ii) It provides mechanisms through which users can *share their activity schedule* in order to infer the colleague's current activity, iii) It provides mechanisms to *automatically group colleagues*, who are involved in an activity or project, and iv) It provides an *availability indicator*, representing the extent to which a colleague could have interest in having an interaction.

3.1   Proving Useful Information about Colleagues

The Start-Smart app provides significant information regarding a colleague's current activity during a project. This information (see Figure 2) refers to the project in which the developer is currently working on, her or his role, current activity, and user status (i.e., online/offline). Is it important to mention that being online only means that the user can be contacted. As seen in Figure 2A, it also includes the city and time zone where the colleague is located. Additionally, the availability indicator, which describes the level of interest that may have the colleague to start the interaction (see Figure 2B), is included. The way in which this indicator is derived will be explained in section 3.4.

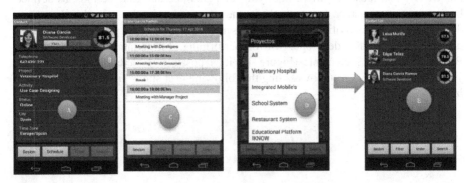

Fig. 2. Contact list (A and B), Activity schedule (C) and Contact list grouped by project (D and E)

3.2   Providing Mechanisms to Share the Activity Schedule

The activity schedule is one way of inferring the task or activity that a potential contact (i.e., colleague) must perform. This is because in the activity schedule, users manage their time through invitations to various events (e.g., meetings, allocation of activities), which allows a user to determine whether to wait for a more appropriate time during which the interaction can be initiated. In Figure 2C the list of events that the contact "Diana Garcia" should attend for a given day is displayed.

3.3   Providing Mechanisms to Automatically Group Colleagues

Through the contact list provided with the Start-Smart app, the user can filter their colleagues by projects (see Figure 2D). This is because the Start-Smart app is connected to the resources of the database project manager, and therefore the app can relate contacts with projects (see Figure 2E).

3.4   Providing an Availability Indicator

To compute the willingness to interact with colleagues, we considered different factors involved in the DSD activity. Firstly, we have to consider if the status of the colleague (online/offline) to initiate a mediated interaction (e.g., call, message). Secondly, it is also important to have the identity of the colleague such as name, role, and time zone. This is important to derive the profile of the developer. Thirdly, it is important to consider the colleague's geographic location, since this can provide information regarding the culture and language, among others that could be inferred. Fourthly, it is important to be aware of the project the colleague is working on (project, activity, task), since this might help determine if the reason to initiate an interaction is appropriate. Fifthly, the activity calendar was considered, as there are events to which the colleague is fully committed to (e.g. workshop, meeting with the client, etc.); with this one can infer what the colleague is (and will be) doing (e.g., eating out). Finally, the sixth consideration, it is important to consider whether the interruption is related to a current project or not.

Taking the previous considerations as a departure point, the availability indicator can be computed, depending a match between the caller and the person to be contacted. The parameters and their values have been determined through the literature (see Table 1). To compute the proposed degree of availability, we propose a set of phases of availability for the Smart Cycle-Interaction. These are:

1. *Availability*: This refers to whether a connection exists, so as to initiate an interaction (online or offline).
2. *Identification*: whether the user has a defined role at that moment based on the activity performed (e.g., programmer, analyst, etc.).
3. *Location*: Time zone and city in which the person to be contacted is located.
4. *Project:* The name of the project in which the colleague is working on at the moment the colleague is to be contacted.
5. *State of Schedule:* Potential spare time of the contact as dictated by the schedule (e.g., no activity, initiate, initiated activity, about to end).
6. *Date Project*: Referred to the validity of the project. In this case, what is considered important is whether the project starts/ends on that day.

The computation consists of six phases, which were defined based on the characteristics of the DSD activities. The phases are proposed considering the level of importance for an interaction during the DSD activities. For this, the following calculations are proposed to identify the range of willingness (0-100%) to interact, as described in [12]. Table 2 shows three columns: Assigned Percentage, in which we describe the total percentage assigned to each item of information. Details of which information elements compose it are presented in the second column. In the third column, the conditions under which the percentage is allocated to achieve the total allocated to each information element.

Table 2. Parameters and conditions to calculate availability level

Assigned Percentage	Information Elements	Conditions	
		YES	NO
10%	Availability		
	Available	10%	0%
10%	Identity	YES	NO
	Has a defined role	10%	5%
10%	Location	YES	NO
	same city	10%	3%
	Same time zone	5%	2%
50%	Project	YES	NO
	Same project	35%	17.50%
	Same activity	15%	7.50%
12%	State of Schedule	YES	NO
	Its about to begin	3%	
	has begun	1%	
	Activity in progress	1%	
	Its about to finish	9%	
	Without activity schedule	12%	
8%	Project Date	YES	NO
	Matches the start date	4%	1%
	Matches the finish date	4%	1%
			<60
TOTAL: 100%			>=60 && < 80
			>=80

It is important to note that at the bottom of the table there are three colors: Red, Yellow and Green, which are used by the Smart-user Interaction app as willingness indicators. A red indicator means that the colleague (i.e., potential contact) is not willing to interact at that time (less than 60% willingness). A yellow indicator means that the contact is somehow willing (between 60 and 79%). Finally, the green indicator means a potential contact with 80% or more of willingness to interact.

As can be seen, the computation and color of the indicator provide a rapid answer to the question: Is it wise to interrupt? In this way, if the indicator shows green means that there are high chances that the potential contact answers and the interruption does not affect considerably, making it a good time to initiate an interaction. Now, if the indicator is yellow, it means that there are chances that the person answers and that the interruption will somehow affect the colleague. In this particular case, the caller must use common sense to determine whether it is really necessary to interrupt at this moment or can wait a little bit more until the indicator becomes green. Moreover, if the indicator is red, it means that it is unlikely that the subject responds to the call and the interruption could really affect her or his colleague's work. This also would result in the type and quality of support requested may not be as expected. Because of this, in these cases, it is not recommended to interrupt a colleague unless there is something that requires urgent attention.

3.5   Implementation Aspects

The architecture of this tool is based on a client-server model (see Figure 3), which consists of three main parts: The Start-Smart client, the application server, and database server. The Start-Smart client is the application on the client side, which was implemented in Android Froyo, and uses the web services connection. This connection provides the functionality required to connect it to an application server (Glassfish 4), to retrieve the list of registered users and data.

The application server receives notifications of a Start-Smart client related to some unit of work of the organization. This server is implemented using Glassfish 4.

The Database server provides the data of Start-Smarts clients based on the current activity (project manager, personal agenda). This server is implemented using MySQL server 5.1. The Start-Smart app was built using the following technologies: Web Services, http protocol and SOAP protocol (Simple Object Access Protocol), which are used for message exchange.

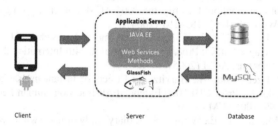

Fig. 3. Architectural diagram

4   Conclusion and Future Work

In this work, we have addressed the lack of timely, adequate opportunities for informal interactions through mobile devices. We have used the literature to identify a set of information elements based on the features of DSD activities. Taking such elements as a departure point, we have developed the in Start-Smart tool implementing the concept of CWSs. The Start-Smart app provides an augmented contact list with elements such an availability indicator. The Start-Smart app was mainly proposed to support the exchange of information between the members of a workgroup who are in different geographical locations in a lightweight and simple manner. Furthermore, the Start-Smart app enables support to initiate an interaction in a selective manner, according to a criterion that we have referred to as "selective availability".

According to our literature analysis, we can infer that Start-Smart can improve the way in which developers initiate interactions through mobile devices in comparison to how this is carried out by using conventional contact lists managers included in smartphones, since the proposed tool helps users to be aware of what each person is working on at each precise moment. Therefore, if one person wishes to start an interaction with another one, the first can check the indicator until s/he determines when the moment is appropriate, preferably when the indicator is green. Hence, it is assumed that the interaction can be less disturbing as this will be related to his/her work at hand. Interactions with the Start-Smart app could consequently be more productive and less disturbing than those made when using traditional contact list tools. Start-Smart was implemented and used according to the specific needs of a DSD group. As future work, Start-Smart will be evaluated in a software factory located in three different cities in northwest Mexico.

Acknowledgement. This work has been funded by P/PIFI-2013-26MSU0023H-06 para el fortalecimiento de la capacidad académica y mejora de la competitividad académica de la DES Navojoa. It is also was supported by Programa de Fortalecimiento y Consolidación de la Investigación (PROFOCI).

References

1. Dabbish, L., Kraut, R.E.: Controlling interruptions: awareness displays and social motivation for coordination. In: Proc. the 2004 ACM Conference on Computer Supported Cooperative Work 2004, pp. 182–191 (2004)
2. Adamczyk, P.D., Iqbal, S.T., Bailey, B.P.: A method, system, and tools for intelligent interruption management. In: Proc. Proceedings of the 4th International Workshop on Task Models and Diagrams. ACM (2005)
3. Cutrell, E.B., Czerwinski, M., Horvitz, E.: Effects of instant messaging interruptions on computing tasks. In: Proc. CHI 2000 Extended Abstracts on Human Factors in Computing Systems, pp. 99–100. ACM (2000)
4. Czerwinski, M., Horvitz, E., Wilhite, S.: A diary study of task switching and interruptions. In: Proc. the SIGCHI Conference on Human Factors in Computing Systems, pp. 175–182 (2004)
5. Mark, G., Gudith, D., Klocke, U.: The cost of interrupted work: more speed and stress. In: Proc. the Twenty-Sixth Annual SIGCHI Conference on Human Factors in Computing Systems, pp. 107–110. ACM Press (2008)
6. Bailey, B.P., Konstan, J.A.: On the need for attention-aware systems: Measuring effects of interruption on task performance, error rate, and affective state. Computers in Human Behavior (2006)
7. Speier, C., Vessey, I., Valacich, J.S.: The Effects of Interruptions, Task Complexity, and Information Presentation on Computer-Supported Decision-Making Performance. Decision Sciences (2003)
8. Ellis, J., Kvavilashvili, L.: Prospective memory in 2000: Past, present and future directions. Applied Cognitive Psychology 14, 1–9 (2000)
9. Mohagheghi, P.: Global Software Development: Issues, Solutions, Challenges. Dept. Computer and Information Science (IDI) (2004)
10. Palacio, R.R., Morán, A.L., González, V.M., Vizcaíno, A.: Selective availability: Coordinating interaction initiation in distributed software development. IET Software 6(3), 185–198 (2012), doi:10.1049/iet-sen.2011.0077
11. Navarro, X.: Manual de Mensajería Instantánea. dbsolutions (2010)
12. Ye, Y.: Supporting software development as knowledge-intensive and collaborative activity. In: Proceedings of the 2006 International Workshop on Workshop on Interdisciplinary Software Engineering Research, Shanghai, China. ACM (2006), doi:http://doi.acm.org/10.1145/1137661.1137666
13. Dabbish, L.A., Kraut, R.E.: Coordinating communication: Awareness displays and interruption. In: Extended Abstracts of the ACM Conference on Human Factors in Computing Systems, pp. 786–787 (2003)

Social Media Collaboration in the Classroom: A Study of Group Collaboration

Liana Razmerita[1] and Kathrin Kirchner[2]

[1] Department of International Business Communication,
Copenhagen Business School, Denmark
lr.ibc@cbs.dk
[2] University Hospital Jena, Germany
kathrin.kirchner@uni-jena.de

Abstract. This article aims to investigate how students use new technology in collaborative group work and tries to measure what factors impact students' satisfaction with overall group collaboration. In particular, this study aims to investigate the following research questions: What are the factors (including challenges) that influence the students' overall satisfaction with collaboration? Does the usage of e-collaboration tools and social media usage influence collaboration satisfaction? The findings of the study are summarized in a model that point towards the main factors influencing student overall group work satisfaction.

Keywords: collaborative work, social media, collaboration, e-collaboration.

1 Introduction

Net generation of students, or digital natives, have grown up with new collaborative technologies and therefore collaboration is natural for them [1]. In the above- mentioned book, it is argued the students are enforcing a change in the model of pedagogy: "from a teacher focused approach based on instruction to a student-focused model based on collaboration". The new generations of students challenge the educators with low tolerance for long lecturing time, have a tendency for multitasking, are very pragmatic, and less patient in class. Therefore designing new learning and teaching methods is an ongoing endeavor for educators [2]. However, we still need an answer to the question what can be gained by using new Information and Communication Technologies (ICT) inside the classroom and to deal with students' attention problem [3].

This article aims to investigate factors that influence collaborative work satisfaction, how the students use the new technology in collaborative group work and also tries to measure its impact on satisfaction with overall group collaboration. The study presented in this article was conducted at Copenhagen Business School in an elective course entitled Web Interaction Design and Communication: New Forms of Knowledge Sharing and Interaction. The course enrolls both Danish bachelor students and

N. Baloian et al. (Eds.): CRIWG 2014, LNCS 8658, pp. 279–286, 2014.

exchange students from universities from all over the world. Within the class, students work in groups on a selected topic of interest and develop their preliminary research ideas in groups and also may collect data in groups. At the end of the course, groups present their research results and receive feedback on their work. This preliminary research in groups represents a springboard for their individual research projects. The aim of group work is both to help students develop their preliminary research related to a selected topic and to foster learning, creativity and innovation of their projects. The course uses Podio as a social media enhanced platform for managing course related materials, communication, sharing information and interactions with the students. Podio is designed to be a "complete work platform for enterprise" and aims to integrate many different work tasks and collaborative work through one application [4].

During the course, students are free to select a topic of interest and how they collaborate either using face-to-face interaction or any e-collaboration tools or social media. Students may form groups based on a common topic of interest and also decide how much they collaborate or cooperate. Some groups just assign tasks to group members in order to collect data and prepare the presentation to be delivered at the end of the semester; other groups collaborate through discussions, brainstorming, sharing of ideas and enter into real collaborative processes that may lead to knowledge building. In a previous study we reported how students form groups in a heterogeneous classroom and proposed a method to assign students to groups based on certain criteria [5].

In this study we aim to investigate the students' perceptions of collaborative work, the factors that impact their satisfaction. On the basis of a questionnaire, this paper aims to address the following research questions: What are the factors (including challenges) that influence the students' overall satisfaction with collaboration? Does the usage of e-collaboration tools and social media usage influence collaboration satisfaction?

2   Related Work

2.1   Collaborative Learning and Factors Influencing Collaboration

Collaborative group work offers many potential advantages for supporting learning, creativity and classwork. However, group work performance, including learning and satisfaction, depend on many variables, including interest in the subject, relations to peers, gender differences, age, individual differences, cultural backgrounds; see among others [5, 6].

Collaborative learning is a teaching method where students work together in small groups to solve a common task [7]. It can improve learner performance when learners discuss a problem and suggest potential solutions [8]. In a meta-analysis of undergraduate courses it was found that group activities improve the attitude toward learning, increase persistence and result in greater academic achievement [9].

In order to understand such team collaboration, group dynamics and group behavior have to be considered [10]. Group dynamics can be described via factors such as:

participation, communication, collaboration, trust and cohesion [11]. Furthermore, group behavior is influenced by team member familiarity. which leads to a positive attitude toward communication and collaboration within the group [12].

A study reported in [13] found that team dynamics, team acquaintance and instructor support have a major influence on teamwork satisfaction. From the students' point of view, establishing team commitment, having clear and frequent communication within the team, using interactive software and synchronous meetings, were all important factors for teamwork satisfaction.

2.2   E-collaboration and Social Media

Online or e-collaboration is the computer-mediated form of in-class collaborative learning, including multilevel interaction, resource sharing and developing competencies in real-world situations [14]. Social media comprising Web 2.0 technologies support the synergetic articulation between personal and collective knowledge which may lead to knowledge creation and innovation within teams and organizations [15].

McConnell [16] defines three aspects of online group collaboration: the process of group work (measured by the ability to develop in-depth discussions, questioning and contributing to group work), social presence (openness between group participants) and outcomes and products of group work. Tseng et al. [13] found that trust among team members and organizational practices are factors that explain satisfaction with online collaboration.

A previous study by [17] aimed at investigating benefits of collaborative learning and challenges of collaborative work using Social Media for foreign language learning. While certain benefits of using new technology for teaching are undeniable, some students felt rather distracted and confused by the new teaching methods, and questioned the use of new technology and social media as compared with traditional teaching methods. Beyond ownership issues, other issues reported in the study were: lack of trust in their capabilities, lack of interest, or students who are not particularly willing to "invade each other's turf" when it comes to correcting, discussing or making changes to language through collaborative editing and group work.

A study of e-collaboration [18] reported that social loafing within teams can diminish team potency assessments, perception of technology usefulness and thus behavioral usage intentions and team performance. Team members loaf if they feel that their contributions are not essential for the end result of the group or if their work is not assessed. Social loafing is defined as a reduction of motivation and effort when individuals work in groups as opposed to when they work individually. Another study of online learning collaboration identified a number of critical challenges, among which: instructor support , well defined instructions, team commitment and clear and frequent communication [10]. In the classroom, social media changes learners from passive content consumers to active participants [19]. A study of more than 600 students found that social media encouraged students to interact with their peers so that they got to know their peers better and developed a positive relationship with them [21]. This is essential to the creation of collaborative learning communities.

Although numerous studies on the topic of collaboration, group work, teamwork, and e-collaboration exist, to our knowledge and according to the literature, there are no studies covering a more comprehensive overview of factors influencing collaboration in the classroom.

3   Research Method and Data Analysis

The study employs a survey research design. Based on the literature review, questionnaire was developed. The survey focused on students' perception of collaboration and e-collaboration. The questionnaire included 13 main questions and covered different aspects of collaboration, including satisfaction, evaluation of end results of collaboration, possible factors influencing collaboration, means of collaboration and e-collaboration. Most of the questionnaire items used a 5-point Likert scale ranging from 1 (strongly agree) to 5 (strongly disagree). In order to collect additional issues a few open questions were included. For each of them, recurring responses were categorized and counted.

The data has been collected using a survey distributed at the end of the students' course within the Web Interaction Design course. Data presented within this study was collected in three semesters: fall 2011, spring 2012 to fall 2012. The course was run as an elective course and it consisted of 8 lectures of three hours.

The survey has been distributed to 140 students, 41 students in Fall 2011, 40 students Spring 2012 and 39 users Fall 2012. A short overview of the purpose of the data collection was provided so that students could understand the underlying objectives of the questionnaire and the study. The students were not given any incentives to fill in the questionnaire. Over the three semesters, 63 valid answers were collected. This accounts for 45% response rate from the total number of students. Out of the 63 respondents 22 were male (35%), 29 were females (46%) and 12 respondents (19%) didn't disclosed their identity and therefore we don't know their gender. In our sample we had 7 responses from Danes and 43 non-Danes and for 13 respondents it was not possible to identify their place of origin.

Data was analyzed using SPSS. Spearman's Rho method was applied for finding correlations between the Likert scale-based ordinal variables. This measure of correlation provides information about the strength and direction of correlation.

Table 1 presents an overview of the different factors measured as ordinal variables considered in the survey. In relation to the collaboration challenges, the mean values suggest that the teams did not experienced major challenges; however, lack of coordination and social loafing seem to be greater challenges for group work than lack of trust, conflict, different backgrounds and cultural differences. According to the results summarized in Table 1, students use e-collaboration tools often and do not perceive major challenges in e-collaboration. In relation with the overall satisfaction of collaboration, 50% of the students evaluate their satisfaction with the overall group collaboration as good and about 27% think it is very good.

Table 1. Overview of Collaboration Factors (Likert scale coded from 1-Strongly agree to 5-Strongly disagree)

Survey	Item	Mean	SD
General collaboration	Enjoy collaboration with peers	1.86	0.780
	Collaboration effect on learning and inspiration	2.3	1.010
	Equal contribution of team members	2.3	1.200
	Evaluation of end result of collaboration	2.43	0.797
	Evaluation of overall satisfaction with collaboration	2.0	0.810
Collaboration challenges	Social loafing	3.95	0.991
	Lack of coordination	3.90	1.043
	Lack of trust	4.56	0.667
	Conflict	4.46	1.010
	Different backgrounds of team members	4.21	0.985
	Cultural differences in the team	4.05	1.007
e-collaboration	Usage of e-collaboration tools	1.85	0.910
	Prefer social interaction	2.39	0.918
	Difficult to use	3.98	0.940
	Not fun	3.75	0.960
	No benefits	4.18	0.866
	No need	4.25	0.830
	Help to advance project ideas	2.46	0.997

Figure 1 presents an overview of e-collaboration tools used in group work during the three semesters (based on a multiple choice question). For collaborating with their peers within their group, students mainly used Podio (which was the platform used in the course) and Facebook for group level e-collaboration. While in fall 2011 email was sometimes used for collaboration, it lost importance in spring 2012 and no longer seemed to play a role in collaboration in fall 2012. According to our data, group collaboration using Facebook grew as a preferred environment for collaboration. One student commented that he used Facebook to learn the names of people in his group and visited their Facebook pages to get to know them better. The preference for Facebook might be because the professor let the students decide what tools they wished to use for collaborating, and Podio was only suggested as a possible option for private group workspace. Previous research has shown that groups prefer to collaborate within private workspaces [8].

Furthermore, 92% of students answered that they would like to use the same e-collaboration tools again if they had access to them in the future. Although social media and e-collaboration in general can support collaboration, some problems still remain open. Even if the use of social media saves time, it does not solve the perceived lack of time of students owing to overlapping course assignments or exams.

For some students Podio seems to be too complex and it was stated that it is: "difficult to get started on a platform with a lot of features."

To answer our research question "What are the factors that influence the students' overall satisfaction with collaboration?" Spearman's Rho was used to correlate the satisfaction with overall group collaboration.

The results are summarized in the model presented in Fig. 2, ** marking a correlation on significance level of 0.01. As shown in Figure 2, the highest influence factor on overall group collaboration satisfaction is the "enjoyment of collaboration with peers" (0.805). Other significant factors positively correlated with the satisfaction are the collaboration effect on learning and inspiration, an equal contribution of team members and the positive evaluation of end results of collaboration.

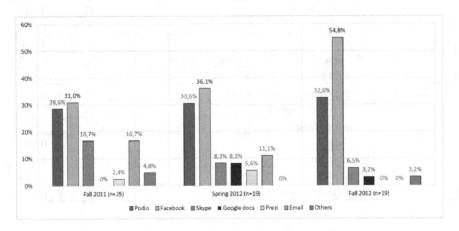

Fig. 1. The usage of e-collaboration tools along the three semesters (n=63)

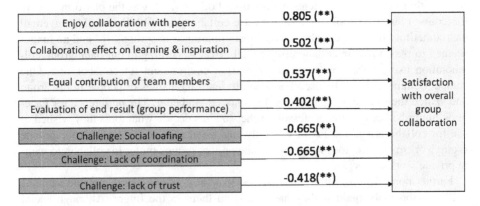

Fig. 2. Significant Influencing Factors on the students' overall group collaboration satisfaction

As shown in Figure 2, the highest influence factor on overall group collaboration satisfaction is the "enjoyment of collaboration with peers" (0.805). Other significant factors positively correlated with the satisfaction are the collaboration effect on learning and inspiration, an equal contribution of team members and the positive evaluation of end results of collaboration. Three major challenges that negatively influence the collaboration satisfaction are: social loafing and lack of coordination and trust. Interestingly, according to our data, e-collaboration does not seem to play a significant role influencing satisfaction.

Students in Denmark are used to working in groups and are not graded based on course participation, group work or presence. They have the freedom to participate in the class or to self-study at home during the semester and, even if they don't attend classes, they are still allowed to take the exam. The course did not prescribe any specific type of collaboration. Although we also found, as reported in [13], that interactive software is useful for collaboration, we have only limited insights into e-collaboration in our study. This is due to the fact that the students could select the type of collaboration themselves. The students in our study used social media for e-collaborating, getting to know their group members better and to coordinate work. Similar benefits of social network applications results were found in [20] and [21].

4   Conclusions and Future Work

The present study aimed to shed light on factors that influence collaboration in the classroom and present how students collaborate using various technologies for group work. Drawing on previous studies and theory, a questionnaire was designed to evaluate collaboration and the usage of collaboration technology for group work. As already pointed out, to our knowledge and according to the literature, not many studies discuss factors influencing collaboration and social media collaboration within the classroom. Despite the fact that students use various collaboration technologies for different purposes and different tools including social media, e-collaboration does not seem to be an influencing factor for satisfaction with collaboration. According to our findings, despite being heterogeneous with different backgrounds (coming from different study programs and countries) digital native students did not experience conflict and seemed to be quite satisfied with their collaboration and collaboration end results. Students do not think that e-collaboration is difficult, but rather fun and that they pretty much benefit from e-collaboration. They also think that e-collaboration is very much needed. The students also agree that group work collaboration has helped them advance their project ideas and evaluate the end-result of their collaboration (in terms of presentation, learning, inspired with new ideas, social interaction) as: really good. More data will be collected and analyzed in future studies. In addition, it would be interesting to compare results across different courses using group work and collaborative technology.

References

1. Tapscott, D.: Grown up digital. McGraw-Hill, New York (2009)
2. Mondahl, M., Rasmussen, J., Razmerita, L.: Web 2.0 Applications, Collaboration and Cognitive Processes in Case-based Foreign Language Learning. In: Lytras, M.D., et al. (eds.) WSKS 2009. LNCS, vol. 5736, pp. 98–107. Springer, Heidelberg (2009)

3. Baloian, N., Pino, J.A., Hoppe, U.H.: Dealing with the Students' Attention Problem in Computer Supported Face-to-Face Lecturing. Journal of Educational Technology & Society 11(2), 192–205 (2008)

4. Razmerita, L.: Collaboration using Social Media: The case of Podio in a voluntary organization. In: Antunes, P., Gerosa, M.A., Sylvester, A., Vassileva, J., de Vreede, G.-J. (eds.) CRIWG 2013. LNCS, vol. 8224, pp. 1–9. Springer, Heidelberg (2013)

5. Razmerita, L., Brun, A.: Collaborative Learning in Heterogeneous Classes: Towards a Group Formation Methodology. In: Proceedings of 3rd International Conference on Computer Supported Education (CSEDU 2011). SciTePress – Science and Technology Publications, Nordwijkerhout (2011)

6. Lou, Y., Abrami, P.C., Spence, J.C., et al.: Within-class grouping: A meta-analysis. Review of Educational Research 66(4), 423–458 (1996)

7. Prince, M.: Does active learning work? A review of the research. Journal of Engineering Education 93(3), 223–231 (2004)

8. Mergendoller, J.R., Maxwell, N.L., Bellisimo, Y.: Comparing problem-based learning and traditional instruction in high school economics. The Journal of Educational Research 93(6), 374–382 (2000)

9. Springer, L., Stanne, M.E., Donovan, S.S.: Effects of small-group learning on undergraduates in science, mathematics, engineering, and technology: A meta-analysis. Review of Educational Research 69(1), 21–51 (1999)

10. Ku, H.-Y., Tseng, H.W., Akarasriworn, C.: Collaboration factors, teamwork satisfaction, and student attitudes toward online collaborative learning. Computers in Human Behavior 29(3), 922–929 (2013)

11. Greenlee, B.J., Karanxha, Z.: A study of group dynamics in educational leadership cohort and non-cohort groups. Journal of Research on Leadership Education 5(11), 357–382 (2010)

12. Janssen, J., Erkens, G., Kirschner, P.A., Kanselaar, G.: Influence of group member familiarity on online collaborative learning. Computers in Human Behavior 25(1), 161–170 (2009)

13. Tseng, H., Ku, H.-Y., Wang, C.-H., Sun, L.: Key Factors in Online Collaboration and Their Relationship to Teamwork Satisfaction. Quarterly Review of Distance Education 10(2), 195–206 (2009)

14. Oliveira, I., Tinoca, L., Pereira, A.: Online group work patterns: How to promote a successful collaboration. Computers & Education 57(1), 1348–1357 (2011)

15. Razmerita, L., Kirchner, K., Nabeth, T.: Social Media In Organizations: Leveraging Personal And Collective Knowledge Processes. Journal of Organizational Computing and Electronic Commerce 24(1), 74–93 (2014)

16. McConnell, D.: E-learning groups and communities. McGraw-Hill International (2006)

17. Mondahl, M., Razmerita, L.: Social Media, Collaboration and Social Learning-a study of Case-based Foreign Language Learning. The Electronic Journal of e-Learning (EJEL) 12, 20 (forthcoming, 2014)

18. Turel, O., Zhang, Y.: Should I e-collaborate with this group? A multilevel model of usage intentions. Information & Management 48(1), 62–68 (2011)

19. McLoughlin, C., Lee, M.J.W.: Personalised and self regulated learning in the Web 2.0 era: International exemplars of innovative pedagogy using social software. Australasian Journal of Educational Technology 26(1), 28–43 (2010)

20. Salaway, G., Caruso, J., Nelson, M.: The ECAR Study of Undergraduate Students and Information Technology, vol. (8). EDUCAUSE Center for Applied Research, Boulder (2008)

21. Rutherford, C.: Using online social media to support preservice student engagement. MERLOT Journal of Online Learning and Teaching 6(4), 703–711 (2010)

An Effort of Communication Measure
for Synchronous Collaborative Search Systems*

Rolando Salazar-Hernández[1], Clarisa Pérez-Jasso[1],
Julio Rodríguez-Cano[2], and Edgar Pérez-Perdomo[2]

[1] Unidad Académica Multidisciplinaria Mante, Universidad Autónoma de Tamaulipas,
Tamaulipas 89840, México
{rsalazar,clperez}@uat.edu.mx
[2] Departamento de Ingeniaría Informática, Universidad Oscar Lucero Moya de Holguín,
Holguín 80100, Cuba
{jcrodriguez,eperezp}@facinf.uho.edu.cu

Abstract. Collaborative work in information seeking and retrieval scenarios is difficult to measure. This paper is about an initial attempt to evaluate the effort of communication, one of the most essential components of collaboration, over synchronous collaborative search (CS) sessions. CS is about explicit collaboration among individuals engaged in a common search task, using groupware technologies to satisfy the shared information needs. Strictly speaking, CS field is not limited to synchronous interaction because some situations are concerned with asynchronous, but explicit collaboration. Nevertheless we consider more attractive and novel the situations that requiring support to coordinate synchronous users' activities in CS systems to measure the effort of communication.

Keywords: communication, collaborative search, groupware, measure.

1 Introducction

What is collaborative search (CS)? Although this phrase means many things to many people, we define it as the field in charge of establishing techniques and methods to satisfy the shared information needs of users associated in a collaborative search session, that work together as a team, extending the information seeking and retrieval process with the knowledge about the user's common goals, the search context, and the explicit collaborative behavior among them. Currently there is not a community consensus about the name of this research field. Some authors prefer, among others less used, the terms "Collaborative Information Retrieval (CIR)", or "Collaborative Information Seeking (CIS)"[1].

Is important to remark that the phrase collaborative information retrieval has been used in the past to refer to many different technologies which support collaboration in the IR process. Much of the early work about collaborative information retrieval

---

* This work has been supported by the Unidad Académica Multidisciplinaria Mante, Universidad Autónoma de Tamaulipas.

[1] http://en.wikipedia.org/wiki/Collaborative_information_seeking

N. Baloian et al. (Eds.): CRIWG 2014, LNCS 8658, pp. 287–293, 2014.

has been concerned with asynchronous, remote collaboration via the reuse of previous search results and processes in Collaborative- Filtering as Recommender Systems (RSs), Collaborative Re-ranking, and Collaborative Footprinting Systems. Asynchronous collaborative information retrieval supports a passive, implicit form of collaboration where the focus is to improve the search process for an individual. RSs are software tools and techniques providing suggestions or recommendations for items to be of use to a user. The suggestions relate to various decision-making processes, such as what items to buy, what music to listen to, or what online news to read [11]. One of the framework approach for RSs, as we already said, is termed Collaborative-Filtering (CF) and its rationale is that if an active user agreed in the past with some users, then the other recommendations coming from these similar users should be relevant as well and of interest to the active user. Please, note that in CF approach, the collaboration above-mentioned is implicit; the active user does not know the rest of the users filtered by the RSs.

The software infrastructures of CS systems are groupware (*software and hardware for shared interactive environments*) technologies conforming. A groupware provides computer support for group work. At a general level, group work includes written and spoken communication, meetings, shared information, and coordinated work. Some group work occurs when people interact whit each other at the same time (synchronously) and in different places (remote or distributed). Face-to-face meetings are an example of people working together at the same time and often in the same place (co-located). People also can work together at different times (asynchronously).

CS represents a significant paradigmatic shift from traditional Information Retrieval (IR) systems, because some researchers have identified many context where the users (e.g., friends, colleagues, classmates) collaborate actively during the search process, such as health-care teams, search-driven software development, digital libraries, and learning environments, among others [1,8,2]. According to Chirag Shah [12], this novel kind of social search owes its popularity in these days for two fundamental reasons. First, using Computer-Supported Cooperative Work (CSCW) frameworks to understand and support collaborative behavior has been around for a while, but it is in the recent years that we have seen more specialized attention given to applying CSCW methods and techniques for IR situations. On the other hand, the field of IR has found the importance of considering social and collective aspects of search and information management.

CS systems may facilitate greatly our information seeking interactions, allowing fluid communication with people who have the same information needs that we have. CS systems and prototypes in use today include *Cerchiamo* [2] that support co-located partners searching online video collections. Each partner has a personal displaying role-specific information, and a shared display shows information relevant to both parties; *iBingo* [13], a collaborative video search system for mobile devices that supports division of labor among users, providing search results to co-located iPod Touch devices; *SearchTogether* and [9], *Coagmento* [5] support remote collaborative Web search, both synchronous and asynchronous, among groups of users. Thus, many CS workshops have recently been held indistinctly at relevant conferences like JCDL, GROUP, ASIS&T, CIKM, and CSCW from the year 2008 to the present.

Traditional IR systems have been evaluated based on recall/precision measures, sometimes with averaging techniques. These metrics assume a single seeker (even if more than one person contributed to the final search results. We need to develop new metrics to assess the contributions of multiple team members to CS processes. Approaches such as averaging by the number of participants do not reflect the reality that marginal rates of finding documents often decrease with the addition more searchers. Furthermore, different team members may contribute in different ways, depending on their roles. Our primary objective in this research is to investigate the effort of communication through the synchronous interactions among users of CS systems.

This paper is organized as follows. We first present a brief overview of related works and place our research in context in Section 2. We then describe in Section 3 our measure of effort of communication. Finally, we dedicate Section 4 to the conclusion remarks.

2 Related Work

Several studies have investigated the effect of communication context on the performance of collaborative teamwork. Some studies have examined the effect of computer-mediated communication. Other studies have compared face-to-face and chat communication, and face-to-face, co-located, and video-mediated communication. Early studies of J. Grudin [6] showed that communication is necessary for system designers of collaborative solutions to understand the costs and benefits related to communication choices under various circumstances. R. González-Ibáñez et al. remark in [4,3] that a common finding in communication research is that physical proximity facilitates communication by increasing spontaneous interactions. Although the literature has shown that communication can facilitate greater exchange of high-quality information, research has also demonstrated negative effects of social interaction on problem solving, generation of ideas, and affective as well as cognitive load. Also, they compare the CS process and products of teams across three different communication contexts (co-located, text chat, audio/text chat) within the theoretical framework of computer-mediated communication. Besides communication needs, introducing support for collaborative search as well as sense-making could help to improve exploratory search experience [9,10].

On the other hand, CS community has proposed some effectiveness evaluation measures for CS systems like viewed precision/recall and selected precision/recall [2], i.e. P_v, the fraction of documents - seen by the group members - that were relevant and P_s, the fraction of documents - judged relevant by the group members - that were marked relevant in the ground truth, and $R_{v|s}$ as their dependent measures. These metrics can measure the performance of the systems in general, but we want to measure our proposal separately, without take in mind all the features of a standard system CS. In such case, we focus in the interaction of the users with a system, another important perspective of systems evaluation, especially in systems with multi-user interfaces.

The majority of CS studies are ethnographic and relies primarily on direct observation and interviewing. J. Kim note in [7] that most studies have been predominantly descriptive and exploratory in nature, and little has been based on existing conceptual or theoretical foundation. It was also found that there is an observable preference toward qualitative studies using observation and interviews. Therefore, the introduction of new approaches in CS is an important area of research that should be better studied by the community.

3   CS Systems Performance Analysis

Even if CS is gaining many popularity for these days, we want to clarify that collaboration is not always useful. Even when collaboration is desired or encouraged, it could induce additional effort that includes the cost to coordinate tasks and teammates. Working in a team may not be beneficial if the participants have conflicts of interest, or they do not trust one another. Therefore, the collaborative search systems are important and necessary only in the situations that really need it. The users of CS systems without doubts, hope that these new kind of search tool should allow them to satisfy their shared information needs with major quality and efficiency. Surely, also they hope to do all of this without a lot effort of communication, or at least, with less effort than using traditional IR systems and some groupware as complement, for instance, instant messaging in chat rooms or email. Otherwise, we would have a great rejection of the CS systems in the online communities. Therefore, our goal is to measure the effort of communication.

3.1   Effort of Communication Measure Description

$$P_s = \frac{|D_{rel} \cap D_{ret} \cap D_{sel}|}{|D_{ret}|} = \frac{|D_{rel} \cap D_{sel}|}{|D_{ret}|}, D_{sel} \subseteq D_{ret} \qquad (1)$$

$$R_s = \frac{|D_{rel} \cap D_{ret} \cap D_{sel}|}{|D_{rel}|} = \frac{|D_{rel} \cap D_{sel}|}{|D_{rel}|}, D_{sel} \subseteq D_{ret} \qquad (2)$$

A good CS system should have high $P_{v|s}$, $R_{v|s}$ (Equations 1 and 2, respectively), and low effort of communication (E). To summarize effectiveness in a single number is convenient to use F measure.

We will propose the effort of communication measure as a comparison point. The effort is the physical activity performed by a person while attempting to accomplish a specific goal, in our case communication in collaborative search tasks. In CS, physical activities could include pressing a key on a search box to type query terms, pressing a mouse button to recommend an relevant item and many others interactions in order to satisfy the shared information needs across communication among group members. Thus, it may be possible to calculate E in CS systems from a weighted sum of explicit recommendations (R), instant messages sent (I), and queries typed that (Q); the most general and illustrative group members activities around the widgets in a CS system interface.

Our definition of E uses continuous functions and is similar with the measure proposed by D. Tamir et al. in [14]. In practice, given its discrete nature, our measure are quantized by converting integrals to sums. Assume that an interactive collaborative search session starts at time t_0, we define E at time t as fallow:

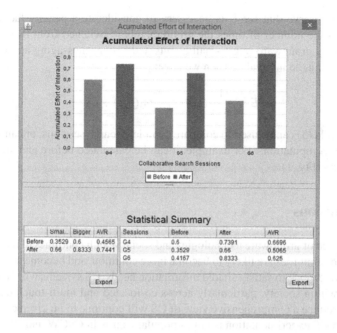

Fig. 1. Effort Accumulated

$$E(t) = \frac{1}{\varphi(t - t_0)} \int_{t_0}^{t} (\omega_1 R(t) + \omega_2 I(t) + \omega_3 Q(t)) dt \qquad (3)$$

Where: $R(t)$, $I(t)$, $Q(t)$ are the number of explicit recommendations, the number of instant messages, and the total of query terms typed by the group members during the time interval $t - t_0$ respectively; φ is the size of the group; and ω_{1-3} are relative constants of every type of explicit interaction (R, I, and Q), which consider in major or minor measurement a weight value for each of these. Finally, we want to remark that $E(t)$ is a monotonically increasing function.

Now, we will consider only static CS interfaces, this means that the multi-user interfaces are determined at design time and cannot be changed. A direction of future research of this work is to consider a dynamic scenario where a CS system can adapts to the group members and enables interfaces improvements at run time. Under the assumption of a static scheme we can ignore the shape of the curve of $E(t)$, and only use accumulated effort of communication at time of completion (τ) of a collaborative search task. Additionally, we assume that $R(t_0) = I(t_0) = Q(t_0) = 0$. Then, in order to derive the relation between $E(t)$ and the static scheme of interfaces, is possible to define the accumulated effort of communication in the following way:

$$E = \frac{1}{\varphi \cdot \tau}((\omega_1 R(\tau) + \omega_2 I(\tau) + \omega_3 Q(\tau)) \qquad (4)$$

For an user group with shared information needs that use a traditional IR systems with instant messaging, email or any other groupware for communication purpose is important to observe that for E is necessary to considerate a penalty factor (ρ) that can measures the number of times the group members switched from IR systems to a complementary groupware (e.g. e-mail clients and instant messaging applications) or vice versa during the interval τ. Then we have:

$$E = \frac{1}{\varphi \cdot \tau}(\omega_2 I(\tau) + \omega_3 Q(\tau) + \rho) \tag{5}$$

This formula (5) can be used to compare complete search sessions, or can also be set intermediate computations. For example, it can be calculated before and after a group gain members (Fig. 1).

4   Conclusions

The acceptance of groupware technologies and social search continues to increase. As a result, CS communities are likely to become more common, and become more accepted as a productive mode to satisfy shared information needs. CS systems will also continue to develop within society, particularly across co-located and multi-touch devices. But studies to produce a comprehensive view of CS will continue to be diverse.

Although CS experimentation is not as popular as it is in CSCW and IR, clearly experiments, simulations and mathematic analysis are further trends that have an important role in understanding the best mechanisms to assist the group members with shared information needs. In this work we propose a measure for the evaluation of communication processes of teams in information seeking and retrieval tasks across different inputs, which may have an impact on the performance analysis of collaborative teams involved in CS tasks.

References

1. Foley, C.: Division of Labour and Sharing of Knowledge for Synchronous Collaborative Information Retrieval. PhD thesis, Dublin City University, School of Computing and Centre for Digital Video Processing (June 2008)
2. Golovchinsky, G., Adcock, J., Pickens, J., Qvarfordt, P., Back, M.: Cerchiamo: a collaborative exploratory search tool. In: CSCW 2008, San Diego, California, USA, pp. 8–12 (2008)
3. González-Ibánez, R., Haseki, M., Shah, C.: Let's search together, but not too close! an analysis of communication and performance in collaborative information seeking. Information Processing and Management (2013)
4. González-Ibánez, R., Shah, C.: A proposal for measuring and implementing group's affective relevance in collaborative information seeking. In: HCIR 2009: Proceedings of the IV Workshop on Human-Computer Interaction and Information Retrieval, New Brunswick, USA (2010)
5. González-Ibáñez, R., Haseki, M., Shah, C.: Let's search together, but not too close! an analysis of communication and performance in collaborative information seeking. Information Processing and Management (2013)
6. Grudin, J.: Groupware and social dynamics: Eight challenges for developers. Communications of the ACM 37(1) (1994)

7. Kim, J.: Collaborative information seeking: A theoretical and methodological critique. In: CIS 2013: 3rd Workshop on Collaborative Information Seeking At the ACM CSCW 2013 Conference, San Antonio, TX, USA (2013)
8. Morris, M.R.: A survey of collaborative web search practices. In: CHI 2008: Proceeding of the Twenty-Sixth Annual SIGCHI Conference on Human Factors in Computing Systems, pp. 1657–1660. ACM, New York (2008)
9. Morris, M.R., Horvitz, E.: Searchtogether: an interface for collaborative web search. In: UIST 2007: Proceedings of the 20th Annual ACM Symposium on User Interface Software and Technology, pp. 3–12. ACM, New York (2007)
10. Morris, M.R., Teevan, J.: Collaborative Search: Who, What, Where, When, Why, and How. Morgan&Claypool Publishers (2010)
11. Ricci, F., Rokach, L., Shapira, B., Kantor, P.B. (eds.): Recommender Systems Handbook. Springer (2011)
12. Shah, C.: Collaborative Information Seeking: The Art and Science of Making the Whole Greater than the Sum of All. Springer (2012)
13. Smeaton, A.F., Foley, C., Byrne, D., Jones, G.J.: iBingo mobile collaborative search. In: CIVR 2008 (2008)
14. Tamir, D., Komogortsev, O.V., Mueller, C.J.: An effort and time based measure of usability. In: WoSQ 2008, Leipzig, Germany (2008)

Using Structural Holes Metrics
from Communication Networks
to Predict Change Dependencies

Igor Scaliante Wiese[2], Rodrigo Takashi Kuroda[1],
Douglas Nassif Roma Junior[2], Reginaldo Ré[2],
Gustavo Ansaldi Oliva[3], and Marco Aurélio Gerosa[3]

[1] PPGI - UTFPR/CP
rodrigokuroda@gmail.com
[2] UTFPR/CM
nassifrroma@gmail.com, {igor,reginaldo}@utfpr.edu.br
[3] Department of Computer Science - IME/USP
{goliva,gerosa}@ime.usp.br

Abstract. Conway's Law describes that software systems are structured according to the communication structures of their developers. These developers when working on a feature or correcting a bug commit together a set of source code artifacts. The analysis of these co-changes makes it possible to identify change dependencies between artifacts. Influenced by Conway's Law, we hypothesize that Structural Hole Metrics (SHM) are able to identify strong and weak change coupling. We used SHM computed from communication networks to predict co-changes among files. Comparing SHM against process metrics using six well-known classification algorithms applied to Rails and Node.js projects, we achieved recall and precision values near 80% in the best cases. Mathews Correlation metric was used to verify if SHM was able to identify strong and weak co-changes. We also extracted rules to provide insights about the metrics using classification tree. To the best of our knowledge, this is the first study that investigated social aspects to predict change dependencies and the results obtained are very promising.

Keywords: structural holes metrics, social network analysis, change dependencies, communication network, Conway's law.

1 Introduction

A good understanding of the impact of communication and cooperation on software evolution can help researchers and practitioners to gain more insights to improve software quality [25,21]. According to some researchers [22,9,18], software modularity and quality can be associated with the interactions among developers. This phenomena was described by Conway, who states that "organizations which design systems are constrained to produce designs which are copies of the communication structures of these organizations" [11]. In this way, Conway's Law assumes a strong association between the software architecture and

N. Baloian et al. (Eds.): CRIWG 2014, LNCS 8658, pp. 294–310, 2014.

the communication structure [2]. During software evolution, developers communicate asynchronously in discussions related to change requests and cooperate around this requests changing files to fix bugs or add new features. The resulting changes may increase the source code complexity and disorganization [8] and are a social product [28]. For example Node.js (Table 2) used the constraintAVG (average of the lack of holes among neighbors) and hierarchySUM (sum of the concentration of constraints to a single node) six times. Considering Rails, neither of these metrics were frequently selected. For Rails (Table 3),effectiveSizeAVG (average of the portion of non-redundant neighbors of a node) and efficiencyAVG (average of the normalization of the effective size by the number of neighbors) – the others two SHM – were selected more often. Ball et al. [1] introduced the concept of co-changes (a.k.a. logical coupling and change dependencies) that involves the notion that some artifacts frequently change together during software development. Previous works have suggested that change dependencies in files influence the quality of software and indicated that entities that changed together in the past are likely to change together in the future [12]. Hence, finding insights of how these co-changes happen in software projects and how their social structures were organized can help developers planning, deciding, accommodating certain types of changes, and tracing through the effects of co-changes in software quality and architecture [20,9].

To understand the impacts of the communication structures, previous studies have been using graph-based analysis [27,4]. In fact, graph-based analysis have shown good results in several different works, for example, to predict defects, to estimate bug severity, to prioritize refactoring efforts, to predict defect-prone releases, and to understand socio-technical aspects [3,6,20,9,25].

Different types of metrics in Social Network Analysis (SNA) have been used to explore the graphs generated from software engineering data [27,3,6,20]. Among them, Structural Holes Metric (SHM) can reflect gaps between nodes in a social network indicating that a developer on either side of the hole have access to different flows of information [7,27]. However, to the best of our knowledge, SHM were studied only as part of social network metrics. Thus, as these metrics were merged with others in previous studies, like centrality and ego measures, their importance remains unclear.

Once that Conway's Law can explain the direct influence of the communication on the quality of software architecture, we expect that it also can explain the indirect relation between communication and co-changes. So, influenced by Conway's Law, we hypothesize that SHM has an important role to identify co-changes and can be explored to provide insights about the organization and the way that source code is modified. We claimed that SHM can describe the social organization during software changes. We used them to build classification models to predict strong and weak change coupling among files, whereas we find a lack of understanding about their prediction importance in previous works. Thus, two main questions were investigated: *RQ1. Does the SHM from communication can predict change dependencies? RQ2. Which is the role of SHM from communication to predict co-changes?*

We used four SHM (effective size, efficiency, constraint, and hierarchy) and three process metrics: sum of number of commits of each file coupled (we called "commits"), the number of updates that the co-change was committed (we called "updates"), and the number of distinct developers that committed each co-change (we called "developers"). To conduct this study we gathered data from two open source projects (Rails and Node.js) hosted on GitHub and built the communication networks using the issues of each project to calculate SHM. We used six well-known classification algorithms comparing SHM against process metrics and against all metrics together.

In summary, the main contributions of this study are:

- We found evidences that SHM computed from communication networks can be used to identify strong and weak co-changes;
- We shown that SHM was better than process metrics in some timeframes of analysis considering the two projects analyzed;
- We extracted a set of rules using classification tree that can provide insights of how communication can be related to co-changes.

The rest of the paper is organized as follows. In Section 2, we present related work about logical coupling and SNA. Section 3 shows the study design. We describe the process to collect the data, build the networks, compute the co-changes, and explain the metrics used in the classification approach. In Section 4, we answered RQ1 comparing the results of classification among each set of metrics using communication network, and RQ2 presenting the classification tree rules to compare the importance of each predictor. Section 5 discusses the threats to validity. Finally, conclusions and future work are presented in Section 6.

2   Related Works

Some related studies are presented in this section. First, we present the works related to logical coupling. Then, we discuss about previous works focused on SNA to build prediction models. We highlight that we only fond studies that predict co-changes without using SNA from communication networks and other set of papers that used SNA to predict only faults instead of co-changes.

2.1   Logical Coupling Studies

Some studies on logical coupling focuses on defects in open source projects [13,17]. D'Ambros et al. [13] performed a study on three large software systems (ArgoUML, Eclipse JDT, and Mylyn) and found that there was higher correlation between change coupling and defects than the one observed only with complexity metrics and defects. In turn, Kouroshfar [17] investigated the impact of specific characteristics of co-change dispersion on software quality. He used statistical regression models to show that co-changes that included files from different subsystems resulted in more bugs than co-changes that include files only from the same subsystem.

Zimmermann et al. [30] developed a tool to try to avoid defects insertion during file changes. They built an Eclipse plug-in that collects information about source code changes from repositories and warns the developers about probable missing co-changes. They used association rules to suggest change coupling among files in method and file level. The authors reported precision values around 30% and recommended that for projects with continuous evolution of file changes the analysis should be made in file-level instead of method-level.

Zhou et al. [29] present a study similar to Zimmermann et al., since they proposed a model do predict co-changes. However, the model is more elaborate because it uses different predictors in Bayesian networks to predict the change coupling between source code entities. They extracted features like static source code dependency, past co-change frequency, age of change, author information, change request, and change candidate. They conducted experiments on two open source projects (Azureus and ArgoUML) and reported precision values around 60% and recall values of 40%. These last two papers are more related to our work. They predicted co-changes, but they did not use social metrics. We also tested more classification algorithm than Zhou et. al.

2.2   Social Network Analysis

Some studies on SNA used different types of networks to build prediction models. Wolf et al. [27] reported results indicating that developer communication plays an important role in software quality. They predicted build failure on IBM's Jazz project yielding recall values between 55% and 75%, and precision values between 50% to 76%. Bicer et al. [5] created models to predict defects on IBM's Jazz project and Drupal. Their results revealed that compared to other metrics such as churn metrics, social network metrics considerably decreased high false alarm rates. We considered that these works are more related to this study, whereas they are the only work that used SHM metrics from communication networks to build prediction models. However, these studies did not report the SHM individual performance and were used to predict fault instead of co-changes.

Other works, like Meneely et al. [19] examined the development network derived from code churn information. They conducted a case study and found a significant correlation between file-based development network metrics and failures. Bird et al. [6] evidenced the influence of combined sociotechnical software networks on the fault-proneness of individual software components. They reported results with precision and recall around 85%.

3   Study Design

In this section, we present how we collected the data, calculated the co-changes and the SHM, and built the classification models and evaluated the results.

3.1 Data Collection

We collected data from two open source projects: Rails and Node.js. We chose these projects because of their influential nature on GitHub[1] ecosystem [16]. We considered the number of "stars" and "forks" on GitHub indicating interesting projects. While Rails have more than 21.000 stars and 7.700 forks, Node.js have more than 28.800 stars and 6.100 forks. Table 1 presents some characteristics of each project collected from the Ohloh.net[2]. We present for each project the number of strong and weak co-changes and the percentage of imbalance in our dataset.

We noticed that the timeframes 2012.1 for Node.js and 2012.2 for Rails have the greater imbalance of our dataset. Class imbalance problem may lead to misclassification of the minority class. The classification algorithms normally are more focused on predicting the majority class [26]. For both projects, weak co-changes were the majority class for all timeframes. We observed that Node.js has more co-changes than Rails. The exception was observed on 2012.1, in which Rails and Node.js have similar number of co-changes considering both classes. Node.js presented a very high number of co-changes in 2011.2 compared to all other timeframes for both projects.

Table 1. Projects characteristics on different timeframes

		2011.1		2011.2		2012.1		2012.2	
		Strong	Weak	Strong	Weak	Strong	Weak	Strong	Weak
Node.js	# Co-changes	221	302	4208	5303	56	157	311	485
	% Imbalance	42	58	44	56	26	74	39	61
Rails	# Co-changes	12	19	47	56	65	138	66	193
	% Imbalance	39	61	46	54	32	68	25	75

We used the GitHubAPI[3] to gather data from the history of the source code and pull requests from the project. We grouped all pull requests and commit metadata considering six months as a timeframe, following the approach of Bird et al. [6]. Besides, we excluded pull requests without commits as we needed commits with at least two files.

To count the number of co-changes, we combined all files of the same pull request in pairs. Going through all pull requests, for each timeframe, we accumulated the number of times that a given pair of files occurred in the history of the project. We removed all pairs with less than five couplings in a timeframe. Five couplings were used as support count to remove a co-change from our analysis, the same approach adopted by Zimmermann et al. [30]. The support count determines the number of occurrences of pair of files in distinct pull requests.

---

[1] https://github.com
[2] http://www.ohloh.net/ accessed on 10/04/2014.
[3] GitHubAPI can be access on: http://developer.github.com/

Since we are interested in using classification algorithms to predict how SHM describe the co-changes of files, we categorized the change dependencies into two classes. The first class represents strong change dependencies based on the amount of co-changes. The second class represents pair of files with weak change dependencies. To split the classes, we used the mean of total co-changes identified in each timeframe. Similar approach was used by Bird et al. [6] to study software defects.

3.2   Communication Networks and Structural Holes

A network is commonly represented as a graph G with a set of edges E and a set of nodes N. This structure is used to represent relationships observed from software development process and evolution [3]. In our network, each developer represents a node. Edges represent the exchange of messages between developers. We weighted the edges counting how many times each developer interacted with the others over the sequence of comments made in all pull requests where the co-changes were identified.

Suppose that developer 1 (d1) submitted a pull request and developer 2 (d2) made a comment. We create nodes d1 and d2 and connect d2 to d1 using a directed edge. Suppose that developer d3 write another message. We create the d3 node and connect it to d1 and d2. Before creating a node, we check if the node already existed. When creating an edge, we check if there is already an edge between two developers. If it is the case, we increment the weight value of this edge.

Burt [7] noticed that SHM concerns the notion of redundancy in networks and the degree to which there are missing links between nodes. SHM denote gaps between nodes in a social network or represents that people on either side of the hole have access to different flows of information, indicating that there is a diversity of information flow in the network. In the following, we describe four metrics proposed by Burt.

- **Effective Size** is the portion of non-redundant neighbors of a node. High values represent that many nodes among the neighbors are not redundant. Low values indicate that there is few non-redundant neighbors.
- **Efficiency** normalizes the effective size by the number of neighbors. High values show that many neighbors are non-redundant. Low values indicate a high redundancy among the neighbors.
- **Constraint** measure the lack of holes among neighbors. This measure is based on the degree of exclusive connection. Low values indicate that there is few alternative to access a single neighbor. High values indicate many alternatives to access the neighbors.
- **Hierarchy** measures the concentration of constraint to a single node. High values indicate that the constraint is concentrated in a single neighbor. Low values indicate that the constraint measure is the same for all neighbors.

Since we are interested in predicting co-changes using SHM obtained from communication networks, we needed to convert the SHM measures computed from each developer for each co-change. To convert the SHM measures, we used the maximum, average, and sum to aggregate them. This approach was adapted from Meneely et al. [19].

3.3 Classification Approach

Classification algorithms learn from training sets composed of a set of features. In this study, the features represent properties of both files and interactions of developers and files. Our algorithms predict the change dependencies of a file, classifying them as "strong" or "weak." The algorithms read the training set and learn which features are most useful to differentiate the classes. We used SHM as features to different machine learning algorithms. We also computed the number of commits, number of updates, and number of developers that committed the file as predictors. Number of commits was computed as the sum of all times that each file was committed. Number of updates means the number of times that the pair was committed together. The number of developers was computed as the number of distinct developers that committed the co-change.

Typically, classification studies assess the predictive power of their model using recall, precision, and F-measure. This is a common way to evaluate the performance of prediction approaches and is widely used in related literature [24]. We calculated recall, precision, F-measure, and MCC for each timeframe and compare the results. We used the Confusion Matrix to compute these metrics. The values returned by the Confusion Matrix are True Positive (TP), True Negative (TN), False Negative (FN), and False Positive (FP) [23]. We calculated recall to identify the proportion of instances of a category that the model could successfully identify. We calculated precision to measure the percentage of correct identification in a class [23]. A good model yields both high precision and high recall. However, increasing one often reduces the other. F-measure returns a balanced score of recall and precision, calculated as the harmonic mean of these two metrics [23].

We also used the Matthews correlation coefficient (MCC) to show the quality of our predictions. MCC is a correlation coefficient between the observed and predicted classifications. This measure takes into account TP, FP, TN, and FN values and is generally regarded as a balanced measure, which can be used even if the classes are unbalanced. This measure returns a value between -1 and +1. A coefficient of +1 represents a perfect prediction, 0 means a random prediction and -1 indicates total disagreement between prediction and observations. We calculated the MCC by using the expression [23]:

$$MCC = \sqrt{\begin{array}{l}(((Recall(ClassStrong) + Recall(ClassWeak)) - 1) * \\ (((Precision(ClassStrong) + Precision(ClassWeak)) - 1)\end{array}}$$

We employed a 10-fold cross validation to evaluate the predictive power of the models. The cross-validation approach split the data, in our case a semester, into 10 sets, each containing 10% of the data. One set is used for testing data, and the remaining nine sets are used as training data. Once we have class imbalanced dataset we used cross validation to generate 10 random folds to train and test our hypothesis. We report the average of the 10-fold cross validation. By building the models and predicting change dependencies, we aimed to investigate the influence of SHM in the prediction. We used the Orange tool [15] to create the learning models and evaluate them. The workflow and raw data can be downloaded from https://github.com/igorwiese/criwg2014.

4 Results

The main aim of this work was to verify if SHM computed from communication networks can be used to predict change dependencies. In the following sections, we discuss the results for each research question.

4.1 RQ1. Can SHM from Communication Networks Predict Change Dependencies?

To study change dependencies, we split them into two classes: strong and weak, following the procedures described in Section 3.1. We used the Orange Toolbox [15] and Minitab Statistical Software to analyze the data.

Following Rahman and Devanbu [24] recommendation, we used more than one classification algorithm to reduce the risk of depending on a particular learning technique. We applied 7 well-known classification approaches, namely: Classification Tree (CT); k-Nearest Neighbors (KNN); Neural Networks (NN); Support Vector Machine (SVM); Regression Logistic (RL); Naïve Bayes (NB); and an Ensemble Bagging (EB) combined with the best algorithm found in each timeframe.

Two experiments were conducted in order to respond RQ1. First, we used all metrics together in a single vector of features. Second, we compared each set of metric splitting the SHM and process metrics to evaluate the performance individually.

Table 2 presents the results for each timeframe for Node.js using SHM and process metrics together. We only reported F-measure, recall, precision, and MCC values for class "strong," because these dependencies are more important since they happen more often during each timeframe of analysis. We highlighted the best values of each evaluation metric in bold. This experiment was performed to check how the whole set of metrics could predict change dependencies.

The results show that we achieved high results of F-measure, precision, and recall at least in three of four periods. Considering 2011.1 using RL, we got 0.75, 0.78, and 0.73 respectively. For 2011.2, we achieved the best results with Neural network. The worst results comparing each semester was in 2012.1. Using NB, we got 0.50, 0.40, and 0.67. Using Bagging+NB, we got better values for precision, but we lost performance to precision. The same happened in 2012.2 using Bagging+KNN. For the last timeframe using KNN, we achieved 0.67, 0.67, and 0.65.

Table 2. Node.js Classification Results

| | Cross Validation - 10 Fold | | | | | | | | | | | | | | | |
| | 2011.1 | | | | 2011.2 | | | | 2012.1 | | | | 2012.2 | | | |
	F1	Prec	Rec	MCC	F1	Prec	Rec	MCC	F1	Prec	Rec	MCC	F1	Prec	Rec	MCC
CT	0.73	0.73	0.74	0.54	0.61	0.65	0.58	0.34	0.30	0.34	0.26	0.09	0.64	0.67	0.62	0.43
KNN	**0.75**	0.75	**0.76**	0.57	0.60	0.63	0.58	0.31	0.47	0.45	0.50	**0.28**	**0.67**	0.67	**0.65**	**0.48**
NN	0.74	0.76	0.72	0.56	**0.65**	**0.70**	**0.61**	**0.41**	0.29	0.39	0.23	0.12	0.62	0.75	0.53	0.45
SVM	0.69	0.68	0.70	0.46	0.57	0.69	0.49	0.34	0.00	0.00	0.00	0.00	0.56	0.57	0.55	0.29
RL	**0.75**	**0.78**	0.73	**0.59**	0.61	0.69	0.54	0.37	0.33	0.45	0.26	0.18	0.60	0.69	0.52	0.40
NB	0.64	0.58	0.71	0.34	0.56	0.68	0.45	0.32	**0.50**	0.40	**0.67**	**0.28**	0.57	0.53	0.62	0.27
EB	**0.75**	**0.78**	0.71	0.58	0.61	0.69	0.55	0.37	0.35	**0.48**	0.28	0.21	0.62	**0.77**	0.52	0.47

We also reported the MCC values. As mentioned in Section 3.3, MCC metric means the correlation of the algorithms and features to predict both classes. For all timeframes, we achieved results ranging from 0.28 to 0.59. We can observe that for 2011.1, 2011.2, and 2012.2 we got good results showing moderate positive correlation to predict both classes. The worst result can be related to the high number of instances and the class imbalance problem noticed and reported in Table 1.

We observed that in each timeframe, we got better results using different classification algorithms. Trying to find the better algorithm for Node.js, we followed D'ambros et al. [14] recommendations to pairwise comparing classification algorithms. We used Mann-Whitney U test (at 95% confidence level) to conduct the comparisons. This test is non-parametric [10] and compare two group of means. The null hypothesis of this test indicates that the mean of the two groups are equal. This way, we grouped the results of F-measure and MCC for each algorithm for all timeframes. After this analysis, we could not reject the null hypothesis at 95% of significance. Thus, there is not a best algorithm because all of them had similar results.

We analyzed the variability of each algorithm using boxplot for both evaluation metrics. Figure 1 shows that NB and CT had the smaller variability among all algorithms. We observed that SVM algorithm presented the highest variability.

We performed the same analysis on Rails and we noticed similar results to the ones found in the Node.js project. Table 3 presents the results for Rails. We obtained the best results of F-measure, precision, recall, and MCC for 2011.1 using NB (0.69, 0.64, 0.75, and 0.47). For the second period, KNN got the best results (0.72, 0.70, 0.71, and 0.47). For 2012.1 and 2012.2, we got the best results using CT. However, we highlight that NB always achieved better values of recall considering all timeframes. The worst results were obtained for 2012.1.

We tried again to find the best algorithm by means of the Mann-Whitney U test (at 95% confidence level) comparing the mean of F-measure and MCC of each algorithm. Once again, we did not reject the null hypothesis.

Figure 2 presents the variability of each evaluation metric of the classification algorithms. Similar to Node.js, we found that CT and NB presented the smaller variability for both evaluation metrics. SVM obtained the highest variability of

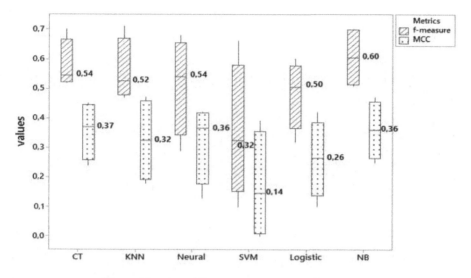

Fig. 1. Boxplot of F-measure and MCC for Node.js

Table 3. Rails Classification Results

| | Cross Validation - 10 Fold | | | | | | | | | | | | | | | |
| | 2011.1 | | | | 2011.2 | | | | 2012.1 | | | | 2012.2 | | | |
	F1	Prec	Rec	MCC	F1	Prec	Rec	MCC	F1	Prec	Rec	MCC	F1	Prec	Rec	MCC
CT	0.52	0.54	0.50	0.24	0.70	0.69	0.72	0.45	**0.53**	0.52	0.55	0.31	**0.56**	0.61	0.53	**0.43**
KNN	0.50	0.50	0.50	0.18	**0.71**	0.70	0.72	**0.47**	0.47	0.47	0.47	0.23	0.55	0.62	0.50	0.42
NN	0.58	0.58	0.58	0.32	0.68	0.68	0.68	0.42	0.29	0.46	0.21	0.13	0.50	**0.70**	0.39	0.41
SVM	0.31	0.42	0.25	0.04	0.66	0.67	0.65	0.39	0.0	0.0	0.0	0.0	0.34	0.57	0.24	0.25
RL	0.50	0.62	0.41	0.28	0.60	0.59	0.61	0.25	0.32	0.41	0.26	0.10	0.51	0.69	0.40	0.42
NB	**0.69**	**0.64**	**0.75**	**0.47**	0.70	0.63	**0.78**	0.41	0.52	0.45	**0.61**	0.25	0.51	0.41	**0.68**	0.31
EB	0.64	0.61	0.66	0.39	0.67	**0.71**	0.63	0.42	0.52	**0.58**	0.47	**0.33**	0.51	0.63	0.43	0.40

F-measure and MCC for Rails. Once we achieved good results for both projects using all metrics together, we performed a more fine-grained analysis to answer RQ1 comparing each set of metrics independently. To perform this analysis, we split SHM and process into two groups of metrics. We also randomly selected only three classification algorithms, since we did not find statistical difference among them.

Comparing the results of F-measure and MCC of each set of metrics, we found that SHM had better results on the last semester (2012.2), and process metrics got better results on the first semester. Considering the second and third timeframes, the results were similar. Due to space constraints, we will not show the complete summarization of these results, but we provide an addendum with the results on our GitHub repository (https://github.com/igorwiese/criwg2014).

To statistically compare the set of metrics, we used Mann-Whitney U test (at 95% confidence level). Once again, we could not rejected the null hypothesis, showing that both set of metrics had similar performance to predict change dependencies. Figure 3 presents the results of each evaluation metric for both

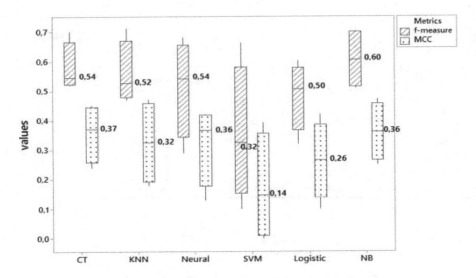

Fig. 2. Boxplot of F-measure and MCC for Rails

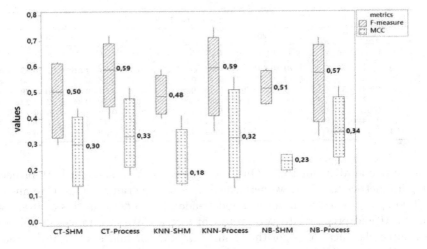

Fig. 3. Comparison among set of metrics and algorithms for Node.js

SHM and process metrics for Node.js. We observed that NB and KNN had the smaller variability using SHM metrics. However, the best result was achieved using KNN+process metrics.

Performing the same analysis for Rails, we achieved good results using SHM on 2011.1. For the other three semesters, we observed a small difference among the set of metrics. We did not find statistical difference between SHM and process metrics. We also present the boxplot to provide insights of how each set of metric works. Figure 4 presents the results for Rails. We can observe that the variability

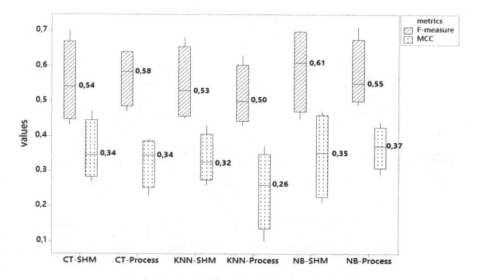

Fig. 4. Comparison among the set of metrics and algorithms for Rails

among the algorithms and set of metrics are very similar. The best results were obtained using CT and NB with SHM metrics.

As we reported in this section, for both projects, SHM and process metrics can be used to predict change dependencies. Together they can improve the results. However, they also can be used as individual predictors without losing predictive power in terms of recall, precision, F-measure, and MCC.

4.2   RQ2. Which Is the Role of SHM from Communication to Predict Co-changes?

To provide insights about how each metric help to predict change dependencies, we used the Orange tool [15] to mine rules from classification tree. We started the analysis building the classification tree for both projects using only SHM, because we did not find statistical difference among the set of metrics and we were interested in investigating the prediction performance of SHM. Then, we analyzed the results combining both set of metrics together.

Figure 5 presents the CT for Node.js for 2011.1 timeframe. The root metric represents the most relevant metric. In this case, constraintAvg was the most relevant metric. Using constraintAvg higher than 0.503 and smaller or equal than 0.503, we split the tree in two parts. The much closer from the root of the tree, the most important the metric is. Red squares means that we found a rule to describe "strong" class. The blue square means that "weak" class was described. For example, we found the rule "IF constraintAVG > 0.503 and hierarchySUM > 2.594 THEN class="strong" to the red square on the third level on the left side. We can observe that using this rule, 27 instances can be predicted as "strong" coupling. Considering the right side, we showed two examples of rules using

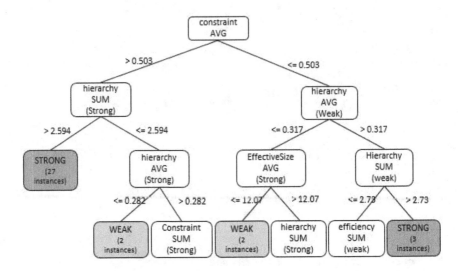

Fig. 5. Classification Tree for Node.js 2011.1

hierarchyAVG on the second and third level consecutively: (i) constraintAVG
<= 0.503 and hierarchyAVG <=0.317 THEN "strong," (ii) constraintAVG >
0.503 and hierarchySUM <= 2.594 and hierarchyAVG > 0.282 THEN "strong."

To rank the metrics we counted the amount of times that each metric appeared on the first four levels of each tree for both projects. For Node.js, we ranked the metrics following this order: constraintAVG (6), hierarchySUM (6), hierarchyAVG (3), efficiencySUM (3), constraintMAX (2), constraintSUM (20), effectiveSizeSUM (2), effectiveSizeAVG (1), and efficiencyAVG (1). Considering Rails, we ranked the metrics following this order: effectiveSizeAVG (3), efficiencyAVG (3), effectiveSizeSUM (2), effectiveSizeMAX (1), efficiencySUM (1), constraintSUM (1), constraintAVG (1), and hierarchyAVG (1).

Conducting these analyses, we found that all SHM were used, however "effective size" were most frequently selected. Comparing Rails with Node.js, we noticed that Rails needed the less amount of SHM to predict change dependencies than Node.js. We also investigated the classification tree using all metrics together. We ranked all metrics following this order to Node.js project: commits (19), updates (14), effectiveSizeSUM (6), efficiencySUM (2), constraintMAX (3), hierarchyMAX (2), efficiencyAVG (2), constraintAVG (2), hierarchyAVG (1), and hierarchySUM (1). For Rails project, we can ranked the metrics following this order: commits (8), updates (3), efficiencyAVG (3), effectiveSizeSUM (3), effectiveSizeAVG (2), constraintSUM (2), efficiencySUM (2), efficiencyMAX (1), hierarchySUM (1), hierarchyAVG (1), and constraintAVG (1).

The root position on the classification tree is the most important metric. We found two process metrics at this position (commits and updates) and one SHM (constraintMAX - 2 times) considering Node.js. For Rails, we observed similar results, finding two process metrics (commits and updates) and two SHM (effectiveSizeAVG and efficiencySUM).

Only to present some examples, we choose three rules mined from classification tree used on Node.js project to predict change dependency. This three rules were mined from communication network for 2011.1 timeframe: (i) IF commits $<=$ 86.0 AND updates $<=$7.5 AND effectiveSizeSUM $<=$115.64 THEN class="weak", (ii) IF efficiencySUM $>$ 1.31 AND hierarchyAVG $<=$0.32 AND updates in (12.5,28.5] THEN class="strong", and (iii) IF efficiencySUM $>$ 1.31 AND hierarchyAVG $>$ 0.32 AND updates in (12.5,28.5] THEN class="strong".

5 Threats to Validity

Co-changes identification: We used five pull requests as support count to remove co-changes of our study. Although this threshold is used in the literature, we still need to evaluate the impact of other values.

Timeframe selection: We used six months to identify the co-change among files. Furthermore, we used "five pull requests" as support to remove pair of files that were committed four or less times among this six months of timeframe. Other approaches like association rules was used to find co-change using a snapshot from source code history. We plan to study the impact of this technique on the GitHub repository to better identify logical coupling and determine what is the best timeframe to group co-changes and build the communication networks. We also planned to run the experiments considering each Release instead of a fix six months timeframe.

Generalizability: This study aimed to explore the use of SHM to predict change dependencies. We used two popular projects from GitHub, but additional studies are necessary to generalize the results to other projects. This is our first effort to predict change dependencies and we plan to add new social network metrics and more projects to investigate the impact of others social metrics over change dependencies.

6 Conclusion and Future Works

During software evolution, developers communicate and cooperate changing files and discussing about change requests. A good understanding of the impact of communication and cooperation can provide insights to improve software quality. Supported by Conway's law that states about the impact of communication over the software design, we used SHM computed from communication networks to study change dependencies. We computed four different SHM, to predict strong and weak change dependencies among files changed in pull request submissions.

We answered RQ1 investigating if all of SHM and process metrics were useful to build classification models, since we did not found previous works that explored social aspects to predict change dependencies. Using seven well-known classification algorithms, we achieved similar results of recall, precision, f-measure, and MCC metrics compared to previous works that used social network metrics to predict bugs and works that predict co-changes using non-social metrics. We did not find statistical difference among algorithms for our two set of metrics (SHM vs process metrics). However, we found that structural metrics for Rails project had a small advantage since it presented smaller variability than process metrics. For Node.js, the results among the two sets were similar. We shown this results presenting boxplots for each project.

Since we found that SHM could be used to predict change dependencies, we explored which of the four metrics were more relevant. We inspected the rules used to build the classification tree and predict strong and weak couplings. We found that Node.js needed more different types of metrics than Rails. These two projects presented very different rules, for example Node.js (Table 2) used the constraintAVG (average of the lack of holes among neighbors) and hierarchySUM (sum of the concentration of constraints to a single node) six times. Considering Rails, neither of these metrics were frequently selected. For Rails (Table 3), effectiveSizeAVG (average of the portion of non-redundant neighbors of a node) and efficiencyAVG (average of the normalization of the effective size by the number of neighbors) – the others two SHM – were selected more often. We observed that different metrics were selected in specific timeframes.

As main contribution, we shown that SHM obtained from communication networks can support the Conway's law and predict change dependencies. Since this first effort to use social metrics presented promising results, we will explore more broadly this social dimension of development process. We want to investigate in future works additional projects and social metrics, and go deeper in providing insights for managers and developers about how their collaboration patterns can influence code quality. Besides, as we are looking into relations of social interaction (communication) and technical work (cooperation), one can investigate the application of these metrics in other collaboration domains.

Acknowledgments. We thank Fundação Araucária, NAWEB and NAPSOL for the financial support. Marco G. receives individual grant from CNPq and FAPESP. Igor W. and Igor S. receive grants from CAPES (Process BEX 2039-13-3).

References

1. Ball, T., Adam, J.K., Harvey, A.P., Siy, P.: If your version control system could talk. In: ICSE Workshop on Process Modeling and Empirical Studies of Software Engineering (March 1997)
2. Betz, S., Mite, D., Fricker, S., Moss, A., Afzal, W., Svahnberg, M., Wohlin, C., Borstler, J., Gorschek, T.: An evolutionary perspective on socio-technical congruence: The rubber band effect. In: RESER 2013, pp. 15–24 (2013)
3. Bhattacharya, P., Iliofotou, M., Neamtiu, I., Faloutsos, M.: Graph-based analysis and prediction for software evolution. In: 2012 34th International Conference on Software Engineering (ICSE), pp. 419–429 (June 2012)
4. Biçer, S., Bener, A.B., Çağlayan, B.: Defect prediction using social network analysis on issue repositories. In: Proceedings of the 2011 International Conference on Software and Systems Process, pp. 63–71 (2011)

5. Bicer, S., Bener, A.B., Cauglayan, B.: Defect prediction using social network analysis on issue repositories. In: Proceedings of the 2011 International Conference on Software and Systems Process, ICSSP 2011 (2011)
6. Bird, C., Nachiappan, N., Harald, G., Brendan, M., Devanbu, P.: Putting it all together: Using socio-technical networks to predict failures. In: Proceedings of the 2009 20th International Symposium on Software Reliability Engineering (2009)
7. Burt, R.: Structural Holes: The Social Structure of Competition. Harvard University Press (1995)
8. Canfora, G., Cerulo, L., Cimitile, M., Di Penta, M.: How changes affect software entropy: an empirical study. Empirical Software Engineering 19(1), 1–38 (2014)
9. Cataldo, M., Mockus, A., Roberts, J.A., Herbsleb, J.D.: Software dependencies, work dependencies, and their impact on failures. IEEE Transactions on Software Engineering 35(6), 864–878 (2009)
10. Conover, W.J.: Practical Nonparametric Statistics, vol. 2. John Wiley and Sons
11. Conway, M.E.: How do committees invent? Datamation (1968)
12. D'Ambros, M., Lanza, M.: Reverse engineering with logical coupling. In: 13th Working Conference on WCRE 2006, pp. 189–198 (2006)
13. D'Ambros, M., Lanza, M., Robbes, R.: On the relationship between change coupling and software defects. In: WCRE 2009, pp. 135–144 (2009)
14. D'Ambros, M., Lanza, M., Robbes, R.: Evaluating defect prediction approaches: a benchmark and an extensive comparison. Empirical Softw. Eng. 17(4-5), 531–577 (2012)
15. Demšar, J., Curk, T., Aleš, E., Gorup, Č., Hočevar, T., Mitar, M., Martin, M., Matija, P., Marko, T., Anže, S., Miha, Š., Lan, U., Lan, Ž., Jure, Ž., Marinka, Ž., Blaž, Z.: Orange: Data mining toolbox in python. Journal of Machine Learning Research 14, 2349–2353 (2013)
16. Ferdian, T., Tegawende, F.B., Lo, D., Jiang, L.: Network structure of social coding in github. In: 15th European Conference on Software Maintenance and Reengineering, pp. 323–326 (2013)
17. Kouroshfar, E.: Studying the effect of co-change dispersion on software quality. In: Proceedings of the 2013 International Conference on Software Engineering, ICSE 2013, pp. 1450–1452 (2013)
18. Kwan, I., Cataldo, M., Damian, D.: Conway's law revisited: The evidence for a task-based perspective. IEEE Software 29(1), 90–93 (2012)
19. Meneely, A., Williams, L., Will, S., Osborne, J.A.: Predicting failures with developer networks and social network analysis. In: SIGSOFT FSE, pp. 13–23 (2008)
20. Meneely, A., Williams, L., Snipes, W., Osborne, J.: Predicting failures with developer networks and social network analysis. In: Proceedings of the 16th ACM SIGSOFT International Symposium on Foundations of Software Engineering, SIGSOFT 2008/FSE-16, pp. 13–23. ACM, New York (2008)
21. Nagappan, N., Murphy, B., Basili, V.R.: The influence of organizational structure on software quality: an empirical case study. In: International Conference on Software Engineering (ICSE), pp. 521–530 (2008)
22. Panichella, S., Canfora, G., Di Penta, M., Oliveto, R.: How the evolution of emerging collaborations relates to code changes: An empirical. In: IEEE International Conference on Program Comprehension (ICPC 2014) (2014)
23. Powers, D.M.W.: Evaluation: From Precision, Recall and F-Factor to ROC, Informedness, Markedness & Correlation. Technical Report SIE-07-001, School of Informatics and Engineering, Flinders University, Adelaide, Australia (2007)
24. Rahman, F., Devanbu, P.T.: How, and why, process metrics are better. In: ICSE, pp. 432–441 (2013)

25. Tsay, J., Dabbish, L.A., Herbsleb, J.D.: Influence of social and technical factors for evaluating contribution in github. In: ICSE (2014)
26. Visa, S.: Issues in mining imbalanced data sets - a review paper. In: Proceedings of the Sixteen Midwest Artificial Intelligence and Cognitive Science Conference, pp. 67–73 (2005)
27. Wolf, T., Schroter, A., Damian, D., Nguyen, T.: Predicting build failures using social network analysis on developer communication. In: Proceedings of the 31st International Conference on Software Engineering, ICSE 2009 (2009)
28. Zanetti, M.S., Sarigöl, E., Scholtes, I., Tessone, C.J., Schweitzer, F.: A quantitative study of social organisation in open source software communities 28, 116–122 (2012)
29. Zhou, Y., Wursch, M., Giger, E., Gall, H., Lu, J.: A bayesian network based approach for change coupling prediction. In: 15th Working Conference on WCRE 2008, pp. 27–36 (2008)
30. Zimmermann, T., Zeller, A., Weissgerber, P., Diehl, S.: Mining version histories to guide software changes. IEEE Transactions on soft Engineering 31(6), 429–445 (2005)

LOST-Map: A Victim-Sourced Rescue Map
of Disaster Areas

André Silva[1], Diogo Marques[1], Carlos Duarte[1],
Maria Ana Viana-Baptista[2], and Luís Carriço[1]

[1] Faculdade de Ciências da Universidade de Lisboa, Lisbon, Portugal
asilva@lasige.di.fc.ul.pt, {dmarques,cad,lmc}@di.fc.ul.pt
[2] Instituto Português do Mar e da Atmosfera, Lisbon, Portugal
mavbaptista@gmail.com

Abstract. In the aftermath of natural disasters, members of the affected communities are often the *de facto* first responders. Local volunteers can respond quickly, are strongly motivated, and have the necessary ground knowledge. However, their search and rescue efforts may be misdirected in the absence of information about the location and status of victims. We propose LOST, a system that gathers data from smartphones in affected areas, even when the regular communication infrastructure fails, and aggregates it in a web interface for visualization. For each individual, LOST-Map shows location traces and activity indicators. The information can be explored by selecting time-frames and/or applying filters over activity indicators. This paper briefly describes the design of LOST, introduces the visualization tool LOST-Map, and reports on a study (n=10) that suggests that it can be effectively used by untrained volunteers.

Keywords: Disaster management, Emergency response, Location services.

1 Introduction

When disasters (like earthquakes, hurricanes or tsunamis) hit populated areas, members of the affected communities often offer themselves to help in the field. While they may not have the necessary knowledge to provide first aid to victims in every situation, these volunteers know the geography and have a better sense of which people are missing. They can be valuable actors in emergency operations, providing immediate response and collecting information useful to other stake-holders, like civil defence efforts. They can report changes in the field, which people need help, etc. In areas that become isolated, local volunteers are sometimes the primary emergency responders for extended periods of time. However, when information flows are chaotic, it is difficult, even for locals, to have an accurate account of the location and status of potential victims. Often the communication infrastructure fails or is overloaded, hindering the dissemination of information.

Technology plays a crucial role in emergency response. Yet, tools that empower local volunteers aren't still common. LOST is a system that tackles the challenge of providing a source of actionable information to volunteer responders. With LOST,

N. Baloian et al. (Eds.): CRIWG 2014, LNCS 8658, pp. 311–318, 2014.
© Springer International Publishing Switzerland 2014

location and activity information is collected automatically from people's smart-phones and then disseminated from peer to peer. When it finally finds its way to an internet connection, it is posted to an online service. Additionally, LOST allows users to write text messages and safe reports, which are also propagated through the net-work. The service that runs on smartphones can be locally activated by the user or remotely triggered by a central authority while there is still communication infrastruc-ture. LOST also includes an online visualization map, LOST-Map, where volunteers can easily navigate and explore this data.

In recent years, several social computing approaches have been applied to disaster management. Agarwal et al. [1] proposes a system which allows people to agree on a secure path to exit a disaster scene, communicating through a Wi-Fi ad-hoc network. Ramesh et al. [6] propose a Bluetooth ad-hoc network formed by the victims' smart-phones that allows exchange of short text messages and location pointers. Although LOST also relies on an peer-to-peer dissemination of information, the approach to communication is different: LOST is targeted at commodity smartphones operating in realistic conditions, and thus cannot rely on Bluetooth (which does not have the nec-essary range) or Wi-Fi ad-hoc mode (which is not available on most off-the-shelf devices). Furthermore, in LOST, when it finds its way to internet, it is made available to support the rescuers' work.

The availability of online maps has also been leveraged in many recent proposals for rescue tools: COORDINATORS [10] aids first responders to coordinate them-selves and share information; WIISARD [2] consists on a mesh network to promote communication and situation updates between first rescuers and command centre; TravelThrough [3] is a system developed for victims and information centres to share and develop a secure path to exit the disaster area. These systems, however, even when they take into account information produced by victims, rely on the existence of an institutional response. LOST is designed to gather information about victims and make it accessible to the local community, as a complement to the institutional re-sponse, or as a substitute when it is not present

Recently, it has also been observed that local, unofficial efforts, are already taking advantage of online collaborative tools. During the Haiti earthquake in 2010 people collaborated to create a local map [4], later used for rescue and planning purposes. LOST-Map displays automatically gathered specific information about battery level, local activity (measured by the accelerometer and number of screen activations), dis-tance travelled (measured by GPS), besides explicit messages. These indicators pro-vide context, helping a rescuer to get a better sense of the situation.

This paper focuses on LOST-Map and its ability to provide information to volunteers. Filters are available that allow users to select a time-frame, to exclude individuals who have been reported safe, and to color markers according to activity information. As a preliminary validation a user study (n=10) was conducted, too as-sess if it can be effectively used by people with no prior knowledge of rescue opera-tions. Findings suggest that users understood the interface and successfully identify potential victims, spending a reasonable amount of time and effort.

2   The LOST Project: Empowering Community Rescuers

The LOST project comprises two main components: LOST-OppNet and LOST-Map. LOST-Map is a web application consisting of a map to display collected data and designed to run in tablet devices and desktop/laptop computers with a standard web browser. LOST-OppNet is responsible for gathering data from the victims using a non-obstructive approach. It consists of an Android application that activates a set of predefined sensors. A dynamic mesh is created with the devices in the area, acting as nodes in an opportunistic network [5]. While the application is intended to be activated by a central remote authority, victims can also manually activate it.

To operate on commodity devices, LOST-OppNet takes advantage of the ability smartphones have to act as Wi-Fi access points. For messages to be disseminated, the devices cycle through: 1) acting as access points, and; 2) trying to connect to access points; 3) trying internet connections. Messages are disseminated for all nodes it can connect. Duplicates are prevented based on authorship and timestamp. Remaining operational details are outside the scope of this paper, but the approach is similar to WiFiOpp [8], except from the internet connection attempts and some optimization parameters.

LOST-OppNet also provides a minimal user interface to victims, allowing them to see system status and send text messages. These allow victims to describe their condition or the status of surroundings with free text. Victims can also mark themselves as safe, if they managed to escape from the disaster.

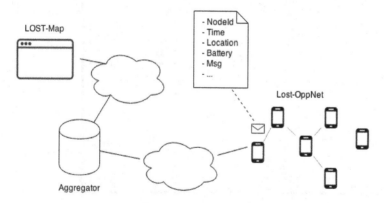

Fig. 1. The LOST architecture, including LOST-Map, LOST-OppNet and the aggregator

LOST-OppNet is a content producer while LOST-Map is a content consumer. Between them, an online repository acts as an aggregator. A LOST-OppNet node that finds itself with an available internet connection deposits in the repository not only its own tracks, but the ones that reached it in the opportunistic network. The LOST-Map webapp is, in essence, a visualization of the data in the aggregator. Fig. 1 summarizes the architecture of LOST.

3   LOST-Map: Visualizing Human Activity

LOST-Map is a web application capable of providing real-time and historical information about a disaster scene. It assumes a target audience that has some knowledge about the rescuing location. It is designed to volunteers as a complement of the rescue operation. LOST-Map allows the visual exploration of:

- Time/location of nodes: the building blocks of location tracks;
- Battery level: if low, we can infer that victim will stop producing data;
- Steps: an indicator of mobility, estimated from accelerometer readings;
- Screen activations: an indicator that an individual is active;
- Text messages: victims can explicitly send messages, readable from the web app;
- Safety status: victims with mobile app can also mark themselves as being safe.

Time and geographical location are used not only to show the last known location, but also a track, which can be an indicator of movement. The remaining information can also inform on victim status, but the inferential step is left to the volunteers.

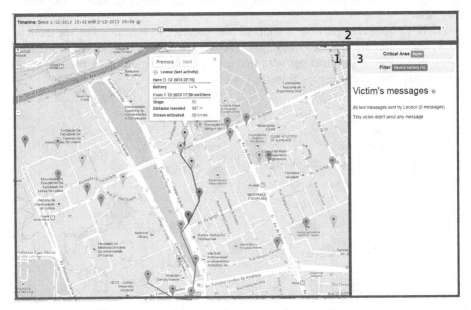

Fig. 2. The user interface split in 3 areas, with all controls applied simultaneously (path for a single victim, timeline and indicators filter)

Figure 2 shows the actual LOST-Map interface. It is divided in three areas. The most prominent is the map, which relies on the Google Maps API[1]. Each marker represents the last known location of a victim. It is possible to show the path for a single victim by clicking the marker (in the figure towards a safe area). Information

---

[1] Google Developers: Maps API – https://developers.google.com/maps/

gathered by smartphone sensors is displayed as a balloon, by clicking any marker in the path (Figure 3). The second area at the top is the time-frame master control. It is possible to select a period of time by adjusting the slider. This acts as a filter to the information shown on the map, removing markers for devices that did not produce information within the interval. The third area on the right is the toolbox. It contains access to filters, shows any posted messages when a marker is selected, and provides space for other tools still being developed.

Fig. 3. Detail of the balloon that is shown when a marker is selected. It displays human activity indicators as well as messages posted by the victim through the smartphone app.

It is also possible to apply a color to markers in order to categorize them. Markers may be colored red, yellow and green according to activity indicator ranges set by the volunteer, using a semaphore analogy. Each indicator has a default scale associated on which a volunteer can control the levels using a slider.

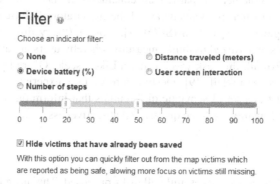

Fig. 4. Detail of the panel to choose an indicator filter and hide victims already safe

As an example, Figure 4 illustrates the selection of a filter over the battery level. In this example, three ranges are selected, from left to right in the scale:

- Lower than 20% is the critical level and is marked with red color;
- From 20% to 50% is the alert level and is marked with yellow color;
- Above 50% is the secure level and is marked with green color.

These levels can be changed by volunteers. Upon applying this filter, markers on the map are colored according to the defined ranges. This filter provides a visual cue to help volunteers to find victims according to a certain criteria. There is also a filter to completely hide individuals that had reported as being safe. This allows the removal of visual clutter on map and helps to pinpoint people in need of assistance.

4   Preliminary Evaluation

To assess if LOST-Map can be a useful and usable tool for untrained volunteers, we conducted a user study. It comprised a number of tasks that simulated the typical use cases that LOST addresses. We recruited 10 volunteers, all being students or research staff. Ages ranged from 22 to 51 (mean=27.3, SD=8.6). All participants had some experience with using Google Maps and good knowledge of the scenario areas.

Participants were given a script containing an introductory explanation about LOST-Map and its relation to LOST. They were asked to complete three. Each task concerned a different disaster scenario. Datasets for each task were generated by computer simulation in advance. Tasks, scenarios details and resulting data are available online[2]. The moderator was responsible for measuring completion times for each task. After each task, participants were asked some questions to understand their comprehension regarding the interface and functionalities. After that, they were asked to answer a between-task survey with a single question. After all tasks were completed, participants were administered a final questionnaire, to assess their overall perception of LOST-Map. All tasks were accomplished using LOST-Map.

We acquired the following measures for each task: 1) Total amount of time to conclude the task; 2) Whether the participant concluded the task without giving up; 3) Number of questions that participants asked the moderator; 4) Ease of use as measured by the Single Ease Question (SEQ) [7], from 1 ("very difficult") to 7 ("very easy"). SEQ is a standardized usability measure, whereby users are asked to complete the statement "Overall, this task was:" using a Likert scale. Overall perceptions were measured with the AttrakDiff [9] questionnaire of user experience, in the ten-item version AttrakDiff is a set of semantic differentials that inform on subjective perceptions of pragmatic quality, hedonic quality and attractiveness.

4.1   Results

All participants concluded successfully all tasks proposed. The mean time to conclude each task is presented in the Table 1. As expected, as tasks were completed in increasing degree of difficulty, participants needed more time to conclude task #3, which was closer to a real world situation, requiring a combination of techniques learnt from the first two tasks. Scores for the SEQ are also presented in Table 1.

---

[2] http://accessible-serv.lasige.di.fc.ul.pt/~astarte/criwg2014

Table 1. Average task completion times, SEQ scores and number of questions for each task

	Avg. task completion time	Avg. SEQ score	Avg. help requests
Task #1	1m48s (SD=37s)	5.50 (SD=1.08)	0.7 (SD=0.7)
Task #2	1m58s (SD=46s)	5.70 (SD=1.34)	0.6 (SD=0.5)
Task #3	2m43s (SD=99s)	4.50 (SD=1.58)	1.0 (SD=1.2)

The answers are in line with the task completion time: the more complex the task is, the less easy users found it. We also measured the number of questions asked by the participants to the moderator. A question to the moderator means that the participant was not able to conclude the current task. When asked, we tended to allow the participant to discover how to continue autonomously. Again, participants asked more of the moderators in the most complex task. An AttrakDiff questionnaire was used to evaluate the full experience provided by LOST-Map. Each semantic differential was given a score ranging from 1 to 7, the latter being the most positive score. The first four differentials are indicators of pragmatic quality and second four are indicators of hedonic quality. All results are between 5 and 7 in the 7 points scale.

4.2   Discussion

Overall, the results validate LOST-Map as an easy-to-use tool for untrained volunteers. All tasks took a reasonable amount of time to conclude. Even the third task which was more realistic, requiring a combination of time control and filters, was completed in less than three minutes. The SEQ scores indicate that participants didn't find the tasks difficult. This is especially encouraging taking into account that participants had no previous training, therefore having to adapt and learn how to work with the tool. From our observations, most of the difficulties were related to using the slider controls. Participants asked for help from the moderator more often than desired, but our observations suggest that these difficulties could be overcome easily.

The responses to AttrakDiff were on the positive side of the differentials, but it seems users' perceived LOST-Map to have greater pragmatic quality than hedonic quality. Our primary focus was the introduction of useful functionalities to filter and understand data gathered from LOST-OppNet. This does not necessarily mean we neglect hedonic aspect of user experience; it was a factor with a slight lower priority.

5   Conclusions

In this paper we presented LOST-Map, part of the LOST project. LOST-Map enables local volunteers in their rescue efforts by augmenting their perception of the disaster area. Volunteers have access to a map and a set of functionalities that allow them to have an overview of the situation. We reported on a study indicating that LOST-Map is usable by novices. The user study also shows that there is room to improve, namely in the filter functionality and the overall hedonic quality. The LOST project is an ongoing effort. We also plan to introduce a native Android version of LOST-Map to have better support on tablet devices. This application in particular will not rely on having an Internet connection, updating the map using data gathered directly from the opportunistic network.

Acknowledgments. This work was funded by FCT through the LaSIGE Strategic Project, ref. PEst-OE/EEI/UI0408/2014 and the EU project ASTARTE - Assessment, STrategy And Risk Reduction for Tsunamis in Europe. Grant 603839, 7th FP (ENV.2013.6.4-3 ENV.2013.6.4-3).

References

1. Agarwal, A., Aggarwal, S., Agrawal, M., Batra, N.: Emergency Relief and Crowd Dispersal using MANET Protocols. cise.ufl.edu, pp. 1–9 (2011)
2. Chipara, O., Griswold, W.G., Plymoth, A.N., Huang, R., Liu, F., Johansson, P., Rao, R., Chan, T., Buono, C.: WIISARD. In: Proc. of 10th Intl Conf. on Mobile Systems, Applications, and Services, MobiSys 2012, p. 407. ACM Press (2012)
3. Gunawan, L.L.T., Fitrianie, S., Yang, Z., Brinkman, W.-P., Neerincx, M.: TravelThrough: a participatory-based guidance system for traveling through disaster areas. In: Proc. 2012 ACM Annu. Conf. Ext. Abstr. Hum. Factors Comput. Syst., CHI EA 2012, pp. 241–250 (2012)
4. Ito World. Mapping the Crisis - OpenStreetMap Response to Haiti Earthquake (2010), http://itoworld.blogspot.com/2010/01/mapping-crisis-openstreetmap-response.html
5. Pelusi, L., Passarella, A., Conti, M.: Opportunistic networking: data forwarding in disconnected mobile ad hoc networks. IEEE Communications Magazine 44, 134–141 (2006)
6. Ramesh, M.V., Jacob, A., Devidas, A.R.: Enhanced emergency communication using mobile sensing and MANET. In: Procs. Intl. Conf. on Advances in Computing, Communications and Informatics, ICACCI 2012, p. 318. ACM Press (2012)
7. Sauro, J., Dumas, J.S.: Comparison of three one-question, post-task usability questionnaires. In: Procs. 27th Intl. Conf. on Human Factors in Computing Systems, CHI 2009, p. 1599. ACM Press (2009)
8. Trifunovic, S., Distl, B., Schatzmann, D., Legendre, F.: WiFi-Opp: Ad-Hoc-less Opportunistic Networking. Perform. Eval., 37–42 (2011)
9. Van Schaik, P., Hassenzahl, M., Ling, J.: User-Experience from an Inference Perspective. ACM Trans. on Computer-Human Interaction 19(2), 1–25 (2012)
10. Wagner, T., Vanriper, R., Phelps, J., Guralnik, V.: COORDINATORS: Coordination Managers for First Responders, pp. 1140–1147 (2004)

Mapping on Surfaces: Supporting Collaborative Work Using Interactive Tabletop

Kanida Sinmai and Peter Andras

School of Computing Science, Newcastle University,
Newcastle upon Tyne NE1 7RU, UK
{kanida.sinmai,peter.andras}@ncl.ac.uk

Abstract. We investigate the usability of our mindmap application using a tabletop integrated with four Android tablets in the context of support for collaborative work. This paper presents two empirical studies that compares the conventional paper-and-pen approach with an interactive touchscreen tabletop system. Our results clearly indicate that the combination of a tabletop and personal devices support and encourage multiple people to work collaboratively. Furthermore, the results confirm earlier results about the usability advantages of the interactive tabletop application. The comparison of the associated emotional attitudes indicates that the interactive tabletop facilitates the active involvement of participants in the group decision making significantly more than the use of the paper-and-pen approach.

Keywords: Tabletop, Collaborative work, Mind Mapping, Personal devices.

1 Introduction

It is becoming increasingly difficult to ignore the benefits of interactive tabletops. The tabletop systems provide a large workspace where people can collaborate in a face-to-face setting simultaneously. Recent developments in the related research of tabletop found that interactive tabletops are enjoyable to use and support group awareness [3],[8],[17],[19, 20]. Tabletop applications have been widely developed for studying and comparing the usability of the interactive surfaces [16],[23, 24]. Although many research studies have investigated around interactive tabletops in the context of collaboration, the tabletop mind map application still needs to be explored [5].

It is a challenge to implement such tools that rely on more advanced technological interfaces, such as interactive multitouch tabletops. We believe that not only multitouch tables enhance the productivity of mind mapping tools, but incorporating the use of personal devices further extends the possibilities for great collaboration.

The main contribution of this paper is to present the results from two empirical studies. On the first experiment we compared the use of tabletop mind map application with conventional paper-and-pen condition. Another experiment is to investigate the use of combination of tabletop and personal devices

N. Baloian et al. (Eds.): CRIWG 2014, LNCS 8658, pp. 319–334, 2014.

(i.e. Android tablets) with the traditional paper-and-pen approach in the context for collaborative work. We made a few assumptions for this study. Firstly, we thought that people would use tablets to perform the task as well as using the tabletop. Secondly, we expected that participants would participate more in our application than the traditional approach. Finally, we expected that users would be satisfied with the use of our system. The results clearly indicate that the combination of a tabletop and personal devices support and encourage multiple people working collaboratively. Furthermore, the results confirm earlier results about the usability advantages of the interactive tabletop application.

2   Related Work

2.1   Mind Mapping: A Tool for Collaborative Work

Mind Mapping is a powerful thinking tool for brainstorming [6]. This valuable method is a graphical diagram for presenting thoughts or ideas relating to the central idea [11]. The mind mapping approach was invented by Tony Buzan [7], and can be applied to every discipline to enhance human performance and to unlock the potential of the brain [7]. While it was initially used for learning and note-taking, recently, mind maps have been widely used in other areas such as planning, presenting, problem solving, decision making and so forth [26].

Mind mapping has several advantages when compared with ordinary note taking. For example, time is saved by just noting down relevant keywords [9], there is increased productivity of 30% and improved understanding of complex issues [14]. Mind mapping empowers creativity, boosts memory, and helps manages information overload [14]. Mind mapping also promotes group collaboration. Shih et al. showed that the mind map structure imposes important framing effect on group dynamics and thought organization during the brainstorming process [25].

2.2   Tabletop Mind Mapping

In the context of tabletop mind mapping, one of the most relevant pieces of research on this area is Tabletop Mind-Maps (TMM) [5]. This project was developed to support collaborative decision making and aimed to compare the usability and usefulness of this approach to the traditional paper-and-pen conditions. A similar approach was presented by Do-Lenh et al. [10]. The researchers evaluated the usage of a tabletop concept mapping tool in cooperative settings. They compared a concept mapping using two different interfaces: a desktop computer using one mouse/keyboard and tabletop to support tangible interaction. In an additional study by [18], they presented Cmate (Concept Mapping at an Adaptive Tabletop for education) to support learners in the form of discussion based on comparing personal understanding as captured in personal concept maps. More recently, Buisine et al. [4] designed an experiment to answer how interactive tabletop systems influence collaboration. This study explored two experiments comparing the use of a tabletop system to the traditional paper-and-pen in two different creative problem solving tools: a Brainpurge on sticky notes and a Mindmap.

3 Our Application

To enhance collaborative work using interactive tabletops, our mind maps not only are built top-down from the root label, but are also created from the user's position (Fig. 1(a)). Each node can be dragged and dropped to connect with their parent. Moreover, for potential participants, we implemented a soft keyboard for each node so that users can work simultaneously (Fig. 1(c)). This section discusses the design for our approach and give implementation details.

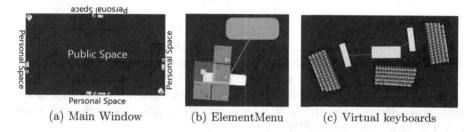

(a) Main Window   (b) ElementMenu   (c) Virtual keyboards

Fig. 1. Our Tabletop Mind Map Application

3.1 Design Goals for Our Mind Map Application

The design of the user experience was guided by the work of Bachl et al. [2] and Scott et al. [22] informed the guidelines for co-located collaboration. We now explain how these were used in the following section.

Support Multi-user. This is one of the most desired goals that several users are able to share the same surface and interact seamlessly together. Our design supports simultaneous and direct touch input using soft keyboards. A keyboard is invoked when a text box gains focus by double-tap interaction. The connection line associates the keyboard with the node as seen in Fig. 1(c). Users may hide the keyboard by double-tapping on the node. The size, orientation and location of each keyboard can be freely changed.

Support Users Arrangement. The tabletop must support a 360° view of the users interaction around the table. Thus, we designed the user interface element so that it can be sized, rotated, and relocated. Therefore, participants are able to view and interact with the application at any point of view around the table.

User-Based Challenges. One of the benefits of multi-touch user interfaces is the possibility to directly interact with the screen using ones fingers for input.

However when objects on the screen are small, some users find it difficult to interact with them. So to support users we included scalable elements.

Support Transitions between Individual and Collaborative Work. We consider the challenge of private activity and collaboration. Users tend to work in parallel[10], so we provide a personal space and use colour to distinguish between users. In the study, we used font colours represent the users who create the nodes, while the background colour of text boxes shows task hierarchy. The background colour can be manually modified.

3.2 Implementation

Our system consisted of two main components: tabletop mind map application and Android mind map application. We implemented the mind map application on a Touchscape table with 47 inch (1190mm) LCD screen diagonal with a display resolution of 1920x1080 full HD pixel. Touchscape is 500 mm high and a table outer dimension is 1300x900 with weight 60 kg [21]. The application code is in C# using the Surface SDK.

The Android application was developed in Java using the Android SDK. Once the device connected to the tabletop, users can see the mind map which is on the table. The application allows users to rotate, size, relocate, create and delete the nodes. Importantly, people can also build a node as a private thought on this device. We have investigated and developed the framework that allows personal tablets to connect to the tabletop using Web service via WiFi connection.

The start of the application places the central idea in the centre of the tabletop display by tapping the start menu. Users can either create the node by tapping the "add a new node" menu in front of them or using the node's ElementMenu. The ElementMenu will always appear when the node is touched. There are four menus at the moment: add a new node, bold, font size, and background colour as seen in Fig. 1(b). The node can be moved, rotated or scaled to support readability from every point of view around the table. Each node has a virtual keyboard. While a node is active the background colour is changed. The node can be freely relocated, rotated, scaled, deleted and also change its parent. Users can drag and drop any unconnected node over a bin provided for deleting. In addition, to change a parent the node is moved over the parent once and then dragged over to the new node. If the node has children nodes, they will also follow their parent automatically.

4   Experiment Design

We investigate the effectiveness and the usability of an interactive tabletop application in the context of support for collaboration work. We designed two controlled laboratory experiments. The first experiment compared the use of our mind map application with paper-and-pen conditions (Fig. 2). The second experiment combined our tabletop mind map application with personal devices and compared this combined application with the conventional approach (See Fig. 4).

4.1 Tasks and Procedure

For the collaborative tasks, we provided two case scenarios: global warming and stress management. The participants were free to choose one of these topics.

Initially, the collaborators were introduced to the purpose of the study and asked to perform two tasks: one using the tabletop and another task using traditional paper-and-pen analysing the same topic. During the experiments, they were sitting around a table. The groups were shown the tabletop application and practised using it with a practice brief for 15 minutes. Participants spent 20 minutes per tasks and after finished performing they had the opportunity to complete a questionnaire about their background, and about their experiences using the applications.

In all cases, the groups were seated around the tables and were free to move chairs. Also they were given an A4-paper and a pen in case they wished to make some notes.

4.2 Data Collection and Analysis

A number of quantitative and qualitative data were gathered from observation of naturally occurring activities. All groups were observed by the researcher, who took notes throughout. The collected data consisted of an observation form, and field notes. The study did not focus on the accuracy of the information presented in the mind-maps. After using the system, each user had to quantify their impression on a 5-point and a 10-point Likert scales for the satisfaction and emotional attitudes sections respectively. They were also particularly prompted to complete with free qualitative comments.

The questionnaires were analysed by the Friedman test for comparing the samples using SPSS version 21. The level of significance was set to 0.05 for these nonparametric statistical tests.

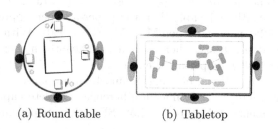

(a) Round table   (b) Tabletop

Fig. 2. Experiment 1 Setup

5   Experiment 1

5.1   Apparatus

The experiment was designed to test the using of the mind map tabletop application. For the tabletop study, we used a Touchscape table and the participants were seated around the table one side each (see Fig. 2(b)).

In the paper-and-pen experiment, the participants were seated around a round table, they used a marker pen to build a mind-map. Each participant used a different coloured marker pen (see Fig. 2(a)).

5.2   Participants

Our experiment involved 40 students (18 male, 22 female) from the local universities' communities. They had no experience with a tabletop although they all had experience with a smart phone or tablet. The age range was from 18 to 45 years old. The participants chosen had a variety of backgrounds, such as economics, business, linguistics, chemical engineering, mechanical engineering, architecture, medical and computer science. The participants were divided into ten groups of four.

5.3   Results

Six groups (60 %) were interested in the global warming and 40% in stress management.

Performance. There were no significant differences appeared in term of performance including number of ideas, completion time and conversation.

Users' Satisfaction. Users' satisfaction results were analysed using the 5-point Likert scale answers. This feedback form contained 10 questions. The significant results are presented in Figure 3.

The results indicate that there was a significant difference in the paper-and-pen approach (M=4.60, SD=0.632) over the tabletop (M=4.25, SD=0.494) with statement the "The device was easy to use". By contrast, the tabletop was rated more significantly enjoyable than the paper-and-pen approach (tabletop: M=4.63, SD=0.628, paper-and-pen:M=3.53, SD=0.960). With regard to the statement "I tried my best", there was a significantly higher rating (tabletop:M=4.70, SD=0.464, paper-and-pen:M=4.38, SD=0.774) for the tabletop compared to the conventional meeting. Similarly, in response to "I had a lot of ideas" statement, there was a significant difference in tabletop approach over the traditional experiments (tabletop: M=4.38, SD=0.774, paper-and-pen:M=4.05, SD=0.876).

However, there were no significant differences between the two approaches with these statements "I was satisfied", "I had high quality ideas", "I was motivated to do well", "The results are important to me" and "I collaborated with

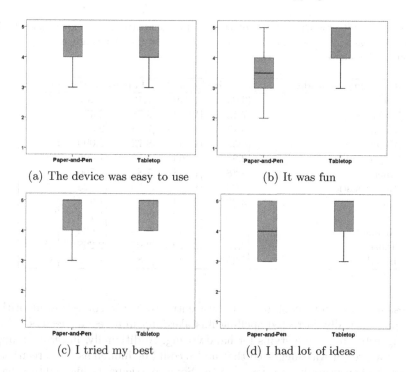

(a) The device was easy to use   (b) It was fun

(c) I tried my best   (d) I had lot of ideas

Fig. 3. Box plots of the distributions of satisfaction rating for the usability of the flip chart and tabletop approaches for the designing of mind maps. The quality of ideas score was based on strongly disagree = 0 and strongly agree = 5. ((3(a)) p=0.004, (3(b)) p<0.001, (3(c)) p=0.029 and (3(d)) p=0.046).

other participants". These statements indicate a preference towards the tabletop over the paper-and-pen approach. While, the statement "It was agreeable", participants ranked the paper-and-pen higher than the tabletop, but again the differences were not statistically significant.

Emotional Attitudes. The emotional attitudes questionnaire contains 12 questions, which users clarified their impression on a 10-point Likert scale. The analysis of the emotional attitudes data is shown in Table 1.

The tabletop was rated significantly higher for alertness, anxiety, energy, enthusiasm, and tiredness. In contrast, the participants felt calmer and more relaxed using the paper-and-pen method.

In response to the remaining emotional questions including happy, depressed, sad, tense and bored, we found that no significant differences appeared between the tabletop and paper-and-pen experiments.

Qualitative Feedback. Qualitative feedback was received through the questionnaire and user comments during the experiments. Some recommendations

Table 1. Results for the ratings of the emotional attitudes for experiment 1. The quality of ideas score was based on strongly disagree = 0 and strongly agree = 10. (NS = Non-significant results, * = significant).

Emotional attitudes	Tabletop		Paper-and-Pen		Sig.
Alert	9.03	1.310	7.10	2.182	*
Anxious	7.13	1.652	3.78	2.475	*
Calm	5.73	2.195	7.83	2.099	*
Happy	8.53	1.567	8.73	2.099	NS
Depressed	2.98	2.465	2.50	1.895	NS
Energetic	8.88	1.488	7.83	1.796	*
Enthusiastic	8.93	1.421	7.90	1.823	*
Tired	3.93	2.664	1.80	1.363	*
Sad	2.23	1.310	2.18	1.130	NS
Relaxed	6.95	1.947	7.73	1.633	*
Tense	3.75	2.570	3.05	2.298	NS
Bored	2.35	1.748	2.83	1.852	NS

suggested an improvement to text data entry by using correction algorithms that automatically corrected spelling mistakes and unnecessary letters. Also it was suggested to have a stylus for handwriting. Additionally, in terms of application features, it was suggested that nodes could be hidden as the screen seems cluttered when users have a lot of ideas. Some participants highlighted that it was very easy to modify the mind map using the tabletop. However, some participants indicated that the reflection of the sunlight may cause issues with the usability of the tabletop. For example, shadows may accidentally interact with objects on the table.

User Observations. Observations with the participants exposed differences between the two experimental conditions. For the paper-and-pen experiment, we found that there was the inequality in participation across the groups. During the discussion phase, in each group, there was a person who was in charge of the task. They invited others to participate and brought the group focus back to the agenda, asked to clarify the questions, restated what others have said, and moved the group forward to reach an agreement. The leader was in charge of writing as well. In some groups there were two different leaders, one was in charge of the session and another was in charge of writing. Some groups drew the mind map during discussion while only two groups discussed the ideas and took notes on an A4 paper before drawing the mind map.

In the tabletop approach, a pattern of parallel work was observed. Four members worked by themselves to type their ideas and create nodes, which they then discussed to reach agreements. In this environment, no one was in charge of writing. During the discussion phase, participants also talked to neighbours about their ideas. Some wrote their ideas down and placed the nodes in their space waiting for others to get ready, then they grouped the ideas together.

We found that after a few minutes of discussion in the paper-and-pen approach, some participants were playing with their mobile phones, and not being fully engaged in the task.

In the tabletop approach, some participants deleted other nodes/ideas by accident. We also observed that some participants enlarged the keyboard as big as the monitor size making it impossible for others to work. However, these did not create major problems as the participants still worked together, and they also laughed after unpredictable behaviour occurred.

6   Experiment 2

6.1   Apparatus

The experiment was designed to test the use of the mind map tabletop application connected to the tablets. We performed the tasks on the Touchscape table. As tablet devices we used three Samsung Galaxy Note 10.1 and a Motorola Xoom. They have 10.1 inches screen with a resolution of 1028x800 pixels.

The participants were seated around the table one side each for the digital study. Each user also has an Android tablet. In the conventional condition, the participants were seated around a round table as same as experiment 1 (Fig 4).

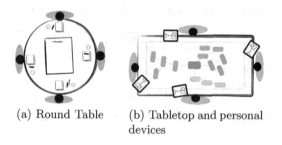

(a) Round Table   (b) Tabletop and personal devices

Fig. 4. Experiment 2 Set up

6.2   Participants

The second experiment involved forty students (16 male and 24 female) from local universities. As with experiment 1, none of the subjects had prior experience using a TouchScape table, but all were experienced of using computers and smart phones. 50% of subjects were in the 18-25 years old age group. The forty participants were divided into ten groups of four. The subjects were randomly formed according to availability. In term of mind mapping, all had experienced drawing a mind map.

6.3   Results

Ten groups of four were formed, all of them worked on both environments. The first five groups worked on the topic of stress management, while the rest worked on the global warming topic. After completing two tasks, participants completed a questionnaire. The first part focused on aspects of user satisfaction. Emotional aspect related questions were asked in the final section. They were also asked to identify their impression of our system and compare the two platforms. Likert scale responses and ranking questions were analysed by the Friedman test using SPSS software. The level of significance was set to 0.05 for these non-parametric statistical tests.

Performance. As Table 2 shows, there is a significant difference between the two considered approaches on duration. We found that when using the tabletop, they spent more time to create a mind map than the paper and pen approach. However, there were no significant differences between both approaches on the number of idea generated and conversation. With regard to the tabletop interaction (e.g. creating, relocating, typing), we found that the participants directly interacted more than via their mobile devices.

Table 2. The table defines mean of duration, number of ideas, conversation, mobile and tabletop interaction in both conditions. (NS = Non-significant results, * p=0.020).

	Tabletop	Paper and pen	Sig.
Duration	22.5	15.2	*
No. of ideas	30.7	30.0	NS
Conversation	33.8	30.5	NS
Mobile interaction	20.8		
Tabletop interaction	76.4		

User Satisfaction. the box plots in Fig. 5 show the significant differences between the two considered conditions for these four statements: "It was fun", "I had a lot of ideas", "I tried my best", and "I was motivated to do well".

According to Fig. 5, the results show the medians of the four statements for the tabletop were higher than the paper-and-pen condition. Furthermore, reference to statement "It was fun", the results indicated that the mean concern for the tabletop (M=4.62, SD=0.627) was significantly greater than the mean concern for the traditional condition (M=3.35, SD=0.833, p<0.001). Similarly, in response to statement "I had a lot of ideas", the mean for the tabletop (M=4.42, SD=0.712) was significantly higher than the mean for the paper-and-pen condition (M=3.80, SD=0.882, p<0.001). Likewise the means for the statement "I tried my best" (tabletop:M=4.70, SD=0.464, Paper-and-Pen:M=4.425,SD=0.742, p=0.003) and "I was motivated to do well" (tabletop:M=4.55, SD=0.552, Paper-and-Pen:M=3.82,SD=0.747, p<0.001).

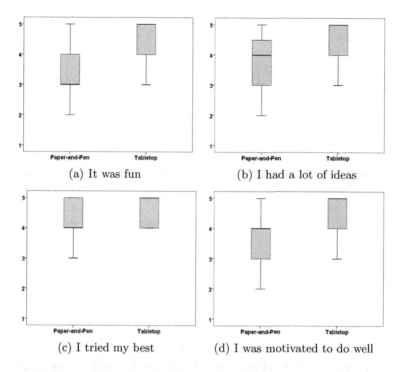

(a) It was fun   (b) I had a lot of ideas

(c) I tried my best   (d) I was motivated to do well

Fig. 5. Box plots of the distributions of satisfaction rating for the usability of the tabletop and paper-and-pent approaches for the designing of mind maps. The quality of ideas score was based on strongly disagree = 0 and strongly agree = 5. (Significant differences 5(a) $p<0.001$).

With regard to user satisfaction on aspects of collaborative work (Figure 6), participants agreed significantly more strongly with the statement "How well did the system support collaborative work?" for the tabletop approach than for traditional conditions. Furthermore, the majority of respondents rated significantly higher with the statement of "How well did the system support you to view the other was talking about?", "How well could you show information to others" and "How well did the system support you to discuss with others" for the tabletop than for paper and pen.

Emotional Attitudes. the analysis of the emotional attitudes data is shown in Table 3.

The tabletop was rated significantly higher for alertness ($p<0.001$), anxiousness ($p<0.001$), happiness ($p=0.004$), energy ($p<0.001$), enthusiasm ($p=0.001$) and tiredness ($p <0.001$). Meanwhile participants felt the paper-and-pen more

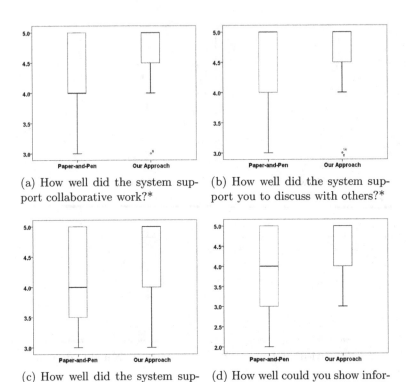

(a) How well did the system support collaborative work?*

(b) How well did the system support you to discuss with others?*

(c) How well did the system support you to view the other was talking about?*

(d) How well could you show information to others?*

Fig. 6. Box plots of the distributions of satisfaction rating for the usability of the tabletop and paper-and-pent approaches for the designing of mind maps. The quality of ideas score was based on strongly disagree = 0 and strongly agree = 5. ((*) indicate significant differences (a) p<0.001, (b) p=0.046, (c) p<0.001, and (d) p<0.001).

calm (p=0.001) and more boring (p=0.004). The results for the remaining emotional questions did not show significant differences.

Qualitative Feedback. Qualitative feedback was received through the questionnaire and users gave some comments during the experiments (N=19). 36.84% of the feedback indicated some recommendations that suggested about input method and also correction algorithms which was similar to experiment 1. 15.79% of the comments highlighted the system was easy to use and learn. 21.05% stated that it was enjoyable and 15.79% indicated that it supported collaboration.

With regard to Android mind map application, users prefer the viewpoints to see the mind map "right way up". Additionally, a scalable feature should be added to allow nodes to be resized to enhance the presentation of the mind map.

Table 3. Results for the ratings of the emotional attitudes for the tabletop and paper-and-pen conditions. The quality of ideas score was based on strongly disagree = 0 and strongly agree = 10. (NS = Non-significant results, * = significant).

Emotional attitudes	Tabletop		Paper-and-Pen		Sig.
Alert	9.075	1.327	6.575	2.448	*
Anxious	6.225	1.954	2.925	2.129	*
Calm	5.425	2.061	7.350	2.423	*
Happy	8.875	1.571	8.250	1.255	*
Depressed	2.575	1.810	2.375	1.674	NS
Energetic	8.875	1.488	7.875	1.712	*
Enthusiastic	8.675	1.542	7.525	1.867	*
Tired	3.550	2.669	1.950	1.319	*
Sad	2.125	1.244	2.100	1.032	NS
Relaxed	7.200	2.102	7.000	1.739	NS
Tense	3.825	2.500	3.600	2.250	NS
Bored	2.300	1.650	3.325	1.817	*

7 Discussion

In this paper, we have explored how the combination of interactive tabletop and mobile devices can support small group collaborative work. We compared the use of our framework with the use of the conventional paper-and-pen in this context. Our results indicate that the use of the interactive tabletop significantly supported the collaborative work within groups. This confirms the results found in a previous study [5]. The finding also showed that the use of the interactive tabletop was more enjoyable than the use of the traditional platform. The table-top approach stimulates users to generate a lot more ideas than the traditional approach. Not only do the soft keyboards help participants pursue their individual creativity, but also the mobile devices encouraged people to generate a lot of ideas, and more importantly, users are enabled to build private thoughts that can be seen in the personal device.

Consequently, the findings support the assumptions for this study. Firstly, users built a mind map using both tabletop and tablets. However, we found that users preferred the tabletop to draw the mind map. Perhaps this is because the tabletop gives more space to the user. Secondly, our results also indicated that users participated more in our application. Finally, they were clearly satisfied that the tabletop mind map application supports collaborative working.

With reference to our first challenge, multiple users shared the same surface and interact seamlessly together, we designed soft keyboards to support multiple users. We found that the soft keyboards can help participants pursue their individual creativity effectively. In addition, the tabletop mind map system supports users arrangements as users can scale, rotate and relocate the elements on the table. As far as user-based challenges are concerned, we also noticed that some users enlarged the soft keyboard for typing, they said that was much easier for them.

With regard to user feelings, it is interesting to note that the tabletop application was mostly driven by user curiosity to explore and play with the tools. It is well established that emotional attitudes are important factors that influence the effectiveness of collaborative work [1]. It has been shown [15] that positive emotions (e.g. happiness, enjoyability) are significantly more influential factors for collaborative working than for individual work. Positive emotions also indicate self-fulfillment of participants in the context of collaborative work [13]. In contrast, negative emotions such as anxiety, sadness, and anger, may be associated with reduced accuracy on tasks and executive functioning by biasing cognitive processing, or may lead to reduction of the production of ideas [12]. Therefore, it is important to consider the emotional attitudes of users in the context of evaluation of techniques aimed to support collaborative work.

The results from the observations show that if one person is in charge the members participate unequally across the groups in the paper and pen condition as found in [4], [20]. On the other hand, we found that in the tabletop application users did participate very actively and more equally while on the traditional approach they merely expressed their ideas to the leader who wrote them down. Therefore, our results suggest that the interactive tabletop supports the empowerment of the participants, and supports them working together in parallel.

The limitations of our study are as follows. First, our participants were students in the university. In the future we aim to add special features to support specific professional groups (e.g. health care professionals). Next, we found that when users put many ideas, the screen seems cluttered. Therefore, to improve this feature, the application should be able to group the nodes.

8   Conclusion and Future Work

The study presented two experiments, first examined the effectiveness and the usability of interactive tabletops for the creation of mind maps, and the second examined the usability of the combination of interactive tabletop and mobile devices. Both situations were compared with traditional paper-and-pen methods in the context of collaborative task.

Our results indicate that the interactive tabletop is more enjoyable to use than paper-and-pen approach. The tabletop appears also to support collaborative work in a small group of people. We also found that the quality and number of ideas generated was higher when participants used the tabletop. Furthermore, the results clearly indicate that the combination of the tabletop and personal devices support and encourage multi users to work collaboratively. The comparison of the associated emotional attitudes indicates that the interactive tabletop facilitates the active involvement of participants in the group decision making significantly more than the use of the paper-and-pen approach.

In the future, we plan to enhance the application with more natural affordances and test on the platform with a wider range of potential users.

Acknowledgements. Special thanks to students for taking part in the study.

References

[1] Ashforth, B.E., Humphrey, R.H.: Emotion in the workplace: A reappraisal. Human Relations 48(2), 97–125 (1995)

[2] Bachl, S., Tomitsch, M., Wimmer, C., Grechenig, T.: Challenges for designing the user experience of multi-touch interfaces. In: Proc. Workshop on Engineering Patterns for Multi-Touch Interfaces (2010)

[3] Baraldi, S., Bimbo, A., Landucci, L.: Natural interaction on tabletops. Multimedia Tools and Applications 38, 385–405 (2008)

[4] Buisine, S., Besacier, G., Aoussat, A., Vernier, F.: How do interactive tabletop systems influence collaboration? Computers in Human Behavior 28(1), 49–59 (2012)

[5] Buisine, S., Besacier, G., Najm, M., Aoussat, A., Vernier, F.: Computer-supported creativity: evaluation of a tabletop mind-map application. In: Harris, D. (ed.) HCII 2007 and EPCE 2007. LNCS (LNAI), vol. 4562, pp. 22–31. Springer, Heidelberg (2007)

[6] Buzan, T.: The mind map book. BBC (2006)

[7] Buzan, T.: Use your head. BBC (2008)

[8] Chi, C., Liao, Q., Pan, Y., Zhao, S., Matthews, T., Moran, T., Zhou, M.X., Millen, D., Lin, C.-Y., Guy, I.: Smarter social collaboration at ibm research. In: Proceedings of the ACM 2011 Conference on Computer Supported Cooperative Work, CSCW 2011, pp. 159–166. ACM, New York (2011)

[9] Chik, V., Plimmer, B., Hosking, J.: Intelligent mind-mapping. In: Proceedings of the 19th Australasian Conference on Computer-Human Interaction: Entertaining User Interfaces, OZCHI 2007, pp. 195–198. ACM, New York (2007)

[10] Do-Lenh, S., Kaplan, F., Dillenbourg, P.: Paper-based concept map: The effects of tabletop on an expressive collaborative learning task. In: Proceedings of the 23rd British HCI Group Annual Conference on People and Computers: Celebrating People and Technology, BCS-HCI 2009, pp. 149–158. British Computer Society, Swinton (2009)

[11] Faste, H., Lin, H.: The untapped promise of digital mind maps. In: Proceedings of the 2012 ACM Annual Conference on Human Factors in Computing Systems, CHI 2012, pp. 1017–1026. ACM, New York (2012)

[12] Fredrickson, B.L.: The role of positive emotions in positive psychology: The broaden-and-build theory of positive emotions. The American Psychologist 56(3), 218 (2001)

[13] Fredrickson, B.L., Losada, M.F.: Positive affect and the complex dynamics of human flourishing. The American Psychologist 60(7), 678 (2005)

[14] Frey, C.: Mind mapping software is an essential tool for todays̗ workers, servey shows (August 2011), http://mindmappingsoftwareblog.com/2011-survey-results-published/

[15] González-Ibáñez, R., Shah, C., Córdova-Rubio, N.: Smile! studying expressivity of happiness as a synergic factor in collaborative information seeking. Proceedings of the American Society for Information Science and Technology 48(1), 1–10 (2011)

[16] Hilliges, O., Terrenghi, L., Boring, S., Kim, D., Richter, H., Butz, A.: Designing for collaborative creative problem solving. In: Proceedings of the 6th ACM SIGCHI Conference on Creativity & Cognition, C&C 2007, pp. 137–146. ACM, New York (2007)

[17] Lee, H., Jeong, H., Lee, J., Yeom, K.-W., Park, J.-H.: Gesture-based interface for connection and control of multi-device in a tabletop display environment. In: Jacko, J.A. (ed.) HCI International 2009, Part II. LNCS, vol. 5611, pp. 216–225. Springer, Heidelberg (2009)

[18] Martínez Maldonado, R., Kay, J., Yacef, K.: Collaborative concept mapping at the tabletop. In: ACM International Conference on Interactive Tabletops and Surfaces, ITS 2010, pp. 207–210. ACM, New York (2010)

[19] North, C., Dwyer, T., Lee, B., Fisher, D., Isenberg, P., Robertson, G., Inkpen, K.: Understanding multi-touch manipulation for surface computing. In: Gross, T., Gulliksen, J., Kotzé, P., Oestreicher, L., Palanque, P., Prates, R.O., Winckler, M. (eds.) INTERACT 2009. LNCS, vol. 5727, pp. 236–249. Springer, Heidelberg (2009)

[20] Pantidi, N., Rogers, Y., Robinson, H.: Is the writing on the wall for tabletops? In: Gross, T., Gulliksen, J., Kotzé, P., Oestreicher, L., Palanque, P., Prates, R.O., Winckler, M. (eds.) INTERACT 2009. LNCS, vol. 5727, pp. 125–137. Springer, Heidelberg (2009)

[21] Ridden, P.: Take a multi-touch break with touchscape's 47-inch coffee table, http://www.gizmag.com/touchscape-multi-touch-coffee-table/17927/

[22] Scott, S.D., Grant, K.D., Mandryk, R.L.: System guidelines for co-located, collaborative work on a tabletop display. In: Proceedings of the Eighth Conference on European Conference on Computer Supported Cooperative Work, ECSCW 2003, pp. 159–178. Kluwer Academic Publishers, Norwell (2003)

[23] Seifert, J., Simeone, A., Schmidt, D., Holleis, P., Reinartz, C., Wagner, M., Gellersen, H., Rukzio, E.: Mobisurf: improving co-located collaboration through integrating mobile devices and interactive surfaces. In: Proceedings of the 2012 ACM International Conference on Interactive Tabletops and Surfaces, ITS 2012, pp. 51–60. ACM, New York (2012)

[24] Shen, C., Everitt, K., Ryall, K.: Ubitable: Impromptu face-to-face collaboration on horizontal interactive surfaces. In: Dey, A.K., Schmidt, A., McCarthy, J.F. (eds.) UbiComp 2003. LNCS, vol. 2864, pp. 281–288. Springer, Heidelberg (2003)

[25] Shih, P.C., Nguyen, D.H., Hirano, S.H., Redmiles, D.F., Hayes, G.R.: Groupmind: supporting idea generation through a collaborative mind-mapping tool. In: Proceedings of the ACM 2009 International Conference on Supporting Group Work, GROUP 2009, pp. 139–148. ACM, New York (2009)

[26] Willis, C.L., Miertschin, S.L.: Mind maps as active learning tools. J. Comput. Sci. Coll. 21(4), 266–272 (2006)

How a Conflict Changes the Way How People Behave on Fandoms

An Investigation of Shipper's Fight in Facebook Groups

Cleyton Souza[1,4], André Rolim[1,3], Jonathas Magalhães[1,4], Evandro Costa[4], Joseana Fechine[1], and Nazareno Andrade[2]

[1] Artificial Intelligence Laboratory - LIA
[2] Distributed Systems Laboratory - LSD
Federal University of Campina Grande - UFCG,
Campina Grande-PB, Brazil
[3] Federal Institute of Education, Science and Technology of Paraíba - IFPB,
Cajazeiras - PB - Brazil
[4] Intelligent, Personalized and Social Technologies Group - TIPS,
Federal University of Alagoas - UFAL,
Maceió-AL, Brazil

Abstract. Shippers are fans of couples. In the virtual sphere, rival Shippers (Shippers of conflicting couples) are obligated to share the same space in order to discuss common interest matters. However, sometimes, conflict emerges between them when the divergence of opinions appears. Our work investigates the implications of these conflicts. In order to propose countermeasures to prevent or reduce the frequency of conflicts, a good understanding of the problem is necessary. We start discussing about why people felt motivated to debate about fictional couples. Next, we report about how this struggle changes the normative sense of people. Finally, we investigate the role of the virtual space in the conflict. Our investigation takes place in a Facebook group called How I Met Your Mother [Brazil]. To support our observations, we collected evidences from posts from the group, interviews with members and answers from a questionnaire.

Keywords: Shippers, Fandoms, Collaborative Systems, Online Communities.

1 Introduction

Shipper is a term used by Fandoms to designate individuals that cheers for a romantic engagement between two characters, be them fictional or real. The word is derived from the English word relationship, and shippers "ship" a couple. Shippers and shipping are a largely unexplored research topic [6]. A notable exception is Scodari and Felder [9], which conducted one of the first studies about this kind of fans. Our research extends their work investigating shipping and shippers' practices and conflicts in a 10,000-member online community called How I Met Your Mother [Brazil].

N. Baloian et al. (Eds.): CRIWG 2014, LNCS 8658, pp. 335–348, 2014.

The sitcom How I Met Your Mother involves a love triangle, what leads fans of the show to split themselves into two major groups of shippers. We observe that the preferences for different couples leads to conflict between shippers of different groups. Nevertheless, to discuss general topics about the sitcom, these conflicting groups share the same virtual space, routinely coping with conflict.

Our analysis focuses on understanding motivations for shipping, how the conflicting groups interact in the community, and how the elements of the virtual space used by the community – a Facebook group – contribute for conflict creation and resolution. To investigate such aspects, we employ a mixed method research design, combining semistructured interviews, privileged ethnography, and questionnaires. To our study, we collected and analyzed several of these discussions. In addition, we also interview ten members of the group to get their perception of the conflict. To support our observations, we used a questionnaire that received answers from 403 members.

Although, our investigation took place in a Facebook group and we focused on Fandoms, conflicts due to opinion divergence between members are not exclusive of this context. Students can argue about the better solution to a problem. Investors can discuss what the best option to invest in the stock market. Readers of a politics news portal may have different opinions depending on their party preferences. With a better understanding of these kind of conflict, we could plan changes (architectural, normative or governmental) to avoid, reduce or incentive this topic or to protect members of its effects or even projecting solutions to somehow affect these conflicts.

This paper is organized as follows. Section 2 presents a brief overview of researches about Shippers found in literature. Section 3 gives a background about the show and about the group itself. Section 4 details our methodology and our data. In Section 5, we discuss the obtained results and some conclusions with proposal for future work are presented in Section 6.

2   Related Work

Jenkins [3] was one of the works that inaugurates the research about Fandoms. In his work, he discuss about how the field evolved, limitations and open questions. Jenkins [3] highlights the mistake made by most researchers that try to explain Fandoms using a religious metaphor, because, unlike Religion, fans can belong to multiple Fandoms.

Regarding studies about Shippers, Scodari and Felder [9] conducted the first research about this specific kind of fan. In their work, they investigate a conflict between X-Files' fans, more precisely, between Shippers (fans who were hoping that the protagonists form a couple) and NoRomos (an acronym of No Romancers that they used to refer to fans who preferred avoid a romantic engagement between the protagonists). In addition of reporting arguments used by the Shippers and by the NoRomos, they traced a theory between the narrative of the show and how it is associated with the conflict. They commented the fact that the show producers picked the side of NoRomos and constantly denied the possibility of engagement between the protagonists, but contradictorily they keep insinuating this in episodes. We made similarly observations in our study.

Actually, our research extends Scodari and Felder's work. However, we explore the role of the virtual space as an environment catalyst to the conflict, while Scodari and Felder focused in the conflict itself and its actors (arguments, behavior, etc). In addition, we investigate Shipper's conflict between Shippers of conflictant couples, instead a fans who are Shipper and who are not. Finally, their investigation takes place in the fandom of the dramatic show X-Files, while ours use a sitcom called How I Met Your Mother. However, we directly compare some of results with Scodari and Felder's observations in X-Files fandom.

More recent research about Shippers concentrate in the study of Slashers (fans who fantasize a gay couple with the characters), e.g. Tosenberger [10] that studies them in the Harry Porter's fandom and dos Santos [8] that studies these fans in the context a British TV show called Sherlock. According Jenkins [4], Slash is an attempt to insert personal aspects inside the object of your admiration. In our research, we have not discussed about this aspect of Shippers, although the topic has emerged during interviews. Our work focuses in a struggle between Shippers of different couples. However, it is worth to say that all interviewee believed that How I Met Your Mother's fandom has no real Slashers and *"when a fan says he cheers for a gay couple either he is just kidding or revealing a fetish"* (LS), as it was said by one of our interviewees. It has no deep emotional connection.

3   Background

How I Met Your Mother (HIMYM) is an American sitcom that premiered on CBS on 2005. The show is about the story of Ted telling his children in 2030 how he met their mother. The story starts back in 2005 and follows the main character and his group of four friends in Manhattan: the couple Marshal & Lily, Ted's friends since college when they all met; Barney, one of Ted's best friends; and Robin that Ted meets in the pilot and by whom he falls in love. Ted & Robin date for two seasons, but she ended not being "The Mother". However, later, in season four, Robin starts date Barney, creating the romantic triangle that splits the fans of the show in two groups of Shippers: the Mosbatsky, those who prefer Ted & Robin, and the Swarkles, those who prefer Barney & Robin. These names are, respectively, combination of last names and nicknames of the characters that compose the couple.

Evidences of this fight can be found on various platforms. However, our investigation take place in a Facebook Group called How I Met Your Mother [Brazil][1] (henceforth HIMYM Brazil), created in 2011, and today with more than ten thousand members. People use the public and common area of the group to share news related to the show, opinions about the episodes, discuss theories about the storyline, etc. Activities like in any other Fandom. In addition, people constantly expose their preferences like favorite episodes. Thus, many discussions occur when there is divergence of opinions, what usually start long debates. One

---

[1] https://www.facebook.com/groups/howimetyourmotherbrasil/

of most recurrent discussion is between Mosbatsky and Swarkles (they debate about several things, for instance, what the best couple is, or who the best Robin's boyfriend was: Ted or Barney, who deserve be happier, etc.).

4   Methodology

Our methodology is one of mixed research presents using a constructive point of view [2]. Three main sources of data are employed: first, one of the authors has been an active member of the group studied for years. As such, he possesses detailed and textured contextual information about the functioning of the group, as well as about its distinguished members and episodes. Incorporating such experience in the research method has been called privileged ethnographic or "aca-fen" [6]. Second, we used semistructured interviews with ten members of the community, discussing perceptions about shippers and their shipping practices. Finally, we performed comprehensive searches for posts in the community discussing shipping, and employed a questionnaire which received answers from 403 respondents. In connecting the three parts of the method, the ethnographic experience and interviews informed generic theories which were further examined using the questionnaire. All the data used in by us are available to download[2]. Also, we highlight that HIMYM Brazil is a group composed mostly by Brazilians, for this reason, questionnaire, answers and interviews are in Portuguese. Translations presented here were performed by the authors.

Recruiting of interviewees was based on responses from a question published in the group in October/2013. The question asked them about their favorite couple. We interviewed ten members, five women and five men. We chose to interview a diversified sample taking in consideration their answers, but also how long the members belong to the group. When talking about results, we will refer to them as interviewees.

In addition, we used the native search engine from Facebook groups to find post where there was some kind of debate among members about what the best couple is. Expressions like "Ted & Robin", "Barney & Robin", "Swarkles", "Mosbatsky" and "Shipper" were used. Results to these queries were individually analyzed; there were post since December/2012 to December/2013 and we select 25, which exemplified the fight between Shippers. When talking about results, we will refer to the authors as posters.

To quantitatively confirm our observations, we also elaborate a questionnaire asking people several things about Shippers and about the living in the group. The questionnaire was answered by 403 people. When talking about results, we will refer to those who answered the questionnaire as respondents.

All direct quotes from a speech are followed by an indexer code that allows us to locate the source of the speech if requested.

---

[2] https://docs.google.com/file/d/0B4ZS_d4fhCZVLUtDZ3lVSTF3UHc

5   Results

The results presented in this study revolve around our two research questions: one which aims to understand the behavior of Shippers inside the group; and a second which investigates the relationship between this behavior and the architecture of the virtual space used by the community. However, before start report our observations, we will briefly describe our respondent's characteristics in next section.

5.1   People Who Answered Our Questionnaire

In Table 1, we show how long people who answered our questionnaire were viewers of the show and how long they belong to HIMYM Brazil.

Table 1. Time following the show vs Time in the group

How long time have you Watched the show?	How long time have you joined the group?					
	One Month	Six Months	One Year	Two Years	More Than Two Years	Sum
One Month	10	0	0	0	0	10
Six Months	26	19	0	0	0	45
One Year	23	52	17	0	0	91
Two Years	13	36	27	5	0	81
More Than Two Years	20	50	74	26	6	176
Sum	92	156	118	31	6	403

Most of our responders are in their first year in the group, but they watched the show longer than that. We also asked if they knew the meaning of the word Shipper; and, after explain what it means, if they believed they were shippers. These results are summarized in Figure 1.

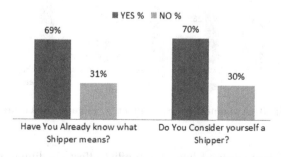

Fig. 1. Know the meaning of "Shipper" (Left) and Consider himself a Shipper (Right)

Fortunately, most of our responders describe themselves as Shipper. Next, we asked what couple they "ship". Based on the answer to this question, people were directed to different questions. We choose keep only these three options to reduce noise in our analysis. Figure 2 shows that Barney & Robin is the favorite couple among our interviewees.

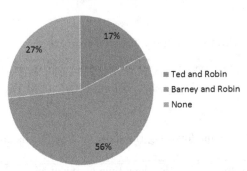

Fig. 2. Favorite Couples of Respondents

Finally, we asked in the questionnaire if they were aware about the conflict, 73% of responders said that they have already seen one of these discussions. In Figure 3, we show that longer the person belongs to the group, the greater the chance he/she had ever witnessed a discussion between Shippers (older members have witnessed the discussion at least once).

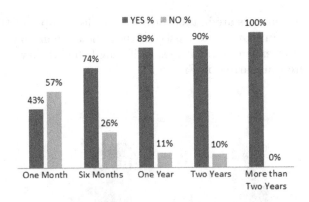

Fig. 3. Awareness of the Conflict × Time in the Group

Next, we will discuss about how this conflict affect the living in the group.

5.2   Shippers: Behavior, Arguments and Perceptions

HIMYM Brazil is a group open to discussions about all topics related to the sitcom. As usually happens in similar groups, there are rules to avoid off-topic posts and spoilers. Regarding shippers' discussions, there are no clear rules, however. In the past, when the group was deciding its operating rules, there were requests to make such discussions forbidden, but this was not done. In contrast with the perception of some that the resulting discussions are not welcome in the group, 65% of the respondents from our questionnaire stated that Shipper's discussion is a recurrent topic in the group.

For all interviewees, it is evident to that there are two main sects of shippers in the community there are Mosbatskty (those who root for Ted & Robin), and there are Swarkles (those who root for Barney & Robin). Moreover, these two sects are usually at conflict, generating recurring and sometimes aggressive discussions. A third portion of the community is those who are indifferent and do not pick a side. We asked respondents of the questionnaire if they agree that there was a split in the HIMYM Fandom. Figure 4 shows the agreement percentage aggregate by favorite couple's choice.

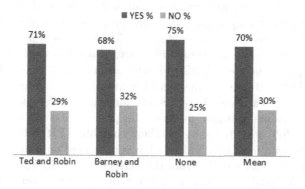

Fig. 4. Agreement Percentage about the Fandom Split (aggregate by favorite couple's choice)

When discussing the origin and motivation for keeping alive the shipping conflict in the community, interviewees seem to perceive that this is a natural consequence of the existence of the love triangle, and of fans turning their support for a couple in part of their identity in the community. Interviewees stated that *"–This struggle between fans is common in any story that has a romantic triangle, especially on TV shows"* (AT), and *"–It's something that happens in almost every show; Shippers struggling is common"* (RL155). Relating to the motivations, interviewees mentioned that *"some fans take the show to serious"* (AT). One said: *"–There are people who raising the flag of a couple and want to defend it as if it were a football team"* (LS). Moreover, on interviewee speculated that this may happen because it is relatively easy to relate to the characters' situation: *"–People identify with characters and start to wish their happiness; for this reason, it is common a feeling of transference, where they imagine themselves in the character place and create expectations about what is better to each one"* (LC272).

In Brazilian culture, it is usual to routinely discuss what the best football team is without reaching any conclusion. Interviewees perceive that some of the community participants follow TV shows the same way others follow football teams. For those people *"the show is a significant part of their lives"* (AT). As their shipping choice becomes part of their identity, it becomes important to defend it. Nevertheless, in spite of having a rationale that is not alien to the community, this behavior is often seen as negative. Some respondents said *"they talk about that because have nothing better to do"* (e.g., GT303, YA232, BS321). In addition, a poster gives an advice to Shippers in general: *"–to everyone who keeping discussing about fictional couples: go get a football team to cheer; it is well better!"* (LV8). HIMYM Brazil, like most online communities, reunites people with a variety of perceptions and cultures.

Regarding identity, people start to saw themselves in the spot of the characters of the show. Then, when someone is arguing about why Robin should end up with Ted, this is just a sign that what he would want to himself. One respondent says *"everyone looks something different in a relationship, that's where the differences come"* (GS168). They keep arguing because *"they wish other people feel the same way"* (JH093). However, one poster states: *"–These things of prefer one or other couple have no argument, we simply like and we identify with different things. It is like everything in life"* (DI22). Identity seems a legit reason to enter in a discussion and these discussions happens because there is difference of identities: each character and each couple represents something different. According one Mosbatsky interviewed, his goal to expose his preference is attract the attention of others "hopeful" with the same thinking that him (PR). He even created a minor group derived from HIMYM Brazil and dedicated only to Shippers of Ted & Robin[3] to protect Mosbatsky from the persecution of Swarkles [1].

When one Mosbatsky comments Swarkles content or vice versa, reactions are almost the same, people start debate and usually they use the same arguments: "Robin is not the one", "Robin and Barney do not match", etc. The aim of this discussion is not change the other's opinion, but convincing the other that his/her opinion is worthwhile. However, like NoRomos in Scodari and Felder's study, the Swarkles are "winning" the Shipper's war. A poster says *"–Mobastsky team lost a long time ago"* (MS4). One of the interviewees compares the *status quo* to the *apartheid*[4], but it is the dominant majority of Swarkles that chases and attacks on Mosbatsky minority in the group (PR). Due to this situation, other Mosbatsky interviewed said that she avoid post content about Ted & Robin, because it would be followed by negative comments by people who disagree of

---

[3] https://www.facebook.com/groups/565654476822300/

[4] Apartheid was a system of racial segregation in South Africa under which the rights of the majority black inhabitants were curtailed by the government compound by the white minority.

her couple's choice (TM). This is a sign that there is a normative sense absorbed by some members that certain types of content are too controversial to be shared in the group feed, even with no explicit rules forbidding. According one of the Swarkles interviewed, there is "hate", and it is reciprocal: *"–Same way Swarkles do not like Mosbatsky content, Mosbatsky do not like Swarkles content"* (AT).

Swarkles are the majority and Mosbatsky minority feels repressed and avoid publishing content related with their ship in order to avoid conflict. A Mosbatsky poster reveals *"–I said that I root for Ted & Robin and they scold me; I even left the group, but decided come back"* (MM11). One Mosbatsky among interviewees said that he often expresses his preference publishing or commenting. He also said that, due to his posture, he is pursued by Swarkles outside the group. *"–I want to be a Mandela to Mosbatsky"* (PR), he said. We tried to confirm this "repression" through our questionnaire, but the behavior of both groups of Shippers was quite similar. However, post analyzes suggests that only a small part of Swarkles (and usually same people) overreact to Mosbatsky's content. In addition to despise Mosbatsky's content; there is a portion of Swarkles that mocks their posts. One of the interviewees reveals a non-verbal way of dispute in Facebook related with the Like option. She says: *"– I* [using Facebook's *like* mechanism] *like the photo to numerically make the couple Barney & Robin win"* (LR). As the support of the community, expressed through Likes, was indicative that who is right in the discussion. This is interesting because, according respondents, "Like" was the Shipper's most supportive reaction to content related to their favorite couple (Mosbatsky: 42%; Swakles: 46%), followed by reading (Mosbatsky: 34%; Swakles: 33%) and commenting (Mosbatsky: 20%; Swakles: 15%).

According most interviewees, the show's producers are guilty for the conflict between fans, because they continue to feed the Mosbatsky group with romantic moments between Ted and Robin, even with the show's authors officially indicating that they will not end up together (AT). One of the Mosbatsky interviewed said in one post, *"–It is not fair they play with the heart of a passionate shipper like me"* (TM5). Another interviewee, with a similar opinion, admitted that there are only four guilts by his preference for Ted & Robin, they are: Carter Bays and Craig Thomas (creators of the show) and Josh Radnor and Cobie Smulders (actors who play Ted and Robin) (PR). He believes that good script and good performances are the main reasons that draw people to prefer this couple. Regardless the accusation of fans, the truth is that producers of the sitcom are aware of this conflict. This even yields a recurring gag on the show, where Marshall and Lily argue about Robin may or may not be the perfect match ("the one") to Ted. Somehow, Marshal represents the hope for Mosbatsky, as a fan declares in a post: *"–Only when Marshal admits he lost the bet, it will be an end to these two"* (RT5).

Mosbatsky themselves recognize that Ted & Robin, probably, is not going to happen. Few still believe that a big plot twist can happen and Ted & Robin end up together; *"–Even though this is almost impossible"* (PR). Mosbatsky are minority in the group and accepted a role of subservient assigned by Swarkles. We explicitly asked respondents who the dominant group was and 81% confirmed

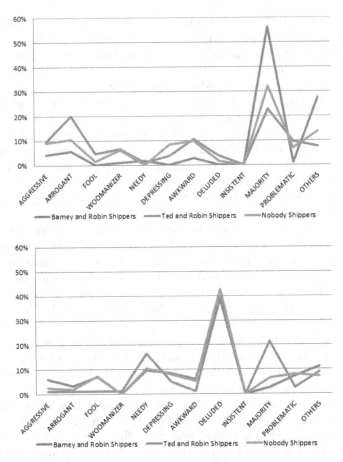

Fig. 5. How interviewees describe Swarkles (Up) and Mosbatsky (Bottom)

that was Swarkles. In addition, we asked respondents how they describe Mosbatsky and Swarkles fans. The results presented in Figure 5 are aggregated by their favorite couple's option.

Figure 5 demonstrates what is inside everyone's mind in the group: Swarkles are majority and Mosbatsky are delusional. Swarkles believe that, since Ted & Robin probably will not end up together, it is nonsense cheer for this couple. The most interest thing is that Mosbatsky themselves think that way. The role assigned to Mosbatsky changes their normative sense and makes most of them behave different from others members. They assume that share Ted & Robin content is prejudicial to the good living in the group (LS, TM). But, why are Mosbtaky "losing" to Swarkles?

A possible explanation could be formulated based on [9]: fans that are supported by the canon[5] are dominant. In Scodari and Felder's study, Shippers were

---

[5] In fiction, canon is the material accepted as "official" in a fictional universe.

powerless, because the runners of the show continuously said that protagonists would never be a couple. Same way, the fact that Mosbatsky need changing the history being told by producers and writers to fit with their own interests makes them weak in the group. One Swarkles poster said: *"–It is funny entering in a debate when you know you gonna win"* (VO11). To confirm this observation, we asked respondents which factors determine if a group is dominant. Figure 6 summarizes their answers.

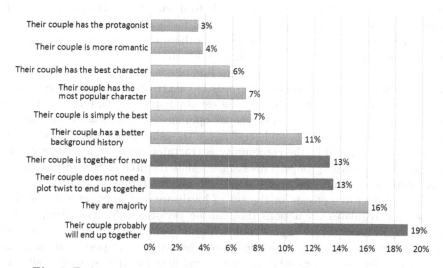

Fig. 6. Factors perceived to determine a group of Shippers as dominant

Reasons related with canon (showed in black in Figure 6) were the top voted options, except by majority that was in 2nd place. Being part of the bigger group is important to be considered dominant. However, mainly, canon defines what group will be dominant and what will be submissive and, consequently, affects the behavior and living of members in the group. This is interesting because means that external factors are able to influence online interactions.

5.3 The Role of the Virtual Space in Shipper's Conflict

Most interviewees state that usually they do not discuss outside the group about what the best couple is. One Mosbastky interviewed said that he does not feel comfortable to discussing this matter with friends (PR), a second prefers only to discuss about the show with people who also prefer Ted & Robin (TM). In addition, other Mosbatsky acknowledges the lack of motivation to discuss this topic, now that the show is in its final season and apparently the couples composition was already established (LS). A couple of Swarkles interviewed said that they do not discuss about "Shipping" with friends: one convinced all her friends that Barney & Robin is the better couple, thus, *"the topic does not comes to light anymore"* (FM); the other said that her friends who also like the show prefer Ted & Robin, thus, to avoid discussion, they usually talk about others aspects of the show (LH). Almost half of the respondents (47%) do not take this kind of discussion out of the group either.

We realize that "Shipping" discussion is not so usual outside the group. Most interviewees also said that they usually do not share Couple's related content in the group's wall. However, when they see a discussion between Shippers, they engage depending on *"the level of absurd commented"* (FM, AT). When we asked respondents about what reasons led people to discuss about their couple's preferences related to fictional character, several motivations were raised: identity with characters (14%), trying to convince others (8%), expose of an opinion (8%), displeasure with story of the show (6%), etc.

A possible explanation for the above findings is that people tend to avoid conflict offline, where discussion and possible aggressive outcomes have a high costs. Even more if the offline discussion usually happens with relatives or close friends. In the virtual space, however, as opinions are broadcast in a common space, it becomes harder to avoid conflict. On the other hand, it is also clear that the conflict generated in the community is tractable, as there are few episodes of aggression between groups. This suggests that the architecture of the virtual space plays a role in conflict generation and administration.

The architecture consists in the set of hardware and software features that defines and regulates the actions that can be performed [5]. Therefore, we believe that the architecture of Facebook groups favors these two kinds of Shippers "find" themselves: there is a public space where users can post content unsupervised (video, photo or text); all content published in that space is visible to everyone in the group; and people can like and comment posts, according their interests. Thereby, Mosbatsky are exposed to Swarkles content and vice versa. The discussions happen because *"some members of the group cannot ignore when they disagree"* (PR, LR). That is when fights start. One interviewee, for example, created a post asking Mosbatsky if they felt pursued in the group (like a virtual bullying) and asked for Swarkles not interact (PR11). Sometime later, Swarkles started to comment the post and verbally attack the author claiming that what was in the collective space may be commented by anyone. In addition, the architecture of communication through Facebook allows misinterpretations and overreactions.

Since these discussions take place in a public space, other members of the group, who identify themselves (identity-based affective commitment) as Shippers, enter and take one side. This way, the amount of people participating in the discussion rises (sometimes, people tag others with the same ideology in the discussion). Preece and Maloney-Krichmar [7] states that people feel empathy for others with similar taste. Thus, supported by the collective power and by the safety of online interaction, people get upset and some users lose the line, starting to offend the rival group (strangers, innocent, deluded, were some of the adjectives used by interviewees to describe people who cheer for a rival couple). In Facebook groups, people and their thoughts are represented by what they write and do; thus, misinterpretations can initiate long ideological discussions. The characteristics of online interaction that facilitate participation, also favor the conflict between Shippers, they are: anonymity, ease of entry and exit and ambiguity of written language [Kraut & Resnick 2010]. Due to these architectural features, Facebook groups become a propitious environment to this discussion happens [7].

We inquiry respondents if they believe that the way how Facebook groups works facilitate that Shippers of different couples find each other to discuss: 92% agreed. However, when we asked if a changing in that way would reduce or eliminate these discussions from the group, 58% disagree. This means that, they believe, even architecture having its effect, these discussions would come to light anyway, regardless the way how the group works.

These are recurring discussions on HIMYM Brazil, one poster jokes *"soon, it will have another topic, with the same content, and almost same comments"* (ML11). According interviewees, a Shipper's discussion only stops when becomes too much exhausting and people get tired of defend their positions. Usually, this begins when the discussion attracts too much attention and the number of people involved rises. Then, comments rate start to increase together with mean size of comments (FM). Finally, people loose focus and change the topic to *"zoeira"* (a Brazilian word that means a state in a discussion where all statements are jokes) (FM, IT). Respondent's answers led to the same observations; "get tired of discuss" (28%) followed by "change the topic to zoeira" (24%) were the two main reasons to end a discussion.

6   Conclusion

In this work, we presented a mixed research studying fans inside a group in Facebook called HIMYM Brazil. The group reunites fans with different perceptions and personalities who continually discuss their preferences regarding who should end with Robin (a character on the show). These fans are known as Shippers. In this research, we reported how the Struggle among conflicting Shippers changes the normative sense of members depending on which group of Shippers they belong. In addition, we discuss the role of the architecture as a catalyst for conflict.

We state three main results in this work: (1) an axiom that considers culture and identity as main reasons that lead people to discuss couples; (2) an axiom about canon establishing a caste of Shippers as dominant and the other as submissive and how being part of a caste change the normative sense; (3) an axiom about how the architecture provides a space where these rival castes can easily find each other and struggle. In addition, we extended Shipper's research initiated by Scodari and Felder [9]. Now that we have a better understanding about this conflict, we could plan changes (architectural, normative or governmental) to avoid, reduce or incentive this topic in Fandoms or to protect members of its effects.

As future work, we plan to investigate the conflict among Shippers in others TV shows, in order to verify if the same observations happen in different contexts. In addition, we aim to analyze the way how the conflict happens in other platforms that are also used by Fandoms to reunite people, e.g. Tumblr, Facebook pages or Wikis. In addition, interviews revealed that fans do not believe that there is Slashers in HIMYM's fans community. Other proposal that deserves further investigation is confirming this statement.

Acknowledgments. We want to thank people who answered our questionnaire and all the people who devoted a few minutes so we could interview them.

References

[1] Butler, B.: When is a Group not a Group: An Empirical Examination of Metaphors for Online Social Structure. Social and Decision Sciences (1999)

[2] Creswell, J.: Research Design: Qualitative, Quantitative, and Mixed Methods Approaches. SAGE Publications (2003),
http://books.google.com.br/books?id=nSVxmN2KWeYC

[3] Jenkins, H.: Textual Poachers: Television Fans & Participatory Culture (Studies in Culture and Communication). Routledge (June 1992),
http://www.amazon.com/exec/obidos/redirect?tag=citeulike07-20&path=
ASIN/0415905729

[4] Jenkins, H.: Fans, bloggers, and gamers: exploring participatory culture/Henry Jenkins. New York University Press, New York (2006), Includes bibliographical references and index

[5] Lessig, L.: Code and Other Laws of Cyberspace. Basic, New York (1999)

[6] Meyer, M.D.E., Tucker, M.H.L.: Textual Poaching and Beyond: Fan Communities and Fandoms in the Age of the Internet. Review of Communication 7(1), 103–116 (2007), http://www.tandfonline.com/doi/abs/10.1080/15358590701211357

[7] Preece, J., Maloney-Krichmar, D.: Online Communities: Focusing on Sociability and Usability. In: The Human-Computer Interaction Handbook, pp. 596–620. L. Erlbaum Associates Inc., Hillsdale (2003),
http://dl.acm.org/citation.cfm?id=772072.772111

[8] dos Santos, P.: I am SherLocked: Afeto e Questões de Gênero no Interior da Comunidade de Fãs da Série Sherlock. Ciberlegenda (28) (2013),
http://www.uff.br/ciberlegenda/ojs/index.php/revista/article/view/624

[9] Scodari, C., Felder, J.L.: Creating a Pocket Universe: Shippers, Fan Fiction, and the X–Files Online. Communication Studies 51(3), 238–257 (2000),
http://www.tandfonline.com/doi/abs/10.1080/10510970009388522

[10] Tosenberger, C.: Homosexuality at the Online Hogwarts: Harry Potter Slash Fanfiction. Children's Literature 36(1), 185–207 (2008),
http://www.worldcat.org/oclc/259648847

Choosing an Appropriate Task to Start with in Open Source Software Communities: A Hard Task

Igor Steinmacher[1] and Marco Aurélio Gerosa[2]

[1] Departamento de Computação – Federal University of Tecnology-Paraná (UTFPR)
[2] Instituto de Matemática e Estatística – University of São Paulo (USP), Brazil
igorfs@utfpr.edu.br, gerosa@ime.usp.br

Abstract. Open Source Software (OSS) projects leverage the contribution of outsiders. Usually these communities do not coordinate the work of the newcomers, who go to the issue trackers and self-select a task to start with. We found that "finding a way to start" is recurrently reported both by the literature and by practitioners as a barrier to onboard to an OSS project. We conducted a qualitative analysis with data obtained from semi-structured interviews with 36 subjects from 14 different projects. We used procedures of Grounded Theory – open and axial coding – to analyze the data. We found that newcomers are not enough confident to choose their initial task and they need information about the tasks or direction from the community.

Keywords: Open Source Software, coordination, task selection, newcomers, onboarding, new developer, joining process.

1 Introduction

Open Source Software projects rely on geographically distributed developers working as a team and incorporating their individual creations into a single, seamless body of source code [13]. Many OSS projects leverage contributions from volunteers and require a continuous influx of newcomers for their survival, long-term success, and continuity. According to Qureshi and Fang [10], it is essential to motivate, engage, and retain new developers to promote a sustainable number of developers in a project.

However, newcomers usually face many difficulties to make their first contribution to an open source project. Therefore, a major challenge for OSS projects is to provide ways to support newcomers' joining. Understanding developer motivation and project attractiveness are well-explored topics in the literature [7, 12, 21]. However, little is known about the barriers that newcomers face when onboarding to a project [20].

In a previous research, we identified 57 barriers faced by newcomers when onboarding to OSS projects from interviews with OSS practitioners. Among these barriers, a barrier called "*difficulty to find a task to start with*" called our attention. In addition, we also evidenced this barrier in other studies conducted [14, 16].

"*Difficulty to find a task to start with*" is a problem inherently related to the coordination of OSS projects. Members of OSS projects usually coordinate their tasks by using issue trackers such as bugzilla, Jira, Rapid Miner, etc. These issue trackers are

N. Baloian et al. (Eds.): CRIWG 2014, LNCS 8658, pp. 349–356, 2014.
© Springer International Publishing Switzerland 2014

systems where any user or developer is free to report, discuss, and choose a task to work on. When newcomers want to contribute, they usually access the issue tracker to choose a bug or a feature request that they can (or want to) handle. However, as reported by a newcomer "*it is a bit frustrating trying to find something I could do or fix.*"

Thus, in this paper, we focus on this barrier, aiming at better understanding it. To achieve this, we analyzed data from interviews conducted with 36 OSS practitioners from 14 different projects. We started our investigation aiming at answering a broad question ("What are the barriers that hinder newcomers' onboarding to OSS projects?") and, as the barrier called our attention, we focused on it by analyzing the existing data and conducting another round of interviews. We did a qualitative analysis based on procedures of grounded theory. The analysis resulted in a model containing 14 concepts that help explaining the barrier.

2   Related Work

Newcomers' onboarding is an issue faced in many online communities. Many studies in the literature deal with newcomers joining process in collective production communities, including studies on online communities [1, 4] and on OSS projects [3, 6, 15, 19].

Von Krogh et al. [6] analyzed interviews with developers, emails, source code repository, and documents and proposed a joining script for developers who want to take part in the project. Nakakoji et al. [8] studied four OSS projects to analyze the evolution of their communities. They presented eight possible roles for the community members and structured them into a model composed of concentric layers, like the layers of an onion, later called the onion patch.

Some researchers tried to understand the barriers that influence the retention of newcomers. Zhou and Mockus [22] worked on identifying the newcomers who are more likely to remain in the project in order to offer active support for them to become long-term contributors. Jensen et al. [5] analyzed if the emails sent by newcomers are quickly answered, if gender and nationality influence the kind of answer received, and if the reception of newcomers is different in users and developers lists. Steinmacher et al. [15] studied how reception influences the retention of newcomers in an OSS project. Park and Jensen [9] show that visualization tools support the first steps of newcomers in an OSS project, helping them to find information more quickly.

Finding the appropriate task is usually classified as a problem because new developers have difficulty to find bugs or features that are of interest, that match their skill sets, are not duplicates, and are important for their future community [19]. Von Krogh et al. [6] found that members of the community encouraged the new participants to find their first tasks themselves. Park and Jensen [9] reported that "... *subjects expressed a need for information specific to newcomers, for instance,... what to contribute to...*"

Regarding support to deal with tasks, Čubranić et al. [2] presented a tool that recommend source code, emails messages, and bug reports to support newcomers. Wang and Sarma [19] presented a tool to enable newcomers to identify similar bugs through synonym-based search. These tools can help newcomers by increasing their knowledge about the tasks and their complexity.

From the literature analysis, we could observe that, from the communities perspective, newcomers should be able to find the most appropriate task themselves, as reported by von Krogh [6]. However, other researches argue that the projects should give special attention to this issue and support the newcomers finding their tasks. The goal of this work is to understand this problem from both perspectives, to enable researchers and community to focus on creating strategies to support newcomers to OSS projects.

3   Research Method

We conducted a qualitative research to understand the difficulties to find an appropriated task and proposed a set of strategies to alleviate the problems caused by this barrier. Qualitative research produces results that cannot be achieved through statistical procedures or similar methods [17]. The results of this kind of approach are richer and more informative, helping to answer questions involving variables that are difficult to quantify, such as human characteristics like motivation and perception [11].

We used semi-structured interviews as data collection method. The participants were recruited primarily through mailing list and forums from 15 different projects and from weblogs postings. We also invited newcomers and project owners directly, identifying them in project pages and by mining and following projects' mailing lists and issue trackers.

We interviewed 36 OSS developers from 14 different projects, including 11 experienced members, 16 newcomers that succeeded, 6 dropout newcomers, and 3 newcomers that were still trying to place their first contributions. Some information about their profile is presented in Table 1. The interviews were conducted using textual based chat tools, like Google Talk. We chose this mean once the participants are used to this kind of tool for their professional and personal activities. Each interview was conducted individually and the data was logged in a local computer. Interviews began with a short and general explanation of the research, followed by some questions to profile the interviewees regarding their technical experience and main occupation.

The first step of the study consisted in a first round of interviews (*It1* in the last column of Table 1) to answer a broad question ("What are the barriers that hinder newcomers' onboarding to OSS projects?"). The interviews of this step focused on identifying the barriers from newcomers and experienced members' perspective. During the analysis of these interviews, we needed to clarify some doubts to better understand some information and conducted few other interviews with some participants (represented as *It2* in the last column of Table 1).

We qualitatively analyzed the data using procedures of Grounded Theory (GT) [17], which is based in three coding steps: open coding, when concepts are identified and their properties and dimensions are discovered in the data; axial coding, when connections between the codes are identified and grouped according to their properties to represent categories; and selective coding, when the core category is identified and described. Although the purpose of the GT method is the construction of substantive theories, its use does not necessarily restricted only to researches with this goal. According to Strauss and Corbin [17], the researcher may use only some of its procedures to meet one's research goals.

During open coding, we assigned codes to sentences, paragraphs, or revisions. This procedure overlapped the axial coding, in which connections between the categories were identified. In practice, open and axial coding were executed several times to refine the emerging codes and categories.

Table 1. Data Collection Summary

	Time spent in OSS per week	First Project?	Profile	Country	Years in the project	Data Collection
P1	less than 5 hours	N	member	France	8	I1
P2	from 5 to 10 hours	Y	member	Germany	3	I1
P3	from 10 to 20 hours	N	member	Germany	3	I1, I2
P4	from 5 to 10 hours	N	member	Canada	10	I1
P5	from 5 to 10 hours	N	member	Germany	15	I1, I3
P6	more than 20 hours	N	member	Hungary	10	I1, I2
P7	more than 20 hours	N	member	Australia	5	I1
P8	more than 20 hours	N	member	Brazil	5	I1
P9	more than 20 hours	N	member	Turkey	8	I1, I3
P10	from 5 to 10 hours	N	member	Brazil	15	I1
P11	less than 5 hours	N	member	Brazil	7	I1
P12	less than 5 hours	Y	newcomer	Germany	0	I1, I3
P13	less than 5 hours	Y	newcomer	Brazil	0	I1
P14	from 5 to 10 hours	Y	newcomer	India	1	I1
P15	from 5 to 10 hours	Y	newcomer	India	0	I1
P16	less than 5 hours	Y	newcomer	Germany	0	I1
P17	less than 5 hours	N	newcomer	USA	0	I1, I2
P18	less than 5 hours	Y	newcomer	USA	0	I1
P19	more than 20 hours	Y	newcomer	Greece	0	I1
P20	less than 5 hours	Y	newcomer	Brazil	0	I1
P21	less than 5 hours	Y	newcomer	Brazil	0	I1, I2
P22	less than 5 hours	Y	newcomer	Brazil	0	I1
P23	N/I	N	newcomer	UK	-	I1
P24	from 10 to 20 hours	N	newcomer	Brazil	1	I1, I2, I3
P25	from 5 to 10 hours	Y	newcomer	Brazil	1	I1
P26	N/I	Y	newcomer	France	0	I1
P27	from 5 to 10 hours	N	newcomer	Germany	0	I1
P28	from 5 to 10 hours	N	dropout	USA	0	I1
P29	less than 5 hours	Y	dropout	India	0	I1
P30	less than 5 hours	N	dropout	Germany	0	I1
P31	less than 5 hours	Y	dropout	Brazil	0	I1
P32	less than 5 hours	Y	dropout	India	0	I1
P33	less than 5 hours	Y	dropout	India	0	I1
P34	less than 5 hours	N	onboarding	China	0	I1
P35	from 10 to 20 hours	Y	onboarding	India	0	I1, I2
P36	less than 5 hours	Y	onboarding	Greece	0	I1, I3

During the analysis of the first set of interviews, we were able to identify 57 different barriers, grouped in six categories, some of them split in subcategories. As our work progressed, we decided to focus on a recurrent and specific category of barriers highlighted in the interviews: *find a way to start*. More specifically, we were interested in better understanding the barrier called *difficulty finding a task to start with*, once the barrier was recurrently evidenced, appearing in 9 interviews. To investigate this specific barrier, we conducted an in-depth analysis of the existing data, and conducted another iteration of interviews with five participants (*It3* in the last column of Table 1). The results of this analysis are presented in the following section.

4   Difficulty to Find a Task to Start With

Our analysis resulted in the emergence of concepts that enabled us to better understand this barrier. These concepts represent a more detailed view on the reasons why newcomers understand the choosing of an appropriate task as a barrier to onboard to the project. The resulting model comprising the concepts and relationships among them is presented as a network in Fig. 1.

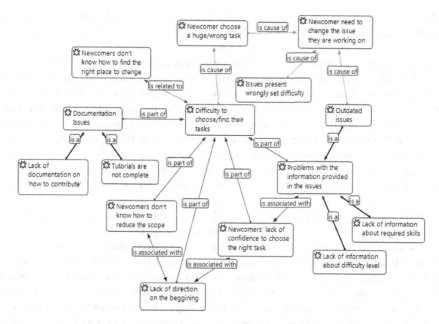

Fig. 1. Result of the qualitative analysis represented as a network

The phenomenon we were trying to understand in this case was the "Difficulty to choose/find their tasks." Five issues emerged as part of the barrier under investigation, and three problems that are consequences of this barrier.

Some developers reported the **lack of direction at the beginning** as part of the explanation why it is difficult to find a task to work on. We found that some newcomers expect that someone indicate them what are the tasks they should start with. To explain this, one of them reported: *"We feel more secure when someone is guiding us. I think this is related to the experience with large projects and Open Source... Less experienced developers always stand on the back foot."*

As it is possible to observe in the previous quote, this issue is related to the **lack of confidence** that newcomers have when choosing a task. The newcomer said, *"I really didn't know which one to pick, which one I was 'authorized' to pick... whether they are important."* It shows he was not sure about what he could do, what would be enough and important. Another newcomer told us a case of a close friend that tried to join the project: *"A college colleague gave up... He did not find a task that he felt confidence to try."* This occurs because newcomers, even when they are aware they can choose a task, they are not sure about what are easy or not, and what they can touch: *"it is frustrating, there is a bunch of issues, but I don't have the proper knowledge to judge what exactly I can touch..."*

On the other side, experienced members reported that **newcomers do not know how to reduce the scope**. However, none of them put it as a newcomer's fault. One said: *"sometimes they want to contribute but don't know how to reduce the scope when starting."* Another practitioner [P6] reported: *"the task chosen [by the newcomer] makes some sense, but is huge, and the newbie thinks she'll be able to implement it in a few days."* The same participant reported the association among the inability of reducing the scope and lack of direction when asked if he had seen some case in which a newcomer gave up without finishing assignment task: *"When they choose a task that is reasonably sized, just we don't give the necessary help."*

Some interviewees also mentioned **documentation issues** as part of the difficulty to find a task to work on. We could identify two related issues from the interviews: **lack of documentation on how to contribute** and **incomplete/outdated tutorials.** These issues were mainly reported by newcomers that recently joined a project. One of them summarized the first one when he said that *"there is no good guide for starter"* [P34]. Two other newcomers, who were onboarding to the project when we conducted the interviews, reported a hard time with the incomplete tutorials. One of them reported: *"A proper up-to-date guide with tutorials would have really helped. As in when I began, I was totally new to open source, but the tutorials skipped over certain instructions and I had no idea what I was doing wrong."* [P33]

Problems with the information provided in the issues were recurrently reported by newcomers and experienced members. We could clearly distinguish three types of issues: **outdated issues; lack of information about required skills;** and **lack of information about difficulty level.** An experienced member [P9] reported that the lack of information frustrated him when he was trying to contribute to a project: *"the issues that I can contribute were not clearly defined so it was my job as a newcomer to find out how to contribute there were no easy or junior bugs as in some projects."* However, defining a task or issue as good for newcomers is sometimes not enough. A newcomer also evidenced that the skills needed to accomplish that task need to be clear: *"the issues should indicate the area of knowledge, like C++, build, shell script"* [P12].

In addition to the issues that helped us understanding the difficulties newcomers face to find an appropriate task, we also evidenced other issues caused by this barrier. We found that, due to these difficulties, **newcomers choose a wrong/huge task**, as per this quote from an experienced member: *"most of them [newcomers] do not know how to start... what can they do first and can choose the wrong task"* [P20]. A newcomer that is contributing for one year to an OSS project reported this problem and accredited the problem to a mistake when the issue was classified as easy: *"Sometimes, I tried to work on a task that was classified as easy hacks and it was too complex that only experienced members could find the solution... the developers who registered these easy hacks, sometimes made mistakes when classifying the difficulty"* [P24]

Another newcomer reported his difficulties finding a task. The coordination mechanism used by the project failed to support his choice. He reported that he had to change tasks twice, but for different reasons: *"when I finally did have something to do, it basically completely changed two times ... one task was too hard and the other was a feature that was already implemented, but the task was not updated"* [P19]

When searching for an issue to work on, newcomers want to be aware of some details about the issues; mainly they want to know whether they will be able or not to handle the task. This links this kind of problem with the **lack of confidence to choose a task.** The way they do this is looking at the issue tracker, aiming at coordinating their contribution with the community. Therefore, newcomers need up-to-date information to support their decision, to make it easier and clearer for them to be aware of what they can do, what they are able to do, and where they can look for support. The main point now is to identify potential strategies to improve this coordination mechanism in a way the newcomers can feel more comfortable in choosing their first task.

5   Threats to Validity

We are aware that each project has its singularities, that the OSS universe is huge, so, the level of support and the barriers can differ according to the project or the ecosystem. Our strategy to consider different projects and different profiles of developers aimed to mitigate this limitation, identifying recurrent barriers from multiple perspectives.

There is also a threat related to the sampling method of the interviews. We sent out invitations to specific development lists and directly to newcomers and project core members, what could introduce some bias.

Another threat to the validity of the results is the subjectivity of the data classification. We used Grounded Theory procedures to mitigate this threat, given that GT requires the entire analysis to be grounded on the data collected. Additionally, the analysis process was discussed along with two other researchers, to encourage a better validation of the interpretations through mutual agreement.

6   Conclusions

Although this study focused on OSS communities, better supporting newcomers is an important issue in many other communities. Many virtual communities count on volunteer contributions. These volunteers can easily leave, since they have no formal relationship with these communities [1]. Moreover, the impoverished awareness information, lack of trust, and the relatively weak interpersonal ties between members in many online groups make it more problematic to attract and retain people than in face-to-face groups [18]. Therefore, studying newcomers and the problems faced by them in virtual communities is a contemporary problem that still needs to be further investigated.

In this paper, we qualitatively analyzed data collected from newcomers, dropouts, and members of OSS projects, aiming at understanding the barriers faced by newcomers to OSS projects. The results of the analysis helped us understanding the "difficulty to find a task to start with" faced by newcomers. It was possible to evidence some of the actual issues that explain this barrier. Lack of confidence to choose a task appeared as a relevant concept in our analysis. Newcomers need the project to provide enough information about the tasks to support their decision about which task is more suitable for them.

The next step of this research encompasses finding suitable solutions and conducting experiments to assess the effectiveness of them in supporting newcomers onboarding to OSS projects.

References

[1] Choi, B., Alexander, K., Kraut, R.E., Levine, J.M.: Socialization Tactics in Wikipedia and Their Effects. In: Proceedings of the 2010 ACM Conference on Computer Supported Cooperative Work, pp. 107–116 (2010)
[2] Cubranic, D., Murphy, G.C., Singer, J., Booth, K.S.: Hipikat: a project memory for software development. IEEE Transactions on Software Engineering 31(6), 446–465 (2005)
[3] Ducheneaut, N.: Socialization in an Open Source Software Community: A Socio-Technical Analysis. Computer Supported Cooperative Work (CSCW) 14(4), 323–368 (2005)

[4] Halfaker, A., Geiger, R.S., Morgan, J., Riedl, J.: The Rise and Decline of an Open Collaboration System: How Wikipedia's reaction to sudden popularity is causing its decline. American Behavioral Scientist 57 (2013)

[5] Jensen, C., King, S., Kuechler, V.: Joining Free/Open Source Software Communities: An Analysis of Newbies' First Interactions on Project Mailing Lists. In: 44th Hawaii Intl. Conf. on System Sciences, pp. 1–10 (2011)

[6] Von Krogh, G., Spaeth, S., Lakhani, K.R.: Community, joining, and specialization in open source software innovation: a case study. Research Policy 32(7), 1217–1241 (2003)

[7] Meirelles, P., Santos, C., Miranda, J., Kon, F., Terceiro, A., Chavez, C.: A study of the relationships between source code metrics and attractiveness in free software projects. In: 2010 Brazilian Symposium on Software Engineering (SBES), pp. 11–20 (2010)

[8] Nakakoji, K., Yamamoto, Y., Nishinaka, Y., Kishida, K., Ye, Y.: Evolution Patterns of Open-source Software Systems and Communities. In: Proceedings of the International Workshop on Principles of Software Evolution, pp. 76–85 (2002)

[9] Park, Y., Jensen, C.: Beyond pretty pictures: Examining the benefits of code visualization for Open Source newcomers. In: 5th Intl. Workshop on Visualizing Software for Understanding and Analysis, pp. 3–10 (2009)

[10] Qureshi, I., Fang, Y.: Socialization in Open Source Software Projects: A Growth Mixture Modeling Approach. Org. Res. Methods 14(1), 208–238 (2011)

[11] Seaman, C.B.: Qualitative methods in empirical studies of software engineering. IEEE Transactions on Software Engineering 25(4), 557–572 (1999)

[12] Shah, S.K.: Motivation, Governance, and the Viability of Hybrid Forms in Open Source Software Development. Manage. Sci. 52(7), 1000–1014 (2006)

[13] Singh, P.V.: The Small-world Effect: The Influence of Macro-level Properties of Developer Collaboration Networks on Open-source Project Success. ACM Trans. Softw. Eng. Methodol. 20(2), 6:1–6:27 (2010)

[14] Steinmacher, I., Silva, M.A.G., Gerosa, M.A.: Systematic review on problems faced by newcomers to open source projects. In: 10th International Conference on Open Source Software, p. 10 (2014)

[15] Steinmacher, I., Wiese, I., Chaves, A.P., Gerosa, M.A.: Why do newcomers abandon open source software projects? In: International Workshop on Cooperative and Human Aspects of Software Engineering (CHASE), pp. 25–32 (2013)

[16] Steinmacher, I., Wiese, I.S., Conte, T., Gerosa, M.A., Redmiles, D.: The Hard Life of Open Source Software Project Newcomers. In: International Workshop on Cooperative and Human Aspects of Software Engineering (CHASE 2014) (2014) (2014)

[17] Strauss, A., Corbin, J.M.: Basics of Qualitative Research: Techniques and Procedures for Developing Grounded Theory. SAGE Publications (1998)

[18] Tidwell, L.C., Walther, J.B.: Computer-Mediated Communication Effects on Disclosure, Impressions, and Interpersonal Evaluations: Getting to Know One Another a Bit at a Time. Human Communication Research 28(3), 317–348 (2002)

[19] Wang, J., Sarma, A.: Which bug should I fix: helping new developers onboard a new project. In: Proceedings of the 4th International Workshop on Cooperative and Human Aspects of Software Engineering, Waikiki, Honolulu, HI, USA, pp. 76–79 (2011)

[20] Wolff-Marting, V., Hannebauer, C., Gruhn, V.: Patterns for tearing down contribution barriers to FLOSS projects. In: 12th Intl. Conf. on Intelligent Software Methodologies, Tools and Techniques, pp. 9–14 (2013)

[21] Ye, Y., Kishida, K.: Toward an Understanding of the Motivation Open Source Software Developers. In: Proceedings of the 25th International Conference on Software Engineering, Portland, Oregon, pp. 419–429 (2003)

[22] Zhou, M., Mockus, A.: What make long term contributors: Willingness and opportunity in OSS community. In: 2012 34th International Conference on Software Engineering (ICSE), pp. 518–528 (June 2012)

Collaborating in the Fog: A Rich Description of Agile Software Development

Diane E. Strode

Whitireia Polytechnic, Wellington, New Zealand
diane.strode@alumni.unimelb.edu.au

Abstract. Collaborative agile software development is inexorably replacing traditional command-and-control project arrangements. To gain a better understanding of collaboration in this context, empirical data was collected from a single co-located agile software development project. Aspects of collaboration in that project are described in a rich description. Collaboration is achieved with an assemblage of collaborative activities, activity sequencing, inter and intragroup interactions, shared artefacts, and other practices. This case description is a first step in informing a theory of collaboration that has potential to contribute to collaboration research and collaboration tool design.

Keywords: Scrum XP hybrid, agile methods, collaboration, qualitative case description.

1 Introduction

Agile software development epitomises collaboration by formalising how project teams and their customer work together to develop a software product [1, 2]. As self-organising agile project teams inexorably replace traditional command-and-control project arrangements [3], and adoption increases worldwide [3], it is useful to refresh our understanding of collaboration in this context. To illustrate how a typical co-located agile software development project organised their complex work to achieve collaboration, this research-in-progress paper presents empirical data from a single case study in a rich description of agile project work. This research asks how collaboration is achieved in typical agile software development projects.

The implications of this proposed research project is a better understanding of the assemblage of activities, artefacts, and interactions that contribute to collaboration in this context. Since collaboration theory in the context of software development (agile or not, distributed or co-located) is not yet available this type of description can be used to begin the process of theory building [4]. Such theory would be useful for collaboration researchers and those creating collaboration tools tailored to this context.

This paper is organised as follows. The most current agile software development literature focusing on collaboration is summarised. There follows the research method and a rich description of selected aspects of collaboration in a single case of a typical agile software development project. Collaboration in the case is discussed followed by the implications for collaboration research.

N. Baloian et al. (Eds.): CRIWG 2014, LNCS 8658, pp. 357–364, 2014.

2   Agile Software Development and Collaboration

Agile software development challenges long-accepted practices in software development and project management [5-7] and has been a focus for research since its inception. Agile software development is a term used to describe the use of any agile method, such as Scrum and Extreme Programming (XP) [7]. World-wide adoption is estimated at 52% to 55% of projects with a hybrid of Scrum and XP the most common mix [3]. Research on agile software development is diverse and has matured over the period from 2000 to 2014. In 2008, Dyba and Dingsoyr systematically reviewed this body of research and found it lacked rigor. Most recently, Chuang, Luor, and Lee [8] have summarised the key areas of agile software development research from 2001 to 2012. Although collaboration is a key characteristic of all agile methods [1], very little of this body of research is centrally concerned with collaboration.

Collaboration has been defined simply *"as joint effort towards a group goal"* [9, p. 122]. Extensive research on collaboration has been carried out on inter and intra organisational collaboration, on virtual and non-virtual groups and teams, and in contexts such as healthcare, tourism, scientific groups, military operations, emergency management, flight crews, and globally distributed software development. Agile methods have been adapted for globally distributed software development, in part, to provide mechanisms to improve collaboration by reducing coordination, control and communication problems [10]. In co-located agile development projects, various physical artefacts and arrangements that support face-to-face collaboration have been identified including story cards used to record requirements, and communal wallboards, discussion spaces, and semi-private cubicles [11]. This small research base focuses on physical artefacts and arrangements. Rich qualitative description explaining how collaboration is achieved in a typical co-located agile software development project is scant. Rich description provides a holistic view of collaboration and illustrates how individuals, groups, artefacts, and the sequencing of activities fit together to achieve collaboration. This paper presents aspects of a single case in the form of a rich description to illustrate how such descriptions might form a basis for theory building work on collaboration in agile software development projects.

3   Research Design

A single positivist case study [12] was selected to explore collaboration in agile software projects. The case study approach is a well-accepted way to explore phenomena where events cannot be controlled and where it is necessary to capture detail in a situation [4, 13, 14]. The guidance of Dubé and Paré [15] on achieving rigor, and addressing validity and reliability in case study research of information systems was followed. The case described in this paper has previously contributed to a multi-case study of coordination [16].

The project was selected because it is an exemplar of the full use of Scrum with some XP practices in a co-located team of 10 developers. Data collection followed a case study protocol [14] and data was collected from a variety of sources. Data was collected over a period of three weeks primarily by semi-structured interview. Five people were selected for interviewing who took a range of roles in the project including the project manager/agile coach, developers, and the tester. Additional material was collected from observations, photographs, sketches, the organisation's annual report and website, and a questionnaire for capturing project information (e.g. project duration, technologies, team experience).

A full description of the case was prepared using a framework developed for the purpose. Eisenhardt [4] recommends this as a first phase of data analysis in the case study method. Using a framework assists in developing a coherent and organised version of events suitable for review by participants. The framework consisted of attributes appropriate for describing a software development project [12] and included details of the organisation, the project, the technologies, the team, the development method, and project problems. To develop the case description, interview transcripts and other data were reviewed to look for evidence for each attribute in the framework. The framework and the data were then written as a detailed case description and sent to one project participant for verification. Factual errors were corrected. For explanations of the various agile practices in the following description see Williams [17].

4   The Case Description

This section describes the project background and the development method used on the project. The method description only includes evidence related to collaboration.

The Project

Project Storm (a pseudonym) took place in a quasi-public sector organisation that provides services to the public and private sectors in New Zealand. In this organisation, a legacy system captured environmental data that was analysed by engineers and disseminated as structured information both locally and internationally. The public and many businesses important to New Zealand's economy rely upon the timeliness and accuracy of this information. Storm was to replace this legacy system.

The project was expected to take two years and data collection took place in the final third of the project. Storm was divided into seven sub-projects. At the time of data collection, two of these sub-projects were successfully completed and two further sub-projects were underway. Storm had a straightforward goal.

"We are an independent contract team...The goal is to replace as much...legacy software and hardware, as possible in the time that it takes me to spend the budget". [CP01]

NOTE: [CPO1, CT01, 2, and 3 are codes uniquely identifying participants]

Only by replacing the 30-year old system could the organisation continue providing effective services because few people in the organisation were able to understand, maintain, or enhance the legacy system. "So that's why they hired in a team of contractors to do it." and "There are three programmers here who are absolute gurus on it, and if they left, yes [the system could not be maintained]." [CT01]

An experienced software project manager was hired on contract. He explained. *"They picked me, and I said I wouldn't take the job unless I can do it agile."* [CP01]. He spent about two months analysing the legacy system and decided that the migration would involve seven sub-projects. *"The seven largest pieces of work that were more tightly coupled and actually running on that main server in the middle."* [CP01].

Storm involved various project stakeholders including 80 engineers (the end-users), four engineers acting as beta testers, a vendor located in South Africa, the engineer's manager, a steering committee, the Central team (supporting the central computer), the Application team, and the Storm development team.

The project manager interviewed and selected the development team at the beginning of the project. He explained his selection criteria.

> *"I figured, if I was going to be working with these people for 18 months I wanted them to be able to gel quickly as an agile team, often most of them with no agile experience.... I needed them to be more than just 'Java smart', but I also needed them to be quite communicative types and with a large dose of pragmatism so that they could handle change well."* [CP01]

The contracted staff consisted of the project manager/agile coach, six software developers, and a test specialist. After some months, a developer from the Application team and a developer from the Central team joined Storm because of their domain and technical knowledge of the legacy system, and so they could gain an understanding of the new system before it was 'switched on'.

The Storm team worked in one room without room divisions. They formed two sub-teams of six and four. Each sub-team worked simultaneously on a different sub-project but these sub-projects had technical interdependencies. The team split into groups when discussing details about their respective sub-projects but worked from a shared wallboard of user stories and tasks. Often their work was closely related. *"[We are all in] one room. ... If there is any design decision everyone just pulls their chairs over and figures it out."* [CP01]

In an agile project, the expectation is that the end-user will 'sit with', and be part of the team. The engineers were not physically part of the Storm team; they were located together in the next room.

The Development Method

Storm used the Scrum method with some practices from XP. The decision on sprint (iterations) duration was defined while the team was learning the agile process. Initially they used a two-week iteration, but they settled on a one-week iteration when the two-week version proved unwieldy.

> *"What we found was, when we then took all of the stories from the backlog that we were working on for two weeks' worth of work and tried to do task breakdown on them, we ended up with too much time to do a task breakdown for that much work. If you can imagine with nine developers, you can get through a lot. So that would become hours and hours to get through that task breakdown and it was a pain."* [CP01]

The 80 engineers worked on three shifts per day on five different cycles. This meant there was no one permanent representative customer to consult about stories and their prioritisation. At unscheduled times, when the engineers were working on a real-time critical task they could not be approached at all. Then, work for the development team would stall, and they would switch to other tasks until the engineers were free to be consulted. The project manager explained:

"...because of the shift work scenario ... we are working into it piece by piece and as we discuss a new function, we ... sit down and discuss [user stories] with them ... and say 'well what happens here?' ...Therefore it is a very, very, slowly moving into a fog, kind of feeling, and the team have to be analysing all the time to see what is coming up next and get a feel from the user whether there is another story hidden inside that function or not." [CP01]

An important practice initiated by the tester was the selection of a small group of four engineers to act as beta testers. These people volunteered for the role, but they were also those that the team felt were more approachable, knowledgeable, or committed to the success of the new system. This group became preferred contacts for the team to approach for assistance.

This section describes the activities at the beginning of a sprint when the Storm team held a stand-up meeting followed by a story breakdown session.

At 9:25 am the stand-up meeting began in front of the Scrum wallboard in the project workroom. In turn, each of the ten team members described his or her progress yesterday and any issues he or she currently faced. The meeting finished at 9:35am.

Next, the sprint burn-down chart was updated to show stories completed in the previous week. Then the stories and tasks on the wallboard were reorganised to reflect the current project status. This involved progressively moving tasks related to a story under development along the board. The team then moved to a 'break-down' session. First, they reviewed the 'pool', a list of prioritised stories (or backlog) and updated it with completed work from the previous sprint. Then five or six user stories with the highest priority on the backlog were printed onto story cards. Then one group returned to their desks to discuss their sub-project while another group moved to a meeting table in the corner to perform task breakdown on each newly selected story card involving their sub-project.

The task breakdown process was to elaborate the story card details and allocate them to tasks. The round-table discussion was wide-ranging. The group gave their opinions on the implications of implementing software tasks. They had to decide if separate new stories were needed and, as discussion progressed, developer tasks and tester tasks were written down on coloured sticky notes. At one point, an approval from an external stakeholder was required before an application could be installed on the engineers' workstations and this meant a task-related decision had to be held over until that group was consulted. Another such decision was held over when a design decision was needed and the engineers needed to be consulted on the details. The tester explained that he used the breakdown session to identify test cases. *"...I assign myself my work in the team discussions."* [CT04]. The group discussed amalgamating stories, they decided if something was 'a story or a task', and they explained things to each other such as specific details about requirements and technical dependencies, for example, the need to *"pull data out of one application for use in another"*. One developer explained to the group how the engineers work and how the system worked for the engineers in a particular scenario. Notes were made about questions, issues, or understandings that the engineers needed to be queried about, and there was talk about the best engineer to approach for specific information. Discussion covered how to test a story or task and the *need* for certain tasks; at one point the tester asked of a task, 'who wants this?' and its relative importance was discussed. The group discussed 'who to talk to'. For example, the tester mentioned that he needed to talk to another group to ensure they carried out appropriate unit tests. Finally, they discussed the 'difficulty' level of one of the tasks.

Once a task was written then it was attached to its story card, and when all the tasks were written, the group allocated story points to each task. Everyone contributed and there was very little speech at this point. Someone would point to a task card, and the others would hold up their fingers to indicate their assessment of the number of 'story points' that the task would take (one finger for one story point, or two fingers for two story points). This was a majority rules voting system. When there was a disagreement on points then each person would explain why they thought the task would be longer or shorter than the majority estimate. This sizing process went smoothly with little disagreement or discussion.

At 11:15 am the meeting ended and stories and tasks were placed on the wallboard.

During a sprint, many activities involved the Storm team working together. Task selection was left to the individual but still involved shared team knowledge.

> *"As each day comes, you are meant to go grab the next highest priority one off the board, but generally we aren't quite that strict about it, we all know … what each person is going to be best tackling. And you just go 'yes' I will do that, and no one argues"* [CT01]

If a team member selected a task that needed elaboration, he would go and sit with an engineer and investigate the details of the task. Engineers were also consulted when requirements emerged during user acceptance testing. For problem solving, people would occasionally pair program.

> *"If there is a sticky problem… if you are blocked, need some help, to tap anyone else on the shoulder and grab them and get them to help unblock you, and so they tend to … pair up … for complicated pieces of work."* [CP01]

Testing and software releases involved the whole team. Story-level tests were written by developers and the tester, and system integration tests involved the team and engineers. Regular software demonstrations to the Central and Application teams involved collaboration in the first months of the project. The Done list was introduced after team negotiation so that it was clear when all tasks for a story were done and a story was complete. *"That is something we only came up with recently because we were finding …you would think some stuff was done, but it hadn't been."* [CT01]

5   Discussion

This qualitative description illustrates how collaboration was achieved in a typical agile software development project using a Scrum XP hybrid method. Collaboration was achieved by combining collaborative activities, sequencing those activities, inter- and intra-group interactions, sharing artefacts, and other practices.

Collaborative activities where the project team worked together included daily stand up meetings, breakdown sessions, designing a done list, and creating stories and tasks. Sequencing of activities was achieved with regular sprints. Wallboard updates, backlog updates, breakdown sessions, and software demonstrations occurred regularly according to the organising structure of the sprint. Interactions during collaborative activities involved discussion, negotiation, sharing understandings, and joint decision-making. Intra-group interaction occurred during pair programing, when sub-groups worked on their sub-projects, and within the team as a whole. Inter-group interaction occurred between the project team and their stakeholder groups. This included negotiating with a vendor in another country, other business units in the organisation, and the large diverse group of end-users. Collaboration was enabled when the team co-opted domain and technical experts from the organisation into the team from the Application and Central teams. This ability to cross-fertilise implies organisational flexibility might be an enabler of effective collaboration. Shared artefacts included those things the team shared such as stories and tasks displayed on a wallboard, a pool of stories, a done list, and software tools (e.g. Subversion, Hudson, and a Wiki) to control software versioning, integrate software, and share documents. Sharp and Robinson [1] reported similar findings regarding shared artefacts supporting collaboration in XP teams. Further support for collaboration was provided by team co-location. Being in a single open plan office enabled ad hoc interactions among the team. Another collaborative practice was the use of the specially formed beta testers group, which improved collaboration with the end users.

The implications of this case description for collaboration research is a better understanding of the assemblage of collaborative practices that contribute to agile software development projects. Since collaboration theory in this context is not yet available, this description might be used to begin the process of theory building.

6   Conclusion

This paper has provided excerpts from a rich description of a single case of co-located agile software development to illustrate the activities, sequencing of activities, inter- and intra-group interactions, shared artefacts and other practices used to achieve collaboration. This qualitative description could form a basis for theory building on collaboration and contribute to a better understanding of collaboration for tool design.

Future work includes an extensive literature review of extant collaboration theories in relevant domains, teamwork literature, and research into decision-making in agile teams. Further additional cases would also increase the validity of any collaboration theory developed to explain agile software development projects.

References

1. Sharp, H., Robinson, H.: Three 'C's of agile practice: collaboration, co-ordination and communication. In: Dingsoyr, T., Dyba, T., Moe, N.B. (eds.) Agile Software Development: Current Research and Future Directions. Springer, Heidelberg (2010)
2. http://www.agilemanifesto.org
3. Stavru, S.: A critical examination of recent industrial surveys on agile method usage. The Journal of Systems and Software (in press, 2014)
4. Eisenhardt, K.M.: Building theories from case study research. The Academy of Management Review 14, 532–550 (1989)
5. Beck, K.: Extreme Programming Explained: Embrace Change. Addison-Wesley, Boston (2000)
6. Schwaber, K., Beedle, M.: Agile Software Development with Scrum. Prentice Hall, Upper Saddle River (2002)
7. Dingsoyr, T., Nerur, S., Balijepally, V., Moe, N.: A decade of agile methodologies: Towards explaining agile software development. The Journal of Systems and Software 85 (2012)
8. Chuang, S.-W., Luor, T., Lu, H.-P.: Assessment of institutions, scholars, and contributions on agile software development (2001-2012). The Journal of Systems and Software (2014)
9. Briggs, R., Kolfschoten, G., Vreede, G.-J., Douglas, D.: Defining key concepts for collaboration engineering. In: Proceedings of the Twelfth Americas Conference on Information Systems, Acapulco, Mexico August 04th-06th AMCIS 2006, paper 17, pp. 121–128 (2006)
10. Holmstrom, H., Fitzgerald, B., Agerfalk, P., OConchuir, E.: Agile practices reduce distance in global software development. Information Systems Management 23, 7–18 (2006)
11. Mishra, D., Mishra, A., Ostrovska, S.: Impact of physical ambiance on communication, collaboration and coordination in agile software development: an empirical evaluation. Information and Software Technology 54, 1067–1078 (2012)
12. Keutel, M., Michalik, B., Richter, J.: Towards mindful case study research in IS: A critical analysis of the past ten years. European Journal of Information Systems Advance Online Publication, 1–17 (October 8, 2013)
13. Pare, G.: Investigating information systems with positivist case study research. Communications of the Association for Information Systems 13, 233–264 (2004)
14. Yin, R.K.: Case Study Research. Sage, Thousand Oaks (2003)
15. Dube, L., Pare, G.: Rigor in information systems positivist case research: Current practice, trends, and recommendations. MIS Quarterly 27, 597–635 (2003)
16. Strode, D.E., Huff, S.L., Hope, B., Link, S.: Coordination in co-located agile software development projects. The Journal of Systems and Software 85, 1222–1238 (2012)
17. Williams, L.: Agile software development methodologies and practices. In: Zelkowitz, M.V. (ed.) Advances in Computers, vol. 80, pp. 1–44. Elsevier, Amsterdam (2010)

Motivating Wiki-Based Collaborative Learning by Increasing Awareness of Task Conflict: A Design Science Approach

Kewen Wu[1], Julita Vassileva[1], Xiaoling Sun[2], and Jie Fang[3]

[1] University of Saskatchewan, Saskatoon, SK, Canada
{kew259,jiv}@usask.ca
[2] Nanjing University of Posts and Telecommunications, Nanjing, JiangSu, China
[3] Urumqi Vocational University, Urumqi, XinJiang, China

Abstract. Wiki system has been deployed in many collaborative learning projects. However, lack of motivation is a serious problem in the collaboration process. The wiki system is originally designed to hide authorship information. Such design may hinder users from being aware of task conflict, resulting in undesired outcomes (e.g. reduced motivation, suppressed knowledge exchange activities). We propose to incorporate two different tools in wiki systems to motivate learners by increasing awareness of task conflict. A field test was executed in two collaborative writing projects. The results from a wide-scale survey and a focus group study confirmed the utility of the new tools and suggested that these tools can help learners develop both extrinsic and intrinsic motivations to contribute. This study has several theoretical and practical implications, it enriched the knowledge of task conflict, proposed a new way to motivate collaborative learning, and provided a low-cost resolution to manage task conflict.

Keywords: wiki, design, collaborative learning, task conflict.

1 Introduction

Wiki, as a kind of collaborative writing system, has been deployed in various collaborative learning contexts to create shared documents. In wiki-based learning project, learners are usually required to jointly write and edit articles through constant negotiation and coordination with their co-learners [1]. Therefore, learners can acquire new knowledge and skills of collaboration. The educational value of wiki has been discussed in past literature [2].

However, deployments of wiki systems to support collaborative learning, as well as public and enterprise wiki collaborations (e.g. Wikipedia) are plagued by lack of motivation [3, 4]. While motivation is considered as a critical factor which determines the success of virtual collaboration [5], how to motivate learners to participate in wiki-based collaborative learning becomes a major issue.

Wiki-based collaborations are often reported by literature regarding the issue of conflict [6–8]. The phenomenon of conflict has been studied for many decades.

N. Baloian et al. (Eds.): CRIWG 2014, LNCS 8658, pp. 365–380, 2014.
© Springer International Publishing Switzerland 2014

And conflict has been categorized into three types: relationship, process, and task conflict [9]. Relationship and process conflict refers to disagreement on interpersonal issues and approaches to the task, respectively, while task conflict only refers to disagreements on ideas and differences of opinions about the task. In wiki pages, a conflict event is more likely to be task-oriented [6], because the most common arguments among wiki users involve opinions about the content [10]. Although conflict (e.g., relationship and process conflict) is often criticized as having a negative effect on group collaboration [11], research evidence shows that task conflict can be beneficial to collaboration. For example, task conflict can increase curiosity, which is an important intrinsic motivation [12]. Therefore, it is possible to enjoy the benefits of task conflict in collaborative learning.

In this paper, we try to motivate learners through designing enhanced tools, and understand how new tools influence learners' motivation. This paper is organized as follows: in Section 2 we summarize some problems encountered in wiki-based collaboration, and provide possible solution. Related studies are reviewed in Section 3. Proposed designs are presented in Section 4. The evaluation process and results are described in Section 5. Finally in Section 6, we draw conclusion and discuss limitations and implications of this study.

2   Problem Statement

Wiki systems have been deployed as a knowledge management tool in many contexts, such as collaborative learning, management of business meta-data, and supporting decision making [13–15]. As a result, much experience about wiki usage has been gained.

The original wiki is designed to hide authorship information. The system presents only the latest version of an article. Such design has its advantages, such as reduced social bias [16]. But it also has significant disadvantages, which may reduce users' motivation to contribute. A summary of problems in wiki-based collaboration is provided in table 1.

The problems mentioned in Table 1 can be attributed to the lack of clues when task conflict occurs. Originally a wiki did not directly show clues like "is there any task conflict issue?", "who has conflicting opinions with me?", "what opinions does he/she hold?" and "when did he/she change my content?" Users who want to know this information should use the "page history" tool to compare every two versions of an article to find the answers. The process can be very time-consuming when the article has a long list of versions.

For issue (A) in Table 1, if users are able to know the details of changes in the content that they are interested in, they would have more confidence in the quality of the information. For issue (B), knowledge exchange activities would be facilitated if the wiki system could give users information about content changes as well as corresponding editors. For issue (C) and (D), the sense of audience can be enhanced by providing peer feedbacks in the form of content changes. For issue (E), making users' social actions and content authorship visible can motivate users to contribute [16, 22].

Table 1. Some wiki-related problems reported in literature

Problems	Context	Demonstration	Source
(A) Generation of content trust not supported well	General	"does not offer ... how the article content has evolved into its most current form...how much the content can be relied upon"	[17]
(B) Knowledge exchange not supported well	General	"users may not be aware of changes of content when the content they contributed is modified by others"	[18]
(C) Limited sense of audience	Education	"having an audience who can comment on what is written directly supports efforts to write clearly and to write well"	[19, 20]
(D) Limited feedback	Enterprise	"...he believed, communicated to staff that their contributions mattered"	[19]
(E) Limited authorship	General/ Education	"this design is less suitable when users are motivated primarily by self-promotion and career-advancement"	[15, 16, 21]

Since each version of a wiki page is confirmed and submitted by a user, for design purpose, we define task conflict in a wiki page as the content difference between two versions of the page. This definition is broader than classic definition of task conflict, which emphasizes opinion differences. Moreover, our definition is similar to *peer feedback*. However, peer feedback is usually a one-to-one collaboration process where one gives reviews to another (a clear receiver), while task conflict in wiki is a many-to-many communication process that everyone edits each other's work to improve the quality of a wiki article.

Regarding the design of enhanced functions, several possibilities are proposed: First, by highlighting modifications between article versions, users can quickly be aware of a conflict event and identify whether this event is related to them. As a result, they may be motivated to express opinions. Second, by providing a paragraph-based revision history, users can focus on a specific part of the content and reduce the time and effort to locate relevant versions from a huge list of versions. Third, by providing a complete edit history of the content, users can generate a sense of community and know exactly the evolution of opinions and which editors to communicate with. Fourth, by providing word-based content authorship, users can quickly identify whether others have modified the content they contributed or not, and fulfill their needs of content ownership.

3 Literature Review

3.1 Task Conflict and Motivation to Participate

The studies on relationship and process conflict have gained consistent results, indicating that these two types of conflict are harmful to group performance [11]. In contrast, the consequences of task conflict seem to be more complex. Groups experiencing task conflict can obtain richer collective knowledge, better decision understanding, quality of decision, and decision commitment, since task conflict encourages a diversity of opinions and positively affects members' relational outcomes (e.g., psychological safety) [23–25]. However, task conflict is also found to have a negative effect on group performance due to the influence of unsolved task issues. Besides, a high level of task conflict can trigger relationship conflict and reduce member satisfaction [26].

Existing conflict theories suggest that there exist four kinds of relationships between task conflict and participation. Firstly, participation can generate task conflict due to opinion differences [27], freedom to participate and to express ones' opinions [28], value diversity [12], different opinions on team goals, importance of task characteristics or actions [29], and perceived informational and value dissimilarity [30]. Secondly, participation could reduce task conflict since information exchange behaviors improves mutual understanding [31, 32], group value congruence, trust, and team spirit [33]. Thirdly, participation can be motivated by task conflict because users' knowledge exchange activities and critical evaluation of conflict issues (or resolution) are stimulated [34, 35]. Finally, task conflict may reduce motivation to participate because it increases stress, tension and dissatisfaction [12, 36].

As summarized above, task conflict can have both positive and negative influence on user's motivation to participate. Therefore, how task conflict influence user's motivation needs to be further investigated. In this study, we tried to evaluate the influence of task conflict in collaborative learning context.

3.2   Designs for Supporting Wiki-Based Collaboration

Many tools have been designed to support wiki-based collaboration. For example, Arazy, et al., designed an embedded tool to show page-level statistics information and help users build their community influence (e.g., a pie chart shows the proportion of contribution from every editor)[16]. Similarly, the Wiki-Dashboard designed by Pirolli, et al., also shows each editor's influence over the article [37]. These tools can reflect information about authorship as it shows the overall influence of each editor, but cannot tell when, where and how the content is changed. A history flow visualization tool designed by Viegas et al. can only reflect the changes of the whole article [38], but cannot facilitate users' communication (almost no user information provided). Moreover, it is difficult to use the tool when there are too many revisions. The same problem also exists in another study conducted by Wattenberg et al. , which uses chromo grams to reflect different edit actions in time series [39]. Ekstrand and Riedl design a history tree visualization tool to give users information about the content evolution [40]. This tool could partially solve the knowledge exchange issue since it points out the difference between users' opinions. However, the tool does not give any information about the actual content changes among revisions, thus it could not help users easily determine the details of conflict opinions.

4 Design of Two Wiki Tools

The article page of the original wiki (wiki-A) provides no information but content text. We proposed two different designs (Figure 1). The first design (wiki-B) is a dialog box triggered by a "view history" hyper-link, which is located at the end of every paragraph in a wiki article page. This tool has two different subfunctions, namely *paragraph-based edit history* and *word-based content authorship*. *Paragraph-based edit history* highlights the added/deleted content between every former and later revision (the revisions can be sorted by revision sequence in ascending and descending orders), and it also shows information about corresponding editors to facilitate further communication (clicking the name of an editor will trigger the navigation to the editor's talk page), and displays the degree of conflict of this paragraph. *Word-based content authorship* shows the author of a word (or sentence, depending on the length).

The second design (wiki-C) assigns different background-colors to words (or sentences) directly in the text of the article based on the computation of severity of task conflict. For example, when a sentence has been modified many times in a certain period, the background color of this sentence will be set to dark red. If the content is not edited by users for certain period of time, the background color of the content will be changed to lighter colors. Unlike the tool in wiki-B, this tool can reflect a direct and detailed view about the distribution of task conflict issues and related information (e.g., last editor, activeness) as soon as users visit the wiki article page.

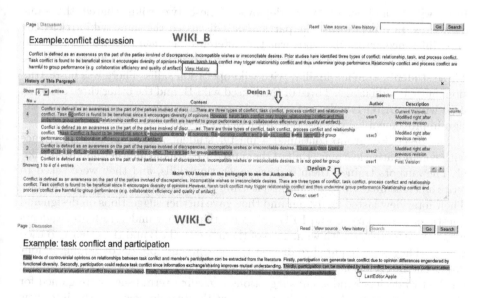

Fig. 1. Two designs of visualization of task conflict in wiki-B and wiki-C

5   Evaluation of Two Wiki Tools

The evaluation of the proposed two designs uses the focus group method, as modified to suit the Design Science Research framework [41]. The basic idea of the modification is the introduction of Exploratory Focus Groups (EFGs) and Confirmatory Focus Groups (CFGs). EFGs are used to iteratively refine the proposed design and the question draft, while CFGs are used to demonstrate the utility of the design in a field setting.

The steps of the focus group method used in this study are: (1) the authors proposed two preliminary designs and issues that need to be addressed, a pilot focus group was organized to help anticipate the issues of managing the focus group interview, including length of interview, generation of the initial questions, and evaluation of the moderators' style. The data gained from the pilot focus group were not used for further analysis. (2) A rolling interview guide [42] was utilized for EFGs. The first EFG was organized to test the designs and give suggestions on survey questions. Based on its outcomes, the quality of design was improved, and the interview guide was refined. Then, the second EFG was organized to re-test the designs and, based on its suggestions, the designs and interview guide were refined until their final version was reached. (3) The field test was conducted, and CFGs were organized to collect learners' feedback about the two designs. (4) The data were analyzed and results were reported.

5.1   Design Refinement and Outcomes of EFGs

Both wiki-B and wiki-C were deployed. These two wikis shared the same databases. In other words, the participants discussed the same wiki article; the only difference across the wikis was the difference between the two designs.

The authors organized a pilot focus group consisting of 6 graduate students. These students were all familiar with wiki operation skills. There was no restriction on their wiki usage. Three of them used wiki-B, and the other three students used wiki-C. After two days of trial and a one-hour interview, suggestions of design improvements and questions draft were gained.

The first EFG group contained 4 master students. They all had experience with collaborative learning. Half of them used wiki-B and another half used wiki-C. The authors trained them in basic operation skills, and they were asked to write a literature review, and discuss it using the modified wiki systems in 4 days. Their interview lasted for 1 hour and they gave further suggestions on design and questions draft. The second EFG group contained four undergraduate students who have experience with wiki. Two of them used wiki-B and the other two used wiki-C. They were required to discuss the benefits and pitfalls of gaming in four days. The interview lasted for half an hour. Almost no suggestions were gained on the two designs, and small suggestions regarding term usage were gained for the interview guide. As a result, the test of the two designs and the interview guide used for wide-scale survey and focus group interview were accomplished.

5.2   Field Test: Collaborative Writing Projects

All three wiki systems (wiki-A, B and C) were deployed in field test (two collaborative writing projects). The procedure is described as follows: (1) before the collaboration starts, there was a 10-minute-long face-to-face instruction session to make the participants fully understand how to use the wiki system. The definition of conflict was further introduced in another 10 minutes of instruction to help the participants distinguish among different types of conflict, since this study only focuses on task conflict. (2) All participants were randomly divided into 2 or 3 groups of almost equal size. Each group was assigned an online discussion topic. The goal was to collaboratively write a high quality discussion paper that is expected to include opinions about the topic from every possible perspective. (3) Participation was anonymous, and each participant used an alias to communicate with the others. Therefore the possibility of generating relationship could be reduced. (4) In each group, one half of the participants used the original wiki, while the others used the modified wiki. (5) Only the basic skills of editing wiki articles were required during the discussion. Thus, the probability of generating process conflict was reduced. Note that the use of the Talk/Discussion tab was still available. (6) All the discussions started simultaneously and lasted for two weeks. (7) The definition of task conflict was re-introduced in order to help the participants with recall.

We conducted the wiki-based collaboration two times. Wiki-A and wiki-B were used for the first round of collaboration (Collaboration-A). Undergraduate students, who were enrolled in a campus-level course named *Computer Ethics*, were invited to participate, and 322 out of 346 students agreed to participate. All students were equally divided into three groups, and the discussion topics selected for these groups were *pirated software, computer related occupational disease*, and *online gaming*. The second round of collaboration (Collaboration-B) used wiki-A and wiki-C; 116 out of 132 undergraduate students who were enrolled in a campus-level course named *Modern Educational Technology* were willing to participate. All students were equally segmented into two groups, and the discussion topics for these two groups were *game-based learning* and *traditional learning versus e-learning*. These two collaborations involved students from two universities separately due to implementation constraints.

5.3   Preliminary Survey and Result

Since focus group studies often face the criticism of a small number of participants, we develop two questions in the preliminary survey that was sent to all participants in the field test. These two questions obtain data about how the students notice conflict, and their willingness to solve conflict.

Questionnaires were sent out immediately after each collaboration ended. 301 out of 322 and 108 out of 116 valid responses were received from Collaboration-A and Collaboration-B respectively. Collaboration-A had 21 invalid responses since 13 students withdrew for personal issues, and 8 students who used the original wiki made an incorrect selection indicating that they noticed the conflict from the

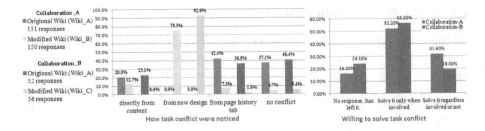

Fig. 2. Summary of wide-scale survey results

new design. Collaboration-B had 8 invalid responses since 3 students withdrew for personal issues, and 5 students who used the original wiki made an incorrect selection indicating that they noticed the conflict from new design.

As is shown in Figure 2, in Collaboration-A, 95 students (62.9%) using wiki-A noticed conflict, while 143 students (95.3%) using wiki-B noticed conflict, including 113 students (75.3%), who noticed conflict from the new design. In Collabroation-B, compared to the 31 (59.6%) students who used wiki-A and noticed conflict, the percentage of being aware of task conflict among students who used wiki-C is much higher (53 students, 94.6%). And 52 students (92.8%) noticed conflict with the new design.

For the question about how participants reacted to the conflict events, in Collaboration-A, 49 students (16.3%) did not want to solve the conflict, 157 students (52.2%) were willing to solve the conflict issues only when they were involved, and the remaining 95 students (31.6%) were willing to solve the conflict issues whether involved or not. In Collaboration-B, 26 students (24.1%) did not want to solve the conflict, 61 students (56.5%) were willing to solve the conflict only when they were involved, and 21 students (19.4%) were willing to solve the conflict issues whether involved or not.

In summary, the results from wide-scale survey suggest that most of the students were willing to solve conflict, and the new designs seemed to be helpful for students to identify conflict.

5.4   Focus Group Interview Process

Focus group method is useful in both exploratory and confirmatory research settings [41]. In this study, we not only want to confirm the utility of out designs, but also want to understand how these tools influence learners' willingness to contribute. Therefore, the focus group method is chosen as a suitable technique. However, the focus groups are based on open discussions or interviews, thus the size of the group and the recruitment of group members was carefully controlled. By considering our field test settings and suggestions from previous study [41], we decided to setup four focus groups (two from wiki-B, and two from wiki-C) consisting of 6-10 participants for the next step in open discussion.

Since we adopted a policy of anonymity during the collaboration, we could only conduct an open call for focus group participation through email in wiki.

We sent out invitation letters (including a description of our interview purpose, style of discussion, time schedule, etc.) to the most active (30%) participants in wiki-B and wiki-C in both Collaboration-A and -B; 62 invitation letters were sent and 57 responses were received. We carefully selected responders based on their profile and telephone conversions. Finally, we selected 20 students who used wiki-B, and 16 students who used wiki-C. These subjects were suitable for our focus group study because they were familiar with the wiki environment and were not too diverse in relationship to the topic of our interests; thus, they had sufficient knowledge to provide data of sufficient depth [41]. We segmented these 36 subjects into four focus groups. Every participant received 20 RMB (3 dollars) plus food and transportation expenses.

We assigned an instructor to lead the conversation in each group. All subjects in each group were seated in a U-shape arrangement to encourage collaboration [41]. Each session of the discussion lasted 60 to 90 minutes. And the discussion were all related to the influence of proposed two designs. For example, the instructors asked the participants to recall their experience with two designs and imagine the situation without design. All conversations were video recorded after gaining permission from the subjects. Notes of key ideas and themes were taken during the discussions by an observer (either a Ph.D student or mater student). A qualitative analysis was carried out to extract the statements.

5.5 Focus Group Interview Results and Discussion

Two categories were extracted from students' feedback: perception of task conflict and motivations triggered by two tools. We further divided motivation factors into two sub-categories: extrinsic and intrinsic motivation. For each motivation factor, we counted the number of students who mentioned the factor.

Perception of Task Conflict. Students said they could notice task conflict by direct view from the content since they had strong impressions about what they wrote. However, if they closed the tools (e.g. the dialog box in wiki-B), they were usually unaware of changes in content contributed by others. Students said they liked the tools that reflected the changes made to the text and that once they got used to these tools, they would not visit the page history any more. Moreover, students generally reported that they were not willing to use Talk/Discussion pages when there were different opinions. They preferred to edit the article content directly, and no one wanted to act as a icebreaker to initiate discussion. One possible reason to explain this phenomenon is that our collaborative writing project did not employ Wikipedia-like mechanisms for students to vote for conflict resolution.

Students reported that the most frequent style of modifications was corrections of typos and punctuation. A content author would accept other's modifications, if the changes were reasonable. Moreover, the phenomenon of self-conflict (someone modified his/her previous contributions) was very pervasive. Students cared more about content written by themselves, after they had no changes to their own content, they began to check and edit others' content. Harsh task conflict (e.g., back-and-forth editing) was rare since students thought the discussion topics were open questions and had no exact answers. Overall, students reported a moderate level of task conflict during collaborative writing.

Extrinsic Motivation Triggered by New Designs. As it can be seen from Table 2, six extrinsic motivation factors were extracted from students' feedback. Five students mentioned that the tools gave them a feeling of being visible to others; sometimes they were motivated to contribute in order to avoid blame for too little contributions from other learners. Eighteen students said that the tools helped them form a sense of group norm of participation (subjective norm). They were motivated because they felt that other learners contributed actively. Since collaborative writing is a group task, and the quality of a group artifact is related to everyone's contribution, a learner at the low-contribution level may be afraid of others' blame, especially when they perceive a strong subjective norm about contribution from system interface (e.g., highlighted modifications occur frequently). In past studies, subjective norm has been proven as a significant predictor of knowledge sharing behavior [28]. But few studies can be found to support our observation about avoiding blame.

Desire for peer recognition was mentioned by 14 students. Since the tools made students' editing activities visible to others, the students could promote themselves in the learning group through frequent contribution. The positive impact of peer recognition on learners' motivation have been found in previous studies [43].

Fifteen students reported that the tools helped them to identify co-learners who held the same opinions, and further helped them to make friends. The desire to increase social capital has been found to maintain users' intention to use a system in previous studies [44].

Nine students felt safe to express opinions when they saw the tools highlighted/listed many modifications and perceived an open style of discussion. As indicated by the Work-Engagement Theory [45], psychological safety is an important factor that predicts job engagement. In our collaborative learning project, different opinions are brought into the open and discussed, which makes students feel safe to express themselves, and encourages learning from each other [24].

Finally, 23 students reported that they generated a sense of belonging with the help of the tools. Students mentioned that they couldn't perceive a community by just looking at the wiki article page. It was the tools that helped students perceive the activeness of the learner group. The value of sense of belonging in collaborative learning has been discussed in previous studies [46].

Intrinsic Motivation Triggered by New Designs. Four intrinsic motivation factors were extracted from students' feedback (Table 3). Thirty-three students reported that they were curious about others' modifications on their own contributions. And the tools increased their frequency of visiting and reading wiki articles. Twenty-nine students felt that the new designs made them feel excited

Table 2. Extrinsic motivation extracted from feedback

Name	Participant quotation	Counts
Avoid blame	"I am afraid that other people think I contribute too little, and they might criticize me"	2(wiki-B); 3(wiki-C)
Subjective norm	I think everyone in my team is very active, and it is unwise not to express my own point of view	8(wiki-B); 10(wiki-C)
Desire for peer recognition	"Once I knew my name was shown in the dialog boxI want to get recognized by other students and become a leader of this team"	9(wiki-B); 5(wiki-C)
Make friends	"The dialog box helped meI think I found someone who has similar opinions with me... We could be friends"	7(wiki-B); 8(wiki-C)
Psychological safety	"When I saw so many red spots in the articleThis kind of open discussion made me express my view freely"	4(wiki-B); 5(wiki-C)
Sense of belonging	"I couldn't feel there is a community working on this page...But the design did show a group of people I feel I am a member of this team, others have not isolated me"	12(wiki-B); 11(wiki-C)

because of the unexpected presence of modification. Moreover, 22 students generated a sense of audience under the influence of new designs. They could feel that their writing has readers. As a result, their satisfaction of participation increased.

Although curiosity to check the modifications did not directly lead to contribution behaviors, it increased the chance of participation in knowledge exchange activities, and the satisfaction of contribution. In the same way, sense of audience and excitement also increases learners' satisfaction. Content modifications and highlights are dynamic; therefore, the sense of curiosity, excitement and sense of audience these actions gave to participants is self-reinforcing. The value of flow experience (curiosity, excitement) has been discussed in previous educational studies [47]. And the positive impact of sense of audience on student's willingness to contribute is discussed in [48].

A few cases of participation caused by unhappiness/annoyance were reported by 6 students. These cases only happened when others' modifications were unreasonable. The reactions (e.g. revert) caused by negative emotions are usually considered as defensive behavior, because people have the need to protect their perceived integrity and worth of self, as suggested by the Self-Affirmation Theory [49].

In summary, the effectiveness of our new tools on motivating learners is strongly connected to the experimental settings. First, we provided an autonomous, supportive style of administration during collaboration that only required basic operation skills, and did not limit students' freedom to use the wiki. Second, the new designs provided up-to-date peer feedback information about content. The positive effects of peer feedback have been discussed extensively in educational studies [50]. Third, since the two designs can directly reflect the interactions of community members, they provide students with a sense of audience and belonging to group. Our findings are in line with the Self-Determination

Table 3. Intrinsic motivation extracted from feedback

Name	Participant quotation	Counts
Curiosity	"I was so curious that I almost go to check if somebody modified my content every once in a while "	18(wiki-B); 15(wiki-C)
Sense of audience	"When I saw the background color of my content change to red, I felt very satisfied, because somebody read my content"	12(wiki-B); 10(wiki-C)
Excitement / Interest	"I think this kind of discussion is interesting, others can modify my content, and this system can unexpectedly show this kind of modification "	15(wiki-B); 14(wiki-C)
Unhappiness/ annoyance	"For myself, when I saw my content was modified, I didn't care if such modification is a kind of extension, but if somebody distorted what I mean, I will feel angry that they changed the content, and revert the content"	2(wiki-B); 4(wiki C)

Theory, which suggests that providing people with senses of autonomy, competence, and relatedness will facilitate internalization of external motivation [51]. Moreover, our collaborative learning project was not bothered by process conflict and relationship conflict since the project did not require students to spend their effort on dealing with process issues, and the generation of relationship conflict was carefully suppressed due to the sample characteristics and the policy of anonymity.

6   Conclusions

6.1   Summary of Findings

The original wiki system has usability problems (it does not well support content trust and knowledge exchange); meanwhile, users in wiki-based collaborations often experience lack of motivation due to a limited sense of audience, limited feedback and limited authorship. These issues can be attributed to the lack of clues of task conflict. In this paper, two wiki tools were introduced to increase user's awareness of task conflict and motivate users to contribute in collaborative learning context. The main evaluation process followed the focus group framework proposed by [41], a method adjusted for Design Science Research.

The effectiveness of the designs was confirmed in two ways. First, the results from the large-scale survey shown in Figure 2 suggest a higher awareness of task conflict with the modified wiki. Second, responses from the CFGs show that our designs, which display clues of task conflict, improve wiki usability and facilitate the generation of a series of extrinsic and intrinsic motivations. The result is in line with our expectations.

6.2   Limitations

Our study has limitations, which may influence the applicability of our findings. First, our field test and focus group interview only used undergraduate students as subjects. Undergraduate students represent the majority of users in collaborative learning. However, disregarding other kinds of users (e.g. middle school students) can cause variance in the final result.

Second, our field test adopted a relative loose policy on user participation, and our focus group interviews only included active participants. Therefore, we cannot address free-riding problems well and we cannot explain free-rider motivations, as well as how our designs might impact their motivations. In future studies, we will focus on evaluating the effects of our designs on the transition from non-active users to active users.

Third, although we did not observe significant negative impact of our design on learner's motivation, the proposed designs can still inhibit contributions behaviors by antagonizing learners. This negative effect is in accordance with the complex consequences of task conflict. We argue that whether task conflict is beneficial to collaborative learning depends on the management style of collaboration (e.g., active and agreeable conflict management style).

Fourth, since our field test was conducted in collaborative learning context, the results from CFGs might not be able to explain the situations in other contexts (e.g., enterprise). Different wiki applications may have very different profiles. For example, the policy of anonymity is usually not allowed in enterprise wikis. Therefore, it may be much harder to mitigate relationship conflict and the generation of motivations factors might be different in other wiki implementations. Since the two designs provide useful tools to let users know about changes of content and provide other task conflict clues, we expect that these designs may be effective in different contexts.

6.3 Implications

This study yielded implications for literature. First, previous studies on task conflict largely focus on exploring the consequences of task conflict, since task conflict can be both beneficial and detrimental to collaboration. This study made an early attempt to use the good side of task conflict. By providing clues of task conflict, problems of lack of motivation caused by wiki usability issues can be addressed. Second, previous studies have not fully explained how task conflict benefits the collaboration process, while this study provided evidence that increasing learner's awareness of task conflict could trigger many motivations which are important for learner's participation–especially for their continual participation (e.g., curiosity, excitement, sense of belonging).

This study also results in practical implications. First, the lack of motivation is partly due to the fact that the collaborative learning platforms cannot generate effective incentive mechanisms [52]. In contrast to the way of using trust and reciprocity to motivate learners, the two tools offer a new approach to achieve the same goal. Second, people in online collaborative work tend to care about their own contributions most. Such behaviors are not compatible with the original objective of collaborative work [15]. The new designs can motivate users to contribute to others' work and generate valuable communication. Third, since the effectiveness of conflict management is based on how well such conflicts can be understood by group members, increasing the awareness of task conflict can help group members to be aware of task conflict at an early stage, track evolution of opinions, and negotiate with each other (active conflict management). Instead of designing complex conflict monitoring and resolution mechanisms, this method could allow conflict to be resolved by people in a self-organized way.

References

1. Larusson, J.A., Alterman, R.: Wikis to support the collaborative part of collaborative learning. International Journal of Computer-Supported Collaborative Learning 4(4), 371–402 (2009)
2. Cress, U., Kimmerle, J.: A systemic and cognitive view on collaborative knowledge building with wikis. International Journal of Computer-Supported Collaborative Learning 3(2), 105–122 (2008)
3. Ebner, M., Kickmeier-Rust, M., Holzinger, A.: Utilizing wiki-systems in higher education classes: a chance for universal access? Universal Access in the Information Society 7(4), 199–207 (2008)
4. Grudin, J., Poole, E.S.: Wikis at work: success factors and challenges for sustainability of enterprise wikis. In: Proceedings of the 6th International Symposium on Wikis and Open Collaboration, p. 5. ACM (2010)
5. Ardichvili, A., Page, V., Wentling, T.: motivation and barriers to participation in virtual knowledge-sharing communities of practice. Journal of Knowledge Management 7(1), 64–77 (2003)
6. Arazy, O., Nov, O., Patterson, R., Yeo, L.: Information quality in wikipedia: The effects of group composition and task conflict. Journal of Management Information Systems 27(4), 71–98 (2011)
7. Collier, B., Bear, J.: Conflict, criticism, or confidence: an empirical examination of the gender gap in wikipedia contributions. In: Proceedings of the ACM 2012 Conference on Computer Supported Cooperative Work, pp. 383–392. ACM (2012)
8. Kittur, A., Kraut, R.E.: Beyond wikipedia: coordination and conflict in online production groups. In: Proceedings of the 2010 ACM Conference on Computer Supported Cooperative Work, pp. 215–224. ACM (2010)
9. Jehn, K.A., Mannix, E.A.: The dynamic nature of conflict: A longitudinal study of intragroup conflict and group performance. The Academy of Management Journal 44(2), 238–251 (2001)
10. Kane, G.C., Fichman, R.G.: The shoemaker's children: Using wikis for information systems teaching, research, and publication. MIS Quarterly 33(1), 1–17 (2009)
11. de Wit, F.R.C., Greer, L.L., Jehn, K.A.: The paradox of intragroup conflict: A meta-analysis. Journal of Applied Psychology 97(2), 360–390 Article (2012)
12. Jehn, K.A.: Enhancing effectiveness: An investigation of advantages and disadvantages of value-based intragroup conflict. International Journal of Conflict Management 5(3), 223–238 (1994)
13. Grace, T.P.L.: Wikis as a knowledge management tool. Journal of Knowledge Management 13(4), 64–74 (2009)
14. Hüner, K.M., Otto, B., Österle, H.: Collaborative management of business metadata. International Journal of Information Management 31(4), 366–373 (2011)
15. Wheeler, S., Yeomans, P., Wheeler, D.: The good, the bad and the wiki: Evaluating student-generated content for collaborative learning. British Journal of Educational Technology 39(6), 987–995 (2008)

16. Arazy, O., Stroulia, E., Ruecker, S., Arias, C., Fiorentino, C., Ganev, V., Yau, T.: Recognizing contributions in wikis: Authorship categories, algorithms, and visualizations. Journal of the American Society for Information Science and Technology 61(6), 1166–1179 (2010)

17. Adler, B.T., Chatterjee, K., De Alfaro, L., Faella, M., Pye, I., Raman, V.: Assigning trust to wikipedia content. In: Proceedings of the 4th International Symposium on Wikis, p. 26. ACM (2008)

18. Wu, K., Vassileva, J., Zhu, Q., Fang, H., Tan, X.: Supporting group collaboration in wiki by increasing the awareness of task conflict. Aslib Proceedings 65, 581–604 (2013)

19. Holtzblatt, L.J., Damianos, L.E., Weiss, D.: Factors impeding wiki use in the enterprise: a case study. In: CHI 2010 Extended Abstracts on Human Factors in Computing Systems, pp. 4661–4676. ACM (2010)

20. Light, D.: Do web 2.0 right. Learning & Leading with Technology 38(5), 10–12 (2011)

21. Chi, E.H.: The social web: Research and opportunities. IEEE Computer 41(9), 88–91 (2008)

22. Preece, J., Shneiderman, B.: The reader-to-leader framework: Motivating technology-mediated social participation. AIS Transactions on Human-Computer Interaction 1(1), 13–32 (2009)

23. Gibson, C., Cohen, S.: Virtual Teams That Work: Creating Conditions for Virtual Team Effectiveness. Wiley (2003)

24. Zhang, X., Chen, Z., Guo, C.: The opening "black box" between conflict and knowledge sharing: A psychological engagement theory perspective. In: 42nd Hawaii International Conference on System Sciences, pp. 1–10. IEEE (2009)

25. Curşeu, P.L., Janssen, S.E., Raab, J.: Connecting the dots: social network structure, conflict, and group cognitive complexity. Higher Education 63(5), 621 (2012)

26. Simons, T.L., Peterson, R.S.: Task conflict and relationship conflict in top management teams: The pivotal role of intragroup trust. Journal of Applied Psychology 85(1), 102–111 (2000)

27. Kankanhalli, A., Tan, B.C.Y., Wei, K.-K.: Conflict and performance in global virtual teams. Journal of Management Information Systems 23(3), 237–274 (2006)

28. Chen, I.Y., Chen, N.S.: Examining the factors influencing participants' knowledge sharing behavior in virtual learning communities. Journal of Educational Technology & Society 12(1) (2009)

29. Weingart, L.R., Todorova, G., Cronin, M.A.: Representational gaps, team integration and team creativity. In: Academy of Management Proceedings, vol. 2008, pp. 1–6. Academy of Management

30. Hobman, E., Bordia, P., Gallois, C.: Consequences of feeling dissimilar from others in a work team. Journal of Business and Psychology 17(3), 301–325 (2003)

31. De Dreu, C.K.: When too little or too much hurts: Evidence for a curvilinear relationship between task conflict and innovation in teams. Journal of Management 32(1), 83–107 (2006)

32. Moye, N.A., Langfred, C.W.: Information sharing and group conflict: Going beyond decision making to understand the effects of information sharing on group performance. International Journal of Conflict Management 15(4), 381–410 (2004)

33. Rose, G.M., Shoham, A.: Interorganizational task and emotional conflict with international channels of distribution. Journal of Business Research 57(9), 942–950 (2004)

34. Liang, T.P., Wu, J.C.H., Jiang, J.J., Klein, G.: The impact of value diversity on information system development projects. International Journal of Project Management 30(6), 731–739 (2012)

35. Son, S., Park, H.: Conflict management in a virtual team. In: 2011 5th International Conference on New Trends in Information Science and Service Science (NISS), vol. 2, pp. 273–276. IEEE (2011)

36. Shaw, J.D., Zhu, J., Duffy, M.K., Scott, K.L., Shih, H.A., Susanto, E.: A contingency model of conflict and team effectiveness. Journal of Applied Psychology 96(2), 391–400 (2011)

37. Pirolli, P., Wollny, E., Suh, B.: So you know you're getting the best possible information: a tool that increases wikipedia credibility. In: Proceedings of the SIGCHI Conference on Human Factors in Computing Systems, p. 1505. ACM (2009)

38. Viégas, F.B., Wattenberg, M., Dave, K.: Studying cooperation and conflict between authors with history flow visualizations. In: Proceedings of the SIGCHI Conference on Human Factors in Computing Systems, pp. 575–582. ACM (2004)

39. Wattenberg, M., Viégas, F.B., Hollenbach, K.: Visualizing Activity on Wikipedia with Chromograms. In: Baranauskas, C., Abascal, J., Barbosa, S.D.J. (eds.) INTERACT 2007. LNCS, vol. 4663, pp. 272–287. Springer, Heidelberg (2007)

40. Ekstrand, M.D., Riedl, J.T.: Rv you're dumb: identifying discarded work in wiki article history. In: Proceedings of the 5th International Symposium on Wikis and Open Collaboration, p. 4. ACM (2009)

41. Tremblay, M., Hevner, A., Berndt, D.: The Use of Focus Groups in Design Science Research. Integrated Series in Information Systems, vol. 22, ch. 10, pp. 121–143. Springer US (2010)

42. Stewart, D.W.: Focus groups: Theory and practice, vol. 20. Sage (2007)

43. Schmidt, J.P., Geith, C., Håklev, S., Thierstein, J.: Peer-to-peer recognition of learning in open education. International Review of Research in Open & Distance Learning 10(5) (2009)

44. Wang, J.C., Chiang, M.J.: Social interaction and continuance intention in online auctions: A social capital perspective. Decision Support Systems 47(4), 466–476 (2009)

45. Kahn, W.A.: Psychological conditions of personal engagement and disengagement at work. The Academy of Management Journal 33(4), 692–724 (1990)

46. So, H.J., Brush, T.A.: Student perceptions of collaborative learning, social presence and satisfaction in a blended learning environment: Relationships and critical factors. Computers & Education 51(1), 318–336 (2008)

47. Liao, L.: A flow theory perspective on learner motivation and behavior in distance education. Distance Education 27(1), 45–62 (2006)

48. Forte, A., Bruckman, A.: From wikipedia to the classroom: exploring online publication and learning. In: Proceedings of the 7th International Conference on Learning Sciences, International Society of the Learning Sciences, pp. 182–188 (2006)

49. Sherman, D.K., Cohen, G.L.: The Psychology of Self-defense: Self-Affirmation Theory, vol. 38, pp. 183–242. Academic Press (2006)

50. Phielix, C., Prins, F.J., Kirschner, P.A.: Awareness of group performance in a cscl-environment: Effects of peer feedback and reflection. Computers in Human Behavior 26(2), 151–161 (2010)

51. Deci, E., Ryan, R.M.: Self-determination theory. In: Handbook of Theories of Social Psychology, p. 416 (2008)

52. Ransbotham, S., Kane, G.C.: Membership turnover and collaboration success in online communities: Explaining rises and falls from grace in wikipedia. Mis Quarterly 35(3), 613–627 (2011)

Author Index